Image Analysis Applications

OPTICAL ENGINEERING

Series Editor

Brian J. Thompson

Provost
University of Rochester
Rochester, New York

Image Analysis Applications

edited by

Rangachar Kasturi
Pennsylvania State University
University Park, Pennsylvania

Mohan M. Trivedi
University of Tennessee
Knoxville, Tennessee

Marcel Dekker, Inc.　　■　　**New York and Basel**

Library of Congress Cataloging-in-Publication Data

Image analysis applications / edited by Rangachar Kasturi, Mohan M. Trivedi.
    p. cm. — (Optical engineering ; 24)
    Includes bibliographic references.
    ISBN 0-8247-8198-8 (alk. paper)
    1. Image processing—Digital techniques. 2. Computer vision. 3. Pattern recognition systems. 4. Robot vision. I. Kasturi, Rangachar. II. Trivedi, Mohan M. III. Series: Optical engineering (Marcel Dekker, Inc.) ; v. 10.
    TA1632.I44   1990
    621.36′7—dc20   90-3065
                                                                        CIP

This book is printed on acid-free paper.

MARCEL DEKKER, INC.
270 Madison Avenue, New York, New York 10016

Current printing (last digit):
10 9 8 7 6 5 4 3 2 1

PRINTED IN THE UNITED STATES OF AMERICA

About the Series

The series came of age with the publication of our twenty-first volume in 1989. The twenty-first volume was entitled *Laser-Induced Plasmas and Applications* and was a multi-authored work involving some twenty contributors and two editors: as such it represents one end of the spectrum of books that range from single-authored texts to multi-authored volumes. However, the philosophy of the series has remained the same: to discuss topics in optical engineering at the level that will be useful to those working in the field or attempting to design subsystems that are based on optical techniques or that have significant optical subsystems. The concept is not to provide detailed monographs on narrow subject areas but to deal with the material at a level that makes it immediately useful to the practicing scientist and engineer. These are not research monographs, although we expect that workers in optical research will find them extremely valuable.

There is no doubt that optical engineering is now established as an important discipline in its own right. The range of topics that can and should be included continues to grow. In the "About the Series" that I wrote for earlier volumes, I noted that the series covers "the topics that have been part of the rapid expansion of optical engineering." I then followed this with a list of such topics which we have already outgrown. I will not repeat that mistake this time! Since the series now exists, the topics that are appropriate are best exemplified by the titles of the volumes listed in the front of this book. More topics and volumes are forthcoming.

Brian J. Thompson
University of Rochester
Rochester, New York

Preface

Visual imagery is one of the most important sensory inputs to the human perceptual system. In our efforts to develop intelligent machines capable of performing useful functions, it is necessary to investigate ways in which one can provide these machines with the capabilities for accurate interpretation of image inputs. In this book we present a collection of papers describing details of a number of such investigations.

Computers have been recognized as an effective and efficient tool for the processing and analysis of high volumes of data ever since their development in the middle of this century. Images, acquired for a variety of applications, while digitized, provide a natural source of large volumes of data. Recognizing this, researchers initiated efforts for computer processing of such images in the 1960s. Earlier research work involved mechanization of the basic image processing tasks such as image encoding, image enhancement, and restoration. Soon it was apparent that computers can effectively be utilized for tasks requiring more complex analysis of image data. These tasks included image segmentation, object/target recognition, motion characterization, 3-D feature analysis, and many other tasks associated with the broad area of scene understanding.

As in most engineering fields, applications play a central role in the development of theories and techniques in the image analysis field. In this book we present a wide spectrum of applications where image analysis has been successfully employed. Although these chapters describe efforts undertaken in diverse application domains, the similarities in the basic image-processing and -analysis tools and techniques will attest to the wide applicability of the general methodologies. Also, it is hoped that a detailed examination of a spectrum of applications within the covers of a single volume will provide the reader with an insight into the relative merits or demerits of a particular technique. We believe that such an insight is quite valuable in developing an effective image-analysis approach for any application.

We have selected a group of distinguished researchers who are highly experienced and very active in the field to present various applications. The contributors have examined applications in document and map data analysis, medical image analysis, satellite/aerial image understanding, robot vision, and fingerprint analysis. In addition, recent theoretical work in mathematical morphology and its importance in image analysis is discussed.

Each of the book's eleven chapters includes a fairly comprehensive list of references dealing with a particular application. The first four chapters deal with the domain of graphics recognition, document analysis, and map data interpretation. In Chapter 1, "Document-Analysis Systems and Techniques," Casey and Wong review algorithms for segmentation of mixed text-graphics-image documents, analysis of graphics data, recognition of text, and determination of the format and representation of documents. In Chapter 2, "A Graphics-Recognition System for Interpretation of Line Drawings," Bow and Kasturi describe a Hough Transform-based approach for the separation of text strings from graphics. The graphics image is further analyzed to generate a line description file which acts as input to an intelligent graphics interpretation system. The final output is a succinct description of the graphical components in the image. In Chapter 3, "Automatic Recognition of Engineering Drawings and Maps," Ejiri et al. review basic techniques to automate inputting of paper-based engineering drawings to CAD systems. Such techniques are expected to result in efficient drawing management systems useful in an industrial environment and to facilitate the merger of paper-based drawings with the state-of-the-art computer-aided design and drafting systems. These techniques are also applied to create a topographical map information system in which data from digitized maps are managed together with other related data and a knowledge base. Such a system would be an important tool for future utility management and planning support systems. Chapter 4, "Image-Analysis Techniques for Geographic Information Systems," by Kasturi, contains a description of a system that attempts to duplicate the process of information extraction from maps by humans. This system contains a query analyzer which converts user queries into appropriate image-processing operations. An image processor, under the control of the query analyzer, extracts relevant information from the digitized images of paper-based maps. A natural language interface is built in to enhance the user-friendliness of the system.

Chapters 5 and 6 describe the applications of medical image analysis. Specifically, applications from the neurosciences and ophthalmology fields are presented. In Chapter 5, "Digital Image Processing and Three-Dimensional

Reconstruction in the Basic Neurosciences," Hibbard describes the impact of image-analysis techniques in autoradiography. Autoradiography is a commonly used method for the study of brain metabolism, blood flow, protein and nucleic acid synthesis, and the neurochemical processing of information. The importance of computer image-processor for 3-D reconstruction of neuronal morphology and connectivity and other ultrastructural studies is also presented. In Chapter 6, "Applying Digital Processing Methods in the Analysis of Retinal Structure," Mitra and Krile describe a PC-based system for the analysis of fundus images to monitor and detect early changes in the retinal structure induced by the onset of diseases such as glaucoma and macular degeneration. Possible techniques of coding physiological images for fast transmission are also discussed in this chapter. In Chapter 7, "Visual Perception Using a Blackboard Architecture," Weymouth and Amini present a vision system for analyzing images of the inner-ear hair cells. The chapter contains a review of the blackboard architectures and a comparative discussion of the goal-directed and data-directed approaches in image analysis. Although the application described deals with the analysis of the hair-cell structure, the methods described have general utility.

Images acquired from airborne or satellite-based platforms offer a unique perspective of the scene. Such images provide valuable information for tasks such as agricultural or forest resource management, weather pattern or oceanographic feature analysis, and military surveillance and reconnaissance. In Chapter 8, "Analysis of High-Resolution Aerial Images," Trivedi presents a discussion of various issues affecting the accurate and efficient analysis of aerial images. The chapter emphasizes use of multispectral imagery and describes an efficient object detection approach utilizing multiresolution copies of an image.

Images, in general, are two-dimensional representations of three-dimensional scenes. Analysis and interpretation of 2-D images to understand the 3-D world are an important consideration in robotic applications. In Chapter 9, " Image Formation and Characterization for Three-Dimensional Vision," Brzakovic and Gonzalez present a vision system that infers depth from texture and two views of a scene. In this system, which is intended for mobile robot guidance, orientations of planar textured surfaces are estimated to determine relative positions of obstacles located in textured surfaces. Exact positions of the obstacles are then determined using two views of the scene.

A unique application of image analysis is described by Krile and Walkup in Chapter 10, "Enhancement of Fingerprints Using Digital and Optical Techniques." Digital processing techniques such as adaptive binarization, homomorphic and adaptive wedge filtering, and coherent optical techniques such as high-pass and Laplacian filters, are applied to degraded fingerprint images. The potential applications of these techniques would have far-reaching implications in criminalistics.

The final chapter, "The Digital Morphological Sampling Theorem," by Haralick et al., presents a methodology which can reduce the number of pixels to be processed while guaranteeing a sufficiently close result to techniques which process all pixels. The sampling theorem presented here has been developed for both binary and gray scale images. The inherent reduction in processing time and cost, resulting from the application of this theorem, makes it extremely attractive for industrial vision applications in which real-time processing of large volumes of data is an important design criterion.

This book is prepared for a wide audience. Researchers, engineers, and scientists who are active in the field will find this selection of chapters useful in updating their reference sources. Those interested in applying image analysis will find this collection helpful in reviewing the state of the art of our field. It has been the experience of the authors that discussion and presentation of real-world applications in the classroom enhance the interest and motivation of students in learning theories and principles. This book contains such material and can be used for reading assignments in image processing, pattern recognition, or computer and robot vision courses.

We are pleased to acknowledge the enthusiastic response of the authors to our invitation for their contributions to this volume and for their efforts in preparing their manuscripts. We are also thankful to the reviewers for their suggestions for improvements. Finally, we would like to thank the editorial and production staff at Marcel Dekker, Inc., for their help and cooperation through all stages of this book's production.

Rangachar Kasturi
Mohan M. Trivedi

Contents

Image Analysis Applications

1

Document-Analysis Systems and Techniques

Richard G. Casey and Kwan Y. Wong

IBM Almaden Research Center
San Jose, California

1.1 INTRODUCTION

Documents used to communicate information in the office can be created, distributed, and stored in paper or in electronic forms. An electronic document, which is created by a computer system, can exist in two types of format: (1) edit format, where special commands are embedded in the file to control the layout of the document when it is being printed or displayed, and (2) print format, where the special commands have been interpreted by a formatter to produce an output ready for printing or displaying the document. Paper documents can be created in a number of ways; e.g., by printing electronic documents with a printer system, by typing on paper, by printing from type-setting machines, or by writing by hand. A paper document can be converted into computer bit-map format by an optical scanner, yielding what we will refer to as an image document.

Persons who receive the printed form of a document ordinarily do not have access to the computer files from which it was generated. Various considerations may motivate the reentry of a paper document into computer-coded form. First, data on the document may be required within a computerized processing and record-keeping system. Second, the document may have to be combined with other information and reprinted, perhaps in edited form. This occurs with submitted manuscripts, proposals, and even correspondence. Finally, conversion to coded representation offers the power of electronic systems for storing information, for retrieving it efficiently, and for distributing or reprinting it as desired. Examples of this type of application are extraction of selected information from magazines or journals for company internal electronic distribution, storing government legislative proceedings in electronic form for retrieval and query, and conversion of drawings (e.g., utility drawings) for retrieval and for merging with existing CAD (computer-aided design) systems.

The process of converting a scanned document from bit-map format into atomic objects that can be interpreted and processed by computers is called *document analysis*. The encoded atomic objects can be text, graphics, or pictures. Here, we restrict ourselves to document-analysis systems that can encode *machine-printed* documents into text, graphic, and pictorial objects. Handwritten text, such as the signature on a letter, will be classified as a graphic object. The accurate interpretation of such material requires further advances in character-recognition methodology.

This chapter provides a survey of different document-analysis systems and techniques. The outline for the remainder of the chapter is as follows: Section 1.2 describes various major components of a document-analysis system and their functions. Section 1.3 discusses resolution requirements for document analysis and some of the commonly used thresholding techniques. Section 1.4 provides a survey of basic block-segmentation and classification techniques. Section 1.5 describes font-recognition techniques with special reference to font capability, format restriction, hardware versus software, user interfaces, and OCR (optical character recognition) status. Section 1.6 deals with graphic recognition and its status. Section 1.7 outlines some of the AI (artificial intelligence) techniques being used in page-structure analysis. Section 1.8 describes general requirements of a document-filing and retrieval system. Section 1.9 provides some examples of current document-analysis systems, both for text and for graphics recognition. Section 1.10 states our conclusions.

1.2   GENERAL ARCHITECTURE OF DOCUMENT-ANALYSIS SYSTEMS

The various major components of a document-analysis system are shown in Fig. 1.1. The tasks of the different components can be summarized as follows:

> Scanning and Thresholding—Scan the document by optical means and convert the signal into digital form (gray-levels) and then threshold the scanned image into bi-level format.

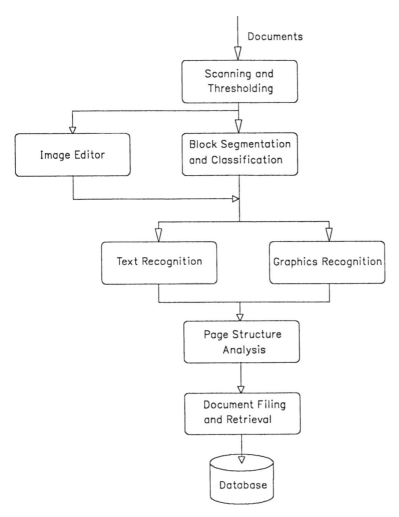

Figure 1.1   Major components of a document-analysis system.

Block Segmentation and Block Classification—Segment an image into blocks automatically and classify each block as text, graphics, or picture.

Image Editor—Edit manually a scanned document into regions of text, graphics, or pictures.

Text Recognition—Process text blocks and recognize individual characters.

Graphics Recognition—Convert block segments of graphics from bitmap format into graphics symbols; e.g., vectors or higher level graphics description.

Page-Structure Analysis—Analyze spatial relationships between blocks and/or textual contents to detect page-structure components such as heading, summary, paragraphs, etc.

Document Filing and Retrieval—Create document indices either manually or automatically for filing and provide a query interface for document retrieval from a database.

Document-analysis systems that exist today may not have all the functional components outlined in Fig. 1.1. In fact, commercial developments indicate that document-analysis systems will tend to evolve incrementally rather than be implemented as full-blown systems. There are important applications that can benefit from access to basic image-handling capabilities only, without the need for block segmentation, text recognition, or graphics recognition. Systems that can simply capture, store, print, and distribute images have widespread utility (Somer/1984). These image-handling capabilities are fundamental to document analysis as well and lay the groundwork for future migration into intelligent processing of the documents.

Optical character recognition is the process by which images of printed letters, numbers, or other symbols are converted to identifying codes such as ASCII (American Standard Code for Information Interchange). Commercially available OCR machines, which recognize textual information only, can be useful in many applications. The lowest priced OCR machines can recognize finite-number fixed-pitch-type fonts, while more sophisticated and expensive OCR machines can recognize fonts from printed journals (*Computer Buyer's Guide*/1986). Commercial OCR machines can be used as a subsystem within a general document-analysis system to perform recognition of documents that have only textual information. The more advanced capabilities of document analysis, such as automatic block segmentation, graphics recognition, etc., can be used to encode complex documents containing a mixture of text, graphics, and pictures.

The current research in document analysis is aimed at automatic segmentation and recognition of documents. Experimental prototype systems are being built by various research organizations and universities to prove feasibility of the concepts. However, there is no commercial system yet available that offers automatic segmentation and recognition of text, graphics, and pictures.

1.3   SCANNING AND THRESHOLDING

An optical scanner for documents generally consists of a transport mechanism plus a sensing device that converts light intensity into gray-levels. Printed documents usually consist of black on a white background. Thus, gray-level signals are converted into binary-image format by comparing the gray-levels either with a prechosen fixed threshold value or with spatially varying threshold values. The latter approach is called *adaptive thresholding*.

Scanner design issues particularly related to spatial sampling and aperture size characteristics are reviewed in (Nagy/1983). The spacing of independent sample points, called *resolution*, greatly influences the accuracy of text and graphics recognition. A discussion of scanning resolution requirements is given in (Srihari/1986).

Our own experience has led us to a rule of thumb for the resolution requirement in OCR. For text recognition, using a template type of approach, the resolution should be sufficient to provide at least two scanned samples across the width of a character stroke and a character height of 15 pixels or more. Feature-based methods can recognize characters at somewhat lower resolution, since features are less sensitive to distortions resulting from coarser sampling. The stroke width of 10- or 12-pitch typed characters, for example, is about 0.2–0.3 mm depending on the font, and the minimum character height is about 1.8 mm for 12-pitch, or 2 mm for 10-pitch. Hence, according to our rule, a scanning resolution of 240 pixels per inch (ppi) (9.5 pixels per mm) is adequate for recognition of typed documents. In journals and magazines, the smallest characters are the footnotes and the references, having stroke widths as small as 0.12 mm. On the other hand, the text in the main body of the IEEE (Institute of Electrical and Electronics Engineers) *Transactions on Information Theory* has lower-case heights of about 1.8 mm and stroke widths of about 0.2 mm. For recognition of such main-body text, our rule of thumb calls for a scanning resolution of at least 300 ppi (12 pixels per mm).

In graphics recognition, a lower resolution can be used for tracking lines than for recognizing symbols, assuming good-quality drafting and adequate spacing between lines. Thus, Toshiba Corporation, in its system for reading circuit diagrams (Okazaki/1985) uses 240 ppi for OCR and symbol recognition, but coarsens the resolution to 80 ppi for line following.

The choice of thresholding method also affects the accuracy of recognition. Improper thresholding can cause lines to break or disappear or small openings in some characters to fill in. For a high-contrast image with uniform background, a prechosen fixed threshold would be satisfactory (Fig. 1.2). However, most documents encountered in practice have a

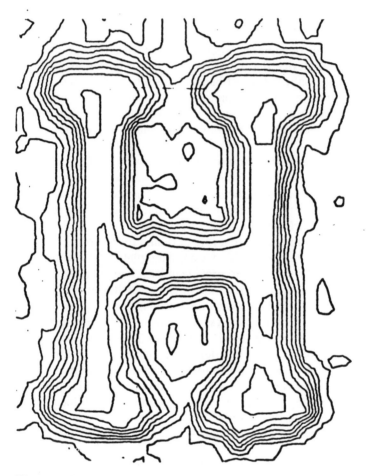

Figure 1.2 Gray-level contours in a scanned letter **H**. This character has been scanned at very fine resolution to show detail of the pattern. For recognition purposes, a sampling of 240–300 ppi (yielding a pattern 20–25 pixels high in this case) would be conventional. The threshold used for converting to black/white representation will affect the thickness of the result and, if poorly chosen, can result in spurious pixel groups (called *noise*).

wide range in contrast and in background levels. For proper image reproduction, an adaptive thresholding method is needed.

Adaptive-thresholding schemes can be either global or local. A global method determines a threshold value based on an analysis of the histogram of image-intensity values (Rosenfeld and Kak/1982). Local adaptive thresholding is done by analyzing local image features to determine a threshold value for each pixel; such a method is sometimes called *dynamic thresholding*.

One dynamic-thresholding method (Morrin/1974) uses an edge gradient to bias the thresholding value from the local average gray-levels to produce a sharper image. The current pixel is set to be black if the current scan value x is less than a threshold value $T(x)$, where

$$T(x) = A + G + b,$$

where A is the average value of the surrounding pixels
   G is the gradient biasing for enhancement
   b is a fixed negative constant to bias background toward white.

A desirable gradient operator in dynamic thresholding is sensitive to character edges but insensitive to random noise and halftone backgrounds. A mini-max corner-sum gradient in a (3×3) matrix and the clustering of scan pixels improves rejection of random noise in a document (Wong/1978). In a dynamic-thresholding algorithm developed for OCR applications (White and Rohrer/1983), two first-order-difference equations are used to update the background levels in both the horizontal and the vertical directions. The thresholding scheme is similar to (Morrin/1974), except that gradient enhancement is not used.

1.4   BLOCK SEGMENTATION AND CLASSIFICATION

There are two basic approaches to block segmentation and classification. The first approach, which may be called a top-down method, segments the document into large blocks and then analyzes the features of these blocks to determine whether they should be classified as text, graphics, or line drawings. The second approach, called a bottom-up method, segments the documents into small blocks (e.g., of the size of a character) and then analyzes the spatial relationship of these blocks to merge them into bigger blocks of text or graphics. Both the top-down and the bottom-up approaches may use the component-labeling technique (described below) to segment an image.

1.4.1   Component Labeling

Component labeling is a well-known image-processing technique that associates adjacent black pixels to detect distinct objects in a field. Component labeling may be done either pixel-by-pixel or using run-lengths. The pixel-by-pixel labeling method is described in (Rosenfeld and Kak/1976) and in (Wahl et al./1982). As the first line is scanned from left to right, the label 1 is assigned to the first black pixel. This label is propagated repeatedly; that is, subsequent adjacent black pixels are labeled with 1s and the first white pixel stops this propagation (see Fig. 1.3). The next black pixel along the line is labeled with 2 and so are its adjacent black pixels. For each black pixel in the second (and succeeding) lines, the neighborhood in the previously labeled line is examined, along with the left neighborhood of the pixel. If a label has already been assigned to a pixel in the neighborhood, the same label is assigned to the current pixel. If a black pixel has no labeled neighborhood, the next label not yet used is assigned to this pixel. This procedure is repeated until the bottom line of the image is reached. Upon completion of this algorithm, some adjacent black regions may be labeled differently.

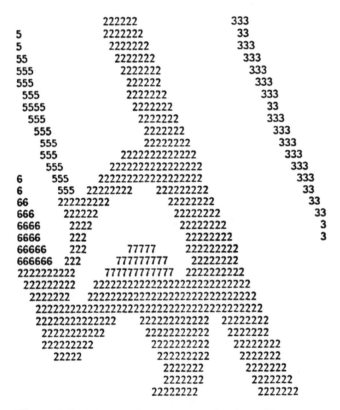

Figure 1.3 Connected-components algorithm. The figure illustrates the labeling of pixels to denote group membership, as the algorithm proceeds row-by-row from top to bottom. When two groups merge (as, for example, 2 and 5), this event is recorded in a merge table, and only one of the labels is carried forward.

The merging of different labels can be detected during the process and the labels combined to form a larger connected object.

The run-length labeling approach has been discussed in (Foith et al./ 1981). It is similar to the pixel-by-pixel labeling process, except that neighborhood adjacency is examined at the end of a black run.

During component labeling, important properties of the connected components can be calculated, e.g., area, perimeter, and maximum x-y dimensions. These properties can be used as features for classifying blocks into different categories. The topic of block classification will be discussed later.

1.4.2 Run-Length Smearing Method

One of the top-down approaches is the run-length smearing (RLS) method (Wong et al./1982), which segments a document into blocks and then classifies them into text, graphics, or more detailed object classes by a statistical feature-classification technique. White background plays an important role in subdividing documents into meaningful blocks. Let us assume that white pixels are represented by 0s and black pixels by 1s. The RLS algorithm takes an input sequence of 0s and 1s and modifies it to obtain an output sequence, y, in which 0s are converted to 1s if the number of adjacent 0s is less than some threshold T_c. For example, with $T_c = 4$, the sequence x is mapped into y as follows:

x : 0 0 0 1 0 0 0 0 0 1 0 1 0 0 0 0 1 0 0 0 0 0 0 0 1 1 0 0 0

y : 1 1 1 1 0 0 0 0 0 1 1 1 1 1 1 1 1 0 0 0 0 0 0 0 1 1 1 1 1

The RLS algorithm is first applied row-by-row (with $T_c = 300$) and then column by column (with $T_c = 500$), yielding two distinct bit-maps, which are combined by a logical AND operation. Some of the results using the RLS method are illustrated in Fig. 1.4 (Wahl et al./1982). An efficient two-pass algorithm for implementation of the RLS algorithm in two dimensions has been proposed in (Wahl and Wong/1982). Component labeling is then used for block segmentation.

1.4.3 Projection-Profile Method

Another top-down approach for segmentation is the projection-profile method (Fig. 1.5). Based on the observation that documents generally have rectangular block structures, a projection of the black pixel counts along the horizontal and vertical directions can be used for block segmentation (Zen and Ozawa/1985; Masuda et al./1985). The projection method is sensitive to document skew with respect to the raster-scanning direction of the scanner. Skew detection is needed, and this can be done by iteratively examining

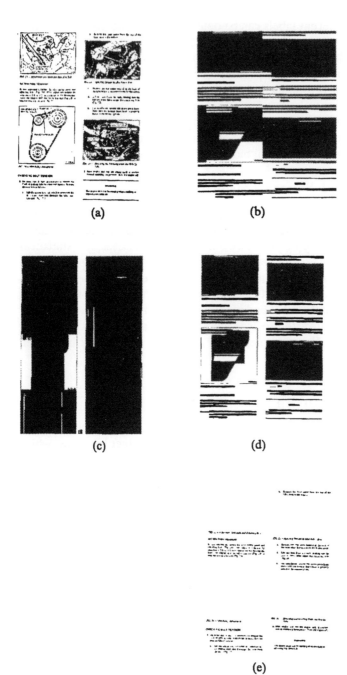

(a)

(b)

(c)

(d)

(e)

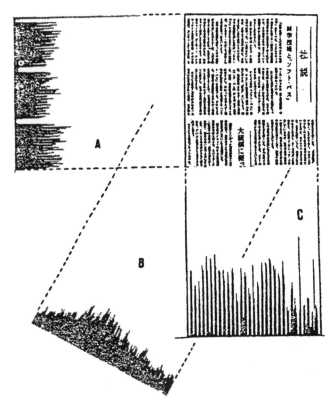

Figure 1.5 Projection profile. This technique uses the fact that lines of text (whether in Western text or Kanji, as here) are conventionally printed in straight rows and/or columns. Thus, skew, if present, can be detected by summing along various directions until a set of alternating peaks and nulls is obtained. Source: (Masuda et al./1985)

Figure 1.4 The RLS process. (a) The original figure, (b) the horizontal RLS map, (c) the vertical RLS map, (d) the superposition and AND of (b) and (c), and (e) the text regions found using connected-component analysis of (d) to define block regions, together with a feature analysis of the document region defined by each block to determine its class (text, image, or line drawing).

small angle deviations from the normal direction to determine which angle gives the steepest variation of the projection profile (Masuda et al./1985).

1.4.4   Block Classification for Top-Down Method

Blocks determined by a segmentation process usually need to be classified into predetermined document object types; e.g., text, graphics, lines, or halftones. We shall describe a few methods that are commonly used for block classifications.

(A)   Feature Classification

During component labeling, certain features of the segmented blocks are calculated, then used to classify the blocks. In (Wong et al./1982), the following features are measured during component labeling: (1) height $H = \Delta y$; (2) eccentricity $E = \frac{\Delta x}{\Delta y}$; (3) ratio of black area B (number of black pixels) to enclosing box area; i.e., $S = \frac{B}{\Delta x \Delta y}$; and (d) R, mean horizontal length of black runs of the original image within each block.

A block is considered to be text if its R and H values are less than some constant multiples of the mean length of black run $\overline{R}$ and mean height $\overline{H}$. The distribution of values in the R-H plane obtained from sample documents are observed to determine the discriminant function. Low R and H values represent regions containing text. A variable, linear, separable classification scheme is used to assign blocks into the following four classes:

Text: if $R < C_{21} \times \overline{R}$ and $H < C_{22} \times \overline{H}$
Horizontal solid black lines: if $R > C_{21} \times \overline{R}$ and $H < C_{22} \times \overline{H}$
Graphics and halftone images: if $E > \frac{1}{C_{23}}$ and $H > C_{22} \times \overline{H}$
Vertical solid black lines: if $E < \frac{1}{C_{23}}$ and $H > C_{22} \times \overline{H}$

The constants C_{ij} are determined heuristically by examining the R-H plane plot of typical documents and the values of $\overline{R}$ and $\overline{H}$.

(B)   Distance-Mapping Shape Measure

A shape measure for determining whether a block is text, graphics, or thresholded gray-level image was proposed by F. Wahl (Wahl/1983). A distance-mapping function based on a border-to-border distance is used for shape discrimination. Given a point (x,y) and an angle ϕ in a connected component of a binary image, the distance d is defined as the length of a line segment that connects (x,y) and two opposite border points (see Fig. 1.6). Thus, for any point (x,y), one can define $D_{\min}(x,y)$ and $D_{\max}(x,y)$, which are the minimum and the maximum line segment lengths at (x,y) over all possible angles ϕ. The eccentricity at a point is defined to be $D_{\text{ecc}}(x,y) = \frac{D_{\max}(x,y)}{D_{\min}(x,y)}$. Likewise, $\bar{d}_{\min}$, $\bar{d}_{\max}$, and $\bar{d}_{\text{ecc}}$ are defined as the average value of $d_{\min}(x,y)$,

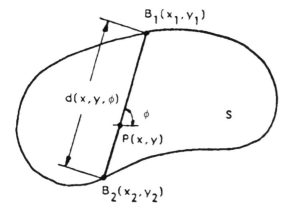

Figure 1.6   Distance-mapping shape measure. The quantity $d(x, y, \phi)$ is the length of a line across the pattern and passing through (x, y). As explained in the text, statistics measured on $d(x, y, \phi)$ as (x, y) and ϕ are varied provide a descriptor for the shape of the pattern.

$d_{max}(x, y)$, and $d_{ecc}(x, y)$ over all the pixels in the connected component. Wahl also defines compact shape factors $f_1 = \frac{area}{d_{min}^2}$ and $f_2 = \frac{area}{d_{max}^2}$. He has shown that $\bar{d}_{min}, \bar{d}_{max}, \bar{d}_{ecc}, f_1,$ and f_2 can be used as features to discriminate text, graphics, and thresholded gray-level pictures. Text has high f_2 value compared to graphics and thresholded gray-level pictures, while thresholded gray-level pictures have low f_1 compared to the other two types of images.

1.4.5 Bottom-Up Approaches

Both the RLS method and the projection method are useful for documents that have horizontal and vertical blocks of text. However, graphic documents such as maps and engineering designs may have line drawings, symbols, or text strings at many orientations. The RLS and projection methods are not able to segment text strings embedded in graphics. We shall describe some bottom-up methods that can separate text from graphics in any orientation. Bottom-up approaches generally require more processing than do top-down approaches.

(A) Hough-Transform Method

A block-segmentation and block-classification method based on the application of the Hough transform has been proposed by L. Fletcher and R. Kasturi (Kasturi et al./1986). The Hough transform is a technique for detecting forms parametrically; e.g., straight lines or other known curves in a noisy image (Pratt/1978). The Hough transform also can be viewed as a projection technique that produces a histogram. Thus, collinear black pixels in a black-white image, when projected onto an orthogonal line, will result in a large accumulator count value.

The algorithms used in (Kasturi et al./1986) can be summarized as follows:

> Group together black pixels by connected component analysis, producing a list of connected components with attributes such as size, black-pixel density, ratio of dimensions, area, position, and center of gravity.

> Detect collinear components by applying the Hough-transform technique to the center of gravity of the connected components at increments of one degree. Because of size and location differences among different characters, the centroids of the connected components of a given text string may not be precisely collinear. An algorithm has been developed to solve the resolution problem of Hough projection so as to avoid either over-grouping or under-grouping.

> To locate words and phrases, the positional relationships among connected components are examined. This is done by examining the intercharacter-gap and interword-gap relations.

> All connected components of the text strings found by Hough-transform analysis are deleted from the overall list of components. The remaining connected components are considered to be graphics.

Examples of this method from (Kasturi et al./1986) are illustrated in Fig. 1.7. A. Rastogi and S. N. Srihari (Rastogi and Srihari/1986) also used the Hough transform to determine text-block structure, and interword and interline spacing by detecting regular spacing of peak-to-valley features in Hough space.

(B)   Rule-Based and Link-Clustering Classifications

Text strings can often be determined by size, using the component-labeling method. However, if characters touch a line, as often occurs in the case of line-drawing graphics, a two-stage segmentation method (Kubota et al./1984) can perform character segmentation. The first segmentation stage, using connected component size, is quite effective as a simple and fast algorithm. The second step is based on a measure called the neighborhood line density (NLD) (Fig. 1.8), which defines the complexity of an image at every black pixel. The two-dimensional distribution of this quantity highlights character patterns appearing in a graphical region, since higher

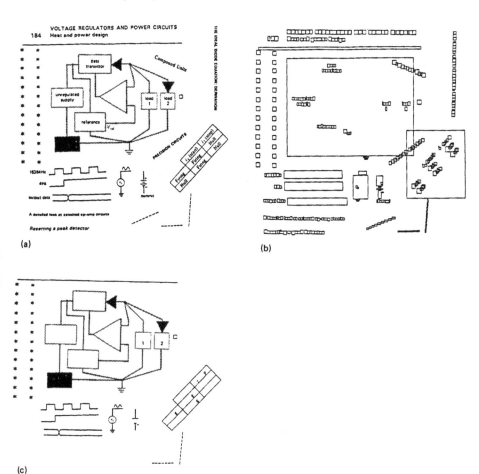

Figure 1.7 Extraction of text from an image using the Hough transform. (a) A scanned line drawing, (b) connected components derived from (a), and (c) the figure after text located by Hough-transform analysis has been removed. The transform facilitates the detection of strings of characters. Note that a character in contact with the drawing does not give rise to a separate connected component and, thus, is not detectable by this method. Source: (Kasturi et al./1986)

NLD peaks appear in the neighborhood of character patterns. A local pattern-analysis procedure then performs the actual segmentation. The NLD method has been used in a prototype document-recognition system (Kida et al./1986).

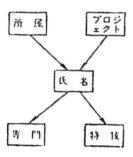

Figure 1.8 Neighborhood line density (NLD). The three-dimensional plot shows NLD for the upper drawing. Complex shapes such as characters give rise to peaks that can be detected in the plot. This property holds even for characters that touch the line drawing, so such patterns can be located using NLD. Source: (Kubota et al./1984)

Another approach for block classification is the link-clustering rule-based procedure. Component labeling is used to do an initial segmentation. A single link (or nearest neighbor) clustering procedure then merges character parts into single characters based on distances between the centroids of surrounding rectangles. In the same manner, characters are linked into words, words into lines, and lines into paragraphs (Doster/1984). In (Meynieux et al./1986), proximity relationships in the horizontal, vertical, and 45° directions are used for the merging of the segments. For each component, internal information (e.g., position, surrounding rectangle, eccentricity, surface) are then evaluated to assign a category.

1.5   TEXT RECOGNITION

Text recognition is the process of mapping an image text block in bit-map format into a machine character-processing format; e.g., ASCII. The term *text recognition* implies that the system uses known properties of language and of document organization in its processing, as opposed to the more conventional OCR, which is based strictly on the shapes of individual character patterns. Text recognition calls for combining OCR with contextual information.

In the following discussion, only the recognition of machine-printed characters in Latin fonts will be considered. Current objectives in document-analysis systems are mainly to read machine printing, although hand-print and script recognition will be included as these techniques improve in accuracy and speed. Recognition of Japanese and Chinese printing, mainly because of the large alphabets involved, requires special techniques that will not be considered here.

Optical character recognition is the most commercially advanced of the functions required for document analysis. However, document analysis presents new challenges. Much broader capability is required in a system that is designed to accept a wide range of character styles. The class of potential users varies from specialists who perform a service for others, to personal users who seek to add information to their own records. Certain trends in automatic text reading are discussed below:

> Increased tolerance to font variations
> Removal of restrictions on print format
> Increased use of software rather than hardware in implementation
> Simplification of the user interface

The following sections will discuss these topics in turn.

1.5.1   Font Capability

The most commonly used typewriter escapement advances a fixed increment at each key stroke and, thus, prints characters spaced at equal intervals. Several manufacturers dominate the typewriter field, and a small number of typing fonts have gained preeminence for commercial and personal communication. These factors permit recognition of typewriter fonts to be done accurately by commercial OCR machines. It is estimated that a machine capable of reading the six most common typing fonts is applicable to about 85% of business correspondence.

In the period before personal computers (PCs) and electronic printing began to replace specialized word processors, typing was a special niche market for optical character recognition. The uniform print spacing and small number of fonts required simplified design of the OCR device. Rough drafts could be created on ordinary electric typewriters and fed through an OCR device to a word processor for final editing. In this way, a word processor could support the output of five typewriters or more, and the overall cost of equipment for a word-processing department might be cut by as much as 50%.

We are only now emerging from this period in the development of OCR readers (Schurman/1982). The commercial market offers primarily machines that read a dozen or so typing fonts. This capability has been expanded in some cases to include output from matrix printers, still regularly spaced, but often of poorer quality than typing.

The few machines that read the typeset fonts used in books, newspapers, and journals are priced much higher than the above devices, currently in the $30,000–$40,000 range. In the near future, however, we can expect to see changes in this area. OCR manufacturers are seeking to enlarge font capability, and the demand for typewriter input to word processors has faded. There is increasing need for systems to build bibliographic databases or to read submitted manuscripts for editing and publication (Baird et al./1986, Kahan et al./1987). With improved techniques, it is undoubtedly with typeset printing that the next generation of OCR machines will operate.

To have wide applicability, the character-recognition capability of document-analysis systems must be flexible. Recent approaches to achieving accurate transcription of printing in arbitrary fonts have attempted to take advantage of the contextual nature of the data. Dictionaries and grammar models can serve as useful tools to help improve recognition rates.

This approach can be carried further when a text document containing a large sample of each font is read. In this case, it has been shown that virtually all the letters (i.e., numbers and punctuation excluded) can be read on the basis of contextual principles alone, with no parameters contained in the OCR system to describe the pattern structure of the different character classes. Since many contextual properties (e.g., statistical distribution of letters and words) are independent of the font and, in fact, vary little from document to document, such a system can encode text regardless of printer, font, scanner, and other variables (Casey/1986).

1.5.2   Format Restrictions

The format, or arrangement of printing on a document, offers two sets of problems for optical character recognition:

> Acceptance—the problem of reading character data in a particular format.

Recognition—the problem of recognizing the format itself and of encoding it along with the text, just as formatting codes are embedded in computer text files during creation of a document.

Examples of the former are the use of underscoring or bold lettering to accentuate text, font changes, and the inclusion of pictorial matter or annotation along with the text. Recent commercial machines detect underscores and are tolerant of bold printing; however, extraneous material on the document tends to interfere with the reading process. This is particularly true of annotations or signatures that may contact or come close to the main text. We have already discussed the discrimination of text from nontext.

Use of slanted italics or of proportional spacing offers another example of difficulties in character arrangement. The implied model of printed text in optical character recognition is a sequence of characters, each occupying a rectangular space. The first task of recognition is to locate these rectangles and to pass them in sequence to a classifier that identifies the contents. Italic fonts tend to violate the model by tilting over into the rectangles of adjacent characters. Proportional fonts introduce even more severe spacing difficulties: because the printing is compacted, adjacent characters may actually touch at the borders, and since character widths vary, there is no a priori rule that will in all cases segment the merged region properly. Humans employ trial segmentation, character recognition, and contextual principles (e.g., word recognition) in combination to read such text. Research has been done along these lines as well (Casey and Wong/1985, Casey/1986). Steps have been taken to recognize formats; i.e., to automatically provide control codes to describe print arrangement. Among the features provided by certain commercial machines are:

Detection of two-column text
Paragraphing
Spacing of lines
Proper alignment of tabular printing
Indented lists

We can expect expansion of these capabilities, as well as the use of contextual principles to identify headings, address fields, subject classification, etc., assuming that the OCR device is given knowledge of the document class.

1.5.3   Hardware Versus Software

The first recognition machines were hard-wired devices. Because electronic circuitry, and particularly scanning hardware, was expensive, throughput rates had to be high to justify the cost, and so extensive parallelism was used.

As the costs of electronics have decreased, the requirement for special circuitry has subsided. Microprocessors, available in a range of speeds and prices, have gradually taken over the control functions and have been incorporated into the recognition process itself. The high-end Palantir OCR system, for example, employs five Motorola 68000 microprocessors to achieve its omnifont capability.

PC-attached document scanners are now sold in the low thousands of dollars. These devices, acquired primarily to support image applications, tend to promote the development of recognition software. Speed requirements are not high in personal applications; a recognition time on the order of a minute per page, corresponding to perhaps 40–50 characters per second, is ordinarily acceptable. Software also offers easy extensibility: as algorithms are improved or new capability is added, existing systems are easily updated. Cost is low, since no additional equipment outlay is required beyond the already existing image system. Finally, if high-level language is used, the OCR function can be implemented in a variety of computers and can remain viable as new computers are introduced.

Hardware-based systems are still in evidence at present. They predominate in systems that employ template matching or complex features, which are highly amenable to parallel processing and at the same time are computation intensive when implemented by general-purpose computers. However, software products with comparable characteristics are appearing.

The attractiveness of software processing can only be enhanced by the trend toward text input from arbitrary print sources. The need to deal with a wide variety of formats, and the constant appearance of new font styles, call for greater flexibility than ever in the reading system. Eventually, AI techniques such as rule-based operation may be adopted to achieve a dynamic structure easily adapted to changes in the nature of the applications.

1.5.4 User Interfaces

Early OCR systems were operated by trained personnel. In fact, hand-print systems were fed input from trained writers! This was a natural effect of the deployment of high-speed, special-purpose, expensive apparatus, and is consistent with other such systems. Early word processors were used only by experienced operators, whereas now editing and formatting programs such as WordStar are sold for office and home use.

In the same way, OCR, especially as applied to personal computing, will be available to unsophisticated users. These users may perhaps be conversant with computers in general, but will desire only to perform particular tasks by means of optical character recognition, and will not want to involve themselves in details of processing. They will react negatively to failures that occur for no readily apparent reason.

It will be the task of the user interface to extract information helpful for recognition in a way that is not painful to the user. There is a history of disconcerting clients due to poor performance on slightly degraded documents, inappropriate output resulting from unfamiliar formats, etc. Present systems show awareness of this history. They stop reading a document when a high frequency of rejected characters is obtained and post a message to the user, in the belief that it is better to acknowledge that the program is in difficulty than to offer garbled output with the same confidence as recognized text.

Character recognition, then, is becoming an interactive process, ideally requiring less and less interaction as software capabilities improve, but always capable of benefiting from user feedback. One interesting area, not explored to date, is feedback between a postprocessing editor and the OCR logic. When the user verifies text and corrects errors in postprocessing, the system has an opportunity to learn where its weaknesses lie. If many s's are being edited to a's, for example, perhaps discrimination can be inclined toward the latter class by an internal change of parameters.

1.5.5 OCR Status

We can summarize current and projected OCR capabilities as follows:

If a printed document is a fixed-pitch, typed original or a clean copy, in a simple paragraph format and in a common typing font, then it can be read with good accuracy (say, 99% or better) by available commercial products. Poor copies or degraded printing will yield less accuracy.

If the document is in a typeset font, then it can be in one of hundreds of possible styles. Even assuming that the characters are well-spaced, so that isolation of characters for classification is reliable, design of a reader to recognize a vast library of fonts is challenging. Currently, only a few expensive systems ($30,000–$40,000) can perform this task. Attempts to achieve low-cost performance are centered on using adaptive systems that employ lexicons and other contextual aids, as well as user assistance, to achieve sufficient accuracy.

Finally, closely spaced printing and the presence of nontext data and noise increase the difficulty of the above processes. Increased use of context and operator assistance is required for these cases.

1.6 GRAPHICS RECOGNITION

Encoding of graphic information is not required in all applications. In text-based systems, graphic fields may simply be stored as bi-level image data for later reprinting, possibly at a different scale or orientation. However, in other applications, the emphasis is on coded graphics; for example, in providing input of maps or engineering drawings to CAD. There are also intermediate problems, such as the creation of databases from chemical catalogs and other literature, in which text and diagrams share equal importance.

In discussing the processing of graphic images, we will be concerned only with line drawings (although the drawings may contain shaded or solidly colored regions). Images such as halftone pictures are not ordinarily subjects for recognition in a document-analysis system. Indeed, one of the objectives of pattern recognition is to eliminate nuance, the very opposite of what is usually desired in image representation. Thus, a document-analysis system for general application will store halftone images as such and will process only line drawings.

Line drawings are encodable at several levels of understanding and may, in fact, progress through these levels in the encoding process. At higher levels, the encoding is appropriate to a particular application, while at the lowest level, a geometric representation is derived, suitable for recreating the original drawing to good approximation or for providing input to a variety of applications.

The low-level encoding provides compaction as well as a primitive representation of the structure of the drawing. For example, an engineering sketch of medium complexity on A4 paper contains about 6 million pixels when scanned at 0.1-mm resolution. Thresholded to a bi-level representation, the drawing can be packed, 8 pixels per character, into 750,000 characters of data. Conventional compression techniques can condense this raster data by a factor of 10–20. However, analysis of the figure into vector-line segments and character codes can yield another fivefold improvement, to a reduction by 50–100 over the original, or on the order of 10K characters. This alone can be an important consideration in a system that stores thousands of such documents or transmits them to remote locations.

Higher levels in the hierarchy of representation call for recognition of certain combinations of lines as graphic symbols, for recognizing the interconnectedness of the symbols, and for associating character data on the drawing with the graphic elements.

In the next section, progress in the automatic encoding of drawings will be discussed. The presentation starts with the low-level vector-encoding process and later considers how this derived information is used in the recognition of graphic symbols.

1.6.1 Vector Encoding

It will be assumed that any character patterns have been filtered out and the pattern region being processed contains only graphic (i.e., noncharacter) information. The problem is to approximate the region as a set of line segments, or vectors.

Solid-colored components of the figure may be detected and separated from the rest in a simple way, by following the boundary of the image of the drawing after a filtering operation to remove thin lines (Harada et al./1985). A closed contour on the thinned image identifies a solid figure. By reference back to the original figure, the contour of the solid region can be extracted and the connectivity to the rest of the drawing ascertained. The solid object itself can be either masked out of the image or ignored in later processing.

The remainder of the image consists of interconnected lines and curves. Some graphic regions may be so complex in shape and interconnection pattern that approximation by vectors will tend to sacrifice detail needed for later recognition of the structure as a whole. (This occurs often when character patterns are represented as vectors, for example.) Such regions, if any, have to be detected and treated by other means.

The image to be analyzed into vectors, then, may be considered to consist of interconnected straight lines and moderately curving arcs with a low density of line crossings. Two main approaches have been defined for obtaining vector approximations of such line patterns: skeletonization and line tracking.

(A)   Vectors from Skeletons

The skeleton of a line pattern is a pixel distribution that is one pixel in thickness and lies wholly within the original line. Ideally, the skeleton lies at the center of the line pattern from which it derives, but the algorithms generally used for skeletonization tend to depart from the ideal at junctions or in noisy regions. Skeletonization is done by iteratively removing edge pixels that can be deleted without destroying connectivity. Line thickness can be measured easily at the same time.

Once having obtained a skeleton, vectors can be derived by segmenting at appropriate points along the figure (Maeda et al./1985, Harada et al./1985). Junctions are usually selected as a subset of the segmentation points. The degree of segmentation determines the accuracy of the resulting vector representation. Topology (i.e., interconnection) is readily obtained by this method. However, errors can occur due to poor quality images (breaks in the scanned version of a continuous line, for example) or to artifacts such as spurious line segments introduced during skeletonizing.

(B)   Line Tracking

This method consists of following the line on the original image, while sequentially determining both points of intersection and points at which to segment the line representation (Fig. 1.9). There are a number of tech-

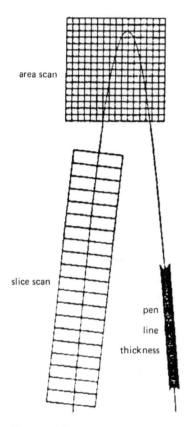

Figure 1.9   Line tracking. The figure shows a grid that is searched for continuations of the curve being tracked. A finer grid is required in regions of sharp curvature
to avoid confusion of the line with neighboring patterns. Source: (Clement/1981)

niques of this class (Maeda et al./1985, Black et al./1981). Some researchers
use an initializing process to establish line direction, then predict ahead to
detect continuations of the line. In this way, not every pixel of the image
need be examined during the vectorizing process, although care must be
taken to detect junctions or sharp corners in the figure being tracked. Once
detected, these occurrences are treated by exception routines.

1.6.2   Higher Level Representation

The vectors obtained by operations of the sort just described may play one
of several roles in the figure being analyzed:

They may combine to form a symbolic element, such as a circuit element or a chemical structure.

They may serve to interconnect symbolic elements; for example, signal lines in a circuit diagram or highway lines between cities on a map.

They may participate in a description or annotation of the drawing proper, as, for example, the dimension lines on an engineering drawing, or an arrowed line relating a textual note to a particular part of the drawing.

Higher level processing consists of identifying these roles and vector combinations in the light of a particular application. The process consists of a series of steps. First, a candidate region is detected on the basis of gross features. Second, the criteria that the vector combination must satisfy in order to belong to each possible symbol category are checked. Finally, a recognized symbol is recorded in the system output tables, along with its interconnection with other drawing components.

An example of this classification process will be seen in the discussion of experimental document-analysis systems.

1.6.3   Status-Graphic Encoding

This technology, driven by the need for input to CAD systems, has begun to reach commercial application. Experimental systems have been developed and applied to special tasks such as circuit diagrams. A few commercial systems are available at high prices.

It appears at present that if care is taken in the preparation of drawings to meet specifications that ease the task of automatic input, then scanner input is accurate and fast. In tests, such a system has doubled the entry rate of manual methods, even taking into account the greater preparation cost. On the other hand, the long-standing problem of converting archives of maps and drawings (many of doubtful quality after frequent manual updates), a task beleaguering the utility and manufacturing industries for many years, can be considered as yet unsolved.

1.7   PAGE-STRUCTURE ANALYSIS

Page-structure analysis is a higher level document understanding than are block segmentation and classification. It attempts to determine an overall block structure of a page; e.g., grouping different segmented image blocks into paragraphs, figures, title, summary, heading, multicolumn printing, diagram, etc. In document printing, we use control commands to lay out a page structure, and the control commands are interpreted by a formatter. Page-structure analysis, in some sense, is the inverse problem of interpreting format-control commands. It attempts to find or understand the control commands that could have been used for laying out image documents.

Document structure varies from one type of document to another. It is not practical, nor easy, to develop a general system to analyze all types of documents *automatically*. Each document-analysis system has to focus on certain chosen types of document that are mostly needed by its application. The a priori knowledge of the target types of document is then entered as decision rules into the system. Examples of specialized applications for document-analysis systems include envelope-address reading (Srihari et al./1985) and Japanese newspaper layout (Inagaki et al./1984).

It should be noted that a human can recognize the structure of a page immediately after it is displayed. For a document-analysis system to cope with general classes of documents, a provision for interactive page-structure identification by the user will be extremely useful and powerful.

Although document structure varies a great deal from one type of document to another, the use of basic block features—such as block height, eccentricity, density over enclosing box area, (x, y) position, centroid, etc., together with a rule-base system—offers a powerful general approach for understanding document structure. For each *specific* document type, the user has to input the a priori knowledge of the document structure via the rule-based system. The rules define the general characteristics expected of the usual components of a document image and their usual relationships; e.g., the title being above the author names, the abstract being above the first paragraph of text, etc. Rules are constructed about the various identified blocks from a given document image that can be used by the inference engine of the rule-based system. This approach has been demonstrated successfully in understanding the structure of mailing envelopes (Srihari et al./1985) and other document types (Niyogi and Srihari/1986).

Another important approach for understanding document structure is the relaxation method, which has been used for image-edge detection and thresholding (Rosenfeld and Kak/1982). The basic relaxation algorithm for document structure understanding is as follows:

1. Specify a set of possible labels (and their associated confidence levels) for each block.
2. Compare related picture blocks and delete labels or adjust confidences to reduce inconsistencies.
3. Iterate step 2 as many times as necessary.

The relaxation method was also used in (Srihari et al./1985) for envelope-structure recognition. One of the difficulties with the relaxation method is to determine the level of consistency for stopping the iteration.

An event-driven relaxation method was used in (Inagaki et al./1984) for newspaper-structure analysis. The relaxation method used differs from the conventional one in the following points:

1. The event detected by a consistency-checking process is not localized to its neighborhood, but propagates its effects to all the labels that it can affect during one iteration cycle of relaxation.
2. Although the relaxation is done on the table of labels, the detection of an event initiates a process that reexamines the original image in detail; e.g., from global to local examination.

The event-propagation technique has been shown to reduce the number of iteration steps remarkably.

1.8 DOCUMENT FILING AND RETRIEVAL

After documents have been created, some sort of file-management facility is needed to allow the user to collect the documents into piles and to put these piles away. There are typically two levels of file organization: file folders and file cabinets. A file folder can contain instances of documents either all of the same type or of different types. A file cabinet contains various file folders. There are many different ways in which documents in a file folder and the file folders in a file cabinet may be organized. A document-management system must allow the user to choose which way he wants to organize his documents.

When a document is filed away in a document-management system, it is necessary to be able to retrieve it at some later time. Typically, two types of retrieval patterns are observed. In one case, the user is not quite sure of what he (or she) is looking for; in which case, he needs to scan or browse a number of documents. On the other hand, he may have an idea of what he is looking for, but tends to be vague when he formulates his request (i.e., fuzzy query). Both of these retrieval methods must be supported in a document-management system (Woo et al./1985).

The design of a complete document-filing and retrieval system is very complex, and it is beyond the scope of this chapter to give a reasonable survey of the techniques used. A recent book devotes several chapters to the subject of document-filing and retrieval systems (Tsichritzis/1985). We shall briefly mention some of the system-design issues here. A document-filing and retrieval system can be considered to have two layers: (1) document-storage subsystems, and (2) user interfaces, as shown in Fig. 1.10. Some of the document-storage-subsystem design issues are: document manager for folders and files, storage-access methods, database, single user or shared system, different levels of document-access control, workflow, communi-

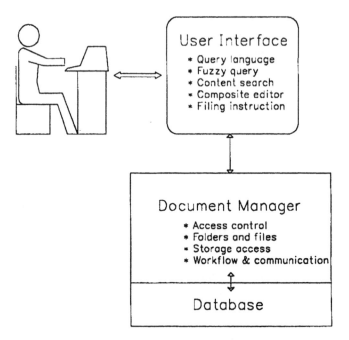

Figure 1.10   Document-filing and retrieval system. The system is divided into a subsystem that responds to a user and a subsystem that deals with the actual data.

cation, etc. The next level above the document-storage subsystem is the user interface, where one addresses issues such as: query languages, fuzzy queries, content search (mainly text), composite editor (for text, graphics, and images), multiple windows, menu-driven screens, document-browse facilities, filing of documents with specific instructions from the user or with interactive suggestions from the systems, etc.

1.9   EXAMPLE SYSTEMS

As examples of document-analysis methods, two experimental systems are presented, one directed at text documents, the other intended for conversion of line drawings to CAD formats.

1.9.1   Text Document Analysis

The document-analysis project of the author and colleagues at IBM Research Laboratory was conceived as a vehicle for assisting in the preparation for publication of submitted manuscripts (Wong et al./1982). It may be compared with other experimental systems intended for similar applications.

All functions after scanning are implemented in software and are essentially independent of scanner resolution and other characteristics. Several different scan sources, having resolutions from 0.125 mm to 0.04 mm, have been used for input. Scanned images are stored as compressed files and processed by a series of analysis routines.

Initial segmentation into text, halftones, or graphic fields can be accomplished in two different ways. Automatic block segmentation and region classification is performed by a method based on smearing the run-length-encoded image (Wahl et al./1982). The classification uses simple features such as average run-length, density of black pixels, etc. A second option is operator selection of portions of the image desired for further processing, using a two-display image editor with joystick control.

At present, graphic fields are merely separated and either stored or discarded. Future plans call for investigation of graphic encoding, probably with respect to a particular application.

Most of the effort to date has focused on text recognition. This is accomplished by means of a pattern-matching scheme that uses an adaptive decision tree as a preclassifier to speed up the matching process (Casey/1984). The system initially groups character patterns into different shape classes (Fig. 1.11); identification of the groups is done in a later step. Cryptographic techniques using a large dictionary are combined with display and manual key input to complete the recognition process (Casey/1986). The method is font-independent, and special combined recognition/segmentation techniques (Fig. 1.12) have been developed to deal with proportionally spaced text (Casey and Nagy/1982).

Practical applications to date have included entering correspondence, resumes, etc., into computer files; restitution of lost computer files from printouts; and reading published computer programs.

1.9.2 Circuit Diagram Input

Automatic encoding of line drawings to a symbolic form for further processing has been demonstrated by, among others, Toshiba Corporation in Japan (Okazaki/1985, Kawamoto et al./1984). The Toshiba system is designed to convert manually prepared circuit diagrams to a CAD format as part of the process of VLSI (very-large-scale integration) design and development (Fig. 1.13). It is an alternative to manual digitization and, thus, offers a comparison between manual and automatic input methods.

Scanning is done at 0.1-mm resolution on a revolving-drum scanner capable of imaging large documents and is followed immediately by connected-

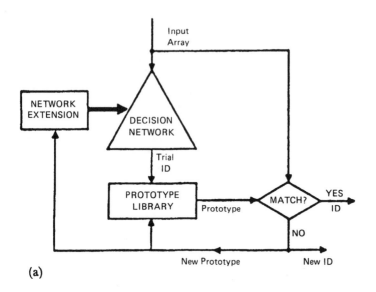

(a)

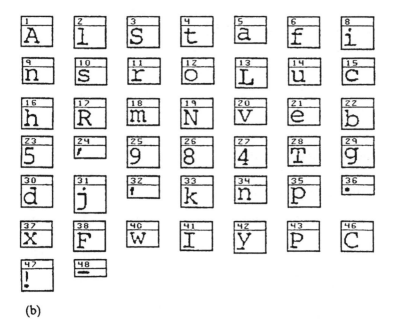

(b)

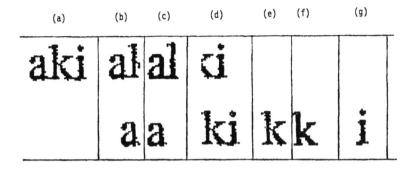

Figure 1.12   Recursive segmentation of touching characters. The technique employs a window whose left edge is fixed at the left of the input pattern (a), while the width is decreased step-by-step. The pattern in the window is matched against the prototype library. In the top row, a partial match results when the left side of the input (b) matches an *al* prototype (c). However, the remainder of the pattern (d) cannot be matched, and so the first match is nullified. Continuation of the process (second row) finds matching *a, k,* and *i* prototypes (*a-g*), after which the entire input has been matched. Below are shown the three prototypes found to be equivalent to the input.

component analysis implemented in special circuitry. Character images are separated from the rest of the line drawing on the basis of the size of the connected component and are recognized by OCR. The connected-component operation is rerun on the inverse image to locate closed contour patterns, which represent gate symbols. A combination of feature analysis and tem-

Figure 1.11   Formation of shape classes by pattern matching. (a) The matching process. Patterns are first seen by a decision network that examines the color of judiciously chosen pixels in order to select a likely matching prototype from the library. If a matching prototype is found, the identity (or ID) of the prototype is output. Otherwise, the unknown pattern is stored in the next vacant position in the library, and the sequence number for this location is output as its ID. The decision network is modified automatically to provide selection logic for a new prototype. (b) A sample prototype library assembled from a single-page typed document.

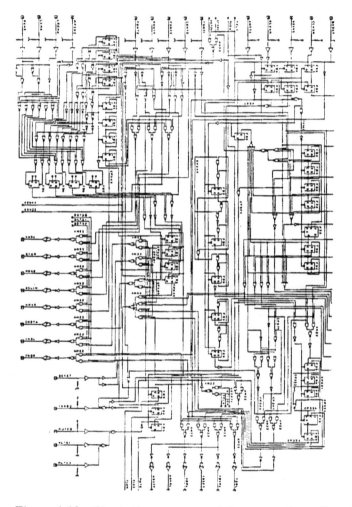

Figure 1.13 Circuit diagram prepared for automatic reading. Certain conventions are followed in the drawing of lines and symbols, but the resulting drawing still presents an imposing task for automatic conversion. Source: (Okazaki et al./1987)

plate matching is used to classify these subimages into various shape categories (Fig. 1.14).

The remainder of the scanned image is processed to locate line intersections, to detect other symbol types such as resistors and capacitors, and to track connecting lines. All classification results plus associated positional data are combined in a postprocess that recognizes gate types, determines

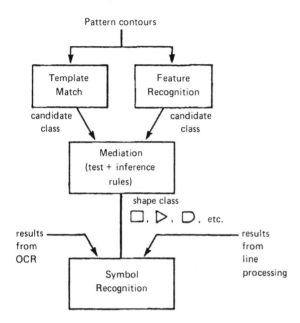

Figure 1.14 Recognition of logic symbols in the Toshiba system. Classification is done by two distinct systems, one using templates, the other using features, and conflicts are resolved by subsidiary tests and inference rules. Source: (Okazaki et al./ 1987)

the interconnection of components, and yields an output in an appropriate CAD format.

The system was tested on hundreds of CAD drawings, yielding over 95% accuracy. Including an editing stage to detect and correct errors, it roughly halved the time required for input compared with a manual procedure using a graphic digitizer.

1.10 CONCLUSIONS

The widespread availability of personal computers and the low cost of current scanning technology have resulted in a rising popularity of image systems for desktop publishing, facsimile transmission, and other applications. By adding intelligent processing functions to these systems, a document-analysis system can be constructed, with new applications possible in, for example, text search and retrieval and interaction with automated design systems.

Three major functions are needed to achieve the potential of document analysis. A segmentation routine initiates the process, partitioning the scanned document into text, graphic, and halftone image areas. Halftone areas are set aside for treatment by conventional image-processing operators, whereas the text region is encoded by OCR methods, and line drawings are analyzed by vector-encoding and symbol-recognition routines.

Applications of pattern recognition have been successful in the past only in narrowly defined problem areas. However, all of the functions required for document analysis have been demonstrated in research studies, and a number of complete system prototypes are currently in development. It remains to be seen whether a document-analysis system having broad function and applicability will attain the performance and acceptance needed to reverse past history.

REFERENCES

Baird, H., et al. (1986). "Components of an Omnifont Page Reader," 8th PRIPS, Paris, pp. 344–348.

Black, W. et al. (1981). A general-purpose follower for line-structured data, *Pattern Recognition, 14, no. 1*: 33.

Casey, R. (1984). "Texture Models for Document Segmentation," CESTA Premier Colloque Image, Biarritz, pp. 571–576.

Casey, R. (1986). "Text Recognition by Solving a Cryptogram," 8th PRIPS, Paris, pp. 349–351.

Casey, R. and Nagy, G. (1982). "Recursive Segmentation and Classification of Composite Character Patterns," 6th PRIPS, Munich, pp. 1023–1026.

Casey, R. and Wong, K. Y. (1985). "Text Recognition Using Adaptive Software," Proceedings of Globecom '85, New Orleans, pp. 353–357.

Clement, T. (1981). The extraction of line-structured data from engineering drawings, *Patt. Rec. 14, no. 1*: 43.

Computer Buyer's Guide and Handbook (1986) "Computer Information Publishing Co.", New York, pp. 53–58.

Doster, W. (1984). "Different States of a Document's Content on Its Way from the Gutenbergian World to the Electronic World," 7th ICPR, Montreal, pp. 872–874.

Foith, J. et al. (1981). Real-time processing of binary images for industrial applications, *Digital Image Processing Systems* (L. Bolc and Z. Kulpa, eds.), Springer-Verlag, New York, pp. 61–168.

Harada, H. et al. (1985). "Recognition of Freehand Drawings in Chemical Plant Engineering," Proceedings of IEEE Workshop on Pattern Architecture and Image Database Management, p. 146.

Huang, T. S. (1977). Coding of two-tone image, *IEEE Trans. on Comm., Com-25, no. 11*: 1406–1424.

Hunter, R. and Robinson, A. H. (1980). International digital facsimile coding standards, *Proc. of the IEEE, 68, no. 7*: 854–867.

Inagaki, K., Kato, T., Hiroshima, T., and Sakai, T. (1984). MACSYM: A hierarchical parallel image processing system for event-driven pattern understanding of documents, *Patt. Rec. 17, no. 1*: 85–108.

Kahan, S., Pavlidis, T., and Baird, H. (1987). On the recognition of printed characters of any font and size, *IEEE Trans. PAMI*: 274–288.

Kasturi, R., Shih, C., and Fletcher, L. (1986). "An Approach for Automatic Recognition of Graphics ," 8th PRIP, Paris, pp. 877–879.

Kawamoto, E. et al. (1984). "Schematic Entry System Equipped with an Automatic Reader," Proceedings of ICCAD '84, pp. 87–89.

Kida, H., Iwaki, O., and Kawada, K. (1986). "Document Recognition System for Office Automation," 8th ICPR, Paris, pp. 446–448.

Kubota, K., Iwaki, O., and Arakawa, H. (1984). "Document Understanding System," 7th ICPR, Montreal, pp. 612–614.

Langdon, G. G., Jr. and Rissanen, J. (1981). Compression of black-white images with arithmetic coding, *IEEE Trans. on Comm., Com. 29, no. 6.*

Maeda, A. et al. (1985). "Application of Automatic Drawing Reader for the Utility Management System," Proceedings of IEEE Workshop on Pattern Architecture and Image Database Management.

Masuda, I., Hagita, N., and Akiyama, T. (1985). "Approach to Smart Document Reader System," Proceedings of the CVPR Conference, San Francisco, pp. 550–557.

Meynieux, E., Seisen, S., and Tombre, K. (1986). "Bilevel Information Recognition and Coding in Office Paper Documents," 8th ICPR, Paris, pp. 442–445.

Morrin, T. H. (1974). A black-white representation of a gray scale picture, *IEEE Trans. on Computers*, vol. 23, p. 184.

Nagy, G. (1983). Optical Scanning Digitizers, *Computer*, pp. 13–24.

Niyogi, D. and Srihari, S. N. (1986). "A Rule-Based System for Document Understanding," Proceedings (AAAI-86) of the Fifth National Artificial Intelligence Conference, Philadelphia, p. 789.

Okazaki, A. et al. (1988). "An automatic circuit diagram reader with loop-structure based symbol recognition," *IEEE Trans PAMI*, May 1988, pp. 331–341.

Pratt, W. K. (1978). *Digital Image Processing*. John Wiley & Sons, New York, pp. 523–525.

Rastogi, A. and Srihari, S. N. (1986). Recognizing textual blocks in document images using the Hough transform, TR 86–01, Dept. of CS, SUNY at Buffalo.

Rosenfeld, A. and Kak, A. C. (1976). *Digital Picture Processing*, Academic Press, New York, pp. 347–348.

Rosenfeld, A. and Kak, A. C. (1982). *Digital Picture Processing* (2nd ed.), Academic Press, vol. 2, chapter 10.

Schurman, J. (1982). "Reading Machines," 6th PRIPS, Munich, pp. 1031–1044.

Somerville, P. (1984). Use of images in commercial and office systems, *IBM Systems J., 23, no. 3*: 281.

Srihari, S., Hull, J., Palumbo, P., Niyogi, D.; and Wang, C. (1985). Address recognition techniques in mail sorting—research directions, Research Report 85-09, Computer Science Dept. University at Buffalo.

Srihari, S. N. (1986). "Document Image Understanding," Proceedings of ACM-IEEE Computer Society '86 Fall Joint Computer Conference, Dallas, Texas.

Tsichritzis, D. (ed.) (1985). *Office Automation*, Springer-Verlag, New York.

Wahl, F. M. (1983). A new distance mapping algorithm and its use for shape measurement on binary patterns, *Computer Vision, Graphics, and Image Processing 23*: 218–226.

Wahl, F. M. and Wong, K. Y. (1982). An efficient method of running a constrained run length algorithm in vertical and horizontal direction on binary image data, IBM Research Report RJ3438.

Wahl, F. M., Wong, K. Y., and Casey, R. G. (1982). Block segmentation and text extraction in mixed text/image documents, *Computer Graphics and Image Processing 2*: 375–390.

White, J. and Rohrer, G. (1984). Image thresholding for optical character recognition and other applications requiring character image extraction, *IBM J. Research and Development, 27, no. 4*, pp. 400–411.

Wong, K. Y., Casey, R. G., and Wahl, F. M. (1982). Document analysis system, *IBM J. of Research and Development, 26, no. 6*: 647–656.

Woo, C. C., Lochovsky, F. H., and Lee, A. (1985). Document management systems, *Office Automation* (D. Tsichritzis, ed.), Springer-Verlag, New York.

Wong, K. Y. (1978). Multi-function auto thresholding algorithm, *IBM Technical Disclosure Bulletin 21, no. 7*: 3001–3003.

Zen, H. and Ozawa, S. (1985). "Extraction of the Fair Document from Mixed Mode Manuscript," Proceedings of the CVPR Conference '85, San Francisco, pp. 544–549.

2

A Graphics-Recognition System for Interpretation of Line Drawings

Sing-tze Bow[*] and Rangachar Kasturi

Pennsylvania State University
University Park, Pennsylvania

2.1 INTRODUCTION

An automatic graphics-recognition system that can generate a succinct description of various graphical primitives and their spatial relationships has many applications. Conversion of paper-based engineering drawings for integration into a CAD/CAM (computer-aided manufacturing) environment, efficient storage and transmission of mixed text/graphics data, and creation of updated versions of existing documents are a few examples of these applications (Bley/1984, Bunke/1982, Chen et al./1980, Ejiri et al./1984, Huang and Tou/1986, Karima et al./1985, Wahl et al./1982). In this chapter, we describe a complete system that has been designed to generate a description of the contents of paper-based line drawings in terms of various graphical primitives, such as polygons, circles, etc., and their interconnections.

A typical graphic diagram consists of lines, polygons, circles, and other higher order curves. These entities have various attributes associated with them. Examples of such attributes are thicknesses of lines, filling details for polygons, etc. Often, these entities overlap with one another and obscure some of the lines. Graphics are typically annotated with text strings.

---

[*]*Present affiliation*: Northern Illinois University, De Kalb, Illinois

Paper-based documents containing mixed text/graphics data are digitized at a resolution of 120 pixels per centimeter. To obtain a graphics-description file of the highest possible level with minimum operator interaction, the following operations are performed on this binary image:

1.   Separation of text strings from graphics
2.   Identification of line segments and their attributes
3.   Recognition of graphical primitives and their attributes
4.   Description of spatial relationships among primitives

A robust algorithm for segmenting the image into text strings and graphics has been developed. This algorithm is described in Sec. 2.2. The text-string image is intended for further processing by a character-recognition system. The graphics image is processed by low-level image-processing routines to identify various types of lines. Thin entities such as lines are represented by their core lines, and thick entities (e.g., polygons that are completely filled) are represented by their boundaries. The line-detection algorithm is described in Sec. 2.3. The line segments are ordered into simple and complex loops using a number of heuristic rules. Simple shapes such as regular polygons are recognized. The complex loops are identified as simple polygons with overlapping objects, whenever possible. Various hatching patterns within closed loops are identified. The spatial relationships among various primitives are found. Large numbers of small primitives, if enclosed by a larger object, are identified as fillings (texture). These processing steps are described in Sec. 2.4. Curved lines are processed to identify circles. Curved lines that are not segments of circles are described as concatenated circular arcs. These steps are explained in Sec. 2.5. The description file generated by the system contains all the information to reconstruct the original data. Clearly, the principal departure of this system from conventional data-compression systems is that it generates a description of the contents of the graphics and not simply data in a compressed form for storage.

2.2   SEPARATION OF TEXT STRINGS FROM GRAPHICS

In this section, an algorithm for the separation of text strings from images of mixed text and graphics is briefly described. The algorithm is robust and separates text strings of various orientations, font styles, and character sizes. This segmentation algorithm is based on grouping collinear connected components of similar size and does not recognize individual characters. The algorithm is described in detail in (Fletcher and Kasturi/1988 and Gattiker/1988). There are five principal steps in this algorithm:

1. Connected-component generation
2. Area and ratio filters
3. Collinear component grouping in Hough domain
4. Separation of components into words and phrases
5. Postprocessing to refine the segmentation

These segmentation steps are described in the following sections.

2.2.1 Connected-Component Generation

The paper-based graphics are digitized at a resolution of 120 pixels per centimeter, resulting in a 2,048 × 2,048-pixel image. The connected components are isolated by raster-scanning this image and growing the components as they are found. This growing is done by a simple algorithm:

> Starting with any pixel in the connected component:

1. If any of the neighboring eight pixels are of the foreground value, then move to that pixel, changing its value to represent the direction of approach.
2. If there are no valid moves (all neighbors either have already been processed by the algorithm or are background), then tag the pixel and backtrack one step.

When the backtracking returns to the original pixel, every pixel of the component is of the tag value. The algorithm also keeps track of the top-, bottom-, left-, and right-most pixels corresponding to the smallest enclosing rectangle of each component and the percentage of pixels within this rectangle that are of the foreground type. The rectangles enclosing the components in Fig. 2.1 are shown in Fig. 2.2. This data is used by the other stages of the segmentation algorithm, thereby minimizing the operations on the large-image array.

2.2.2 Area and Ratio Filters

The area filter is designed to identify the components that are very large compared to the average size of the connected components in the image. In a mixed text/graphics image, such components are likely to be graphics. By obtaining a histogram of the relative frequency of occurrence of components as a function of their area, an area threshold that broadly separates the larger graphics from the text components is chosen. A priori knowledge about the types of documents being processed (such as the amount of text data compared to graphics), if available, is helpful in determining this threshold. For documents that contain a substantial amount of text and some large graphics, the most populated area will, in general, represent mostly text components or, at least, small graphics. By ensuring that

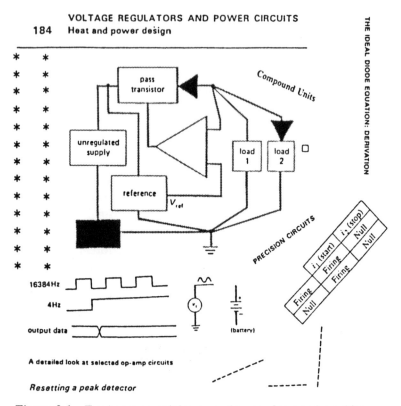

Figure 2.1   Test image containing several types of text and graphics.

the threshold is set above the most populated area, A_{mp}, the possibility of discarding members of the text character set is avoided. This alone is not adequate to process images that contain text strings of different sizes, since it is likely that the most populated area corresponds to the smallest characters. Thus, a second parameter, the average area A_{avg}, is computed. The area threshold is then set at five times the larger of the two parameters A_{mp} and A_{avg}. The histogram is searched to locate components that are larger than this threshold. Since text characters of the same size may have different areas, the histogram is blurred such that neighboring areas are grouped together to determine the most populated area.

Connected components that are enclosed by rectangles with a length-to-width ratio larger than 10 are not likely to be text characters. Such components (e.g., long horizontal or vertical straight lines) are marked as graphics and are eliminated from further consideration.

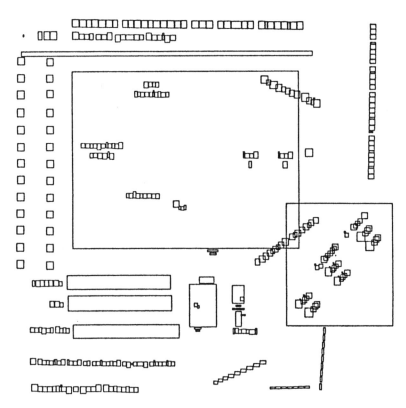

Figure 2.2   Smallest enclosing rectangles around connected components of image in Fig. 2.1.

The results of these filters applied to the components illustrated in Fig. 2.2 can be seen in Fig. 2.3. Notice that the components corresponding to the horizontal lines at the top of the image have been removed by the ratio filter, and the large central graphic, as well as the component representing the table, among others, has been removed by the area filter.

2.2.3   Collinear Component Grouping

We define a text string as a group of at least three characters that are collinear and satisfy certain proximity criteria. This part of the algorithm identifies collinear characters by applying the Hough transform to the centroids of the connected components. In our implementation, the angular resolution θ in the Hough domain is set at one degree, whereas the spatial resolution, ρ, is a variable that is set to $0.2 \times A_h$, where A_h is the average

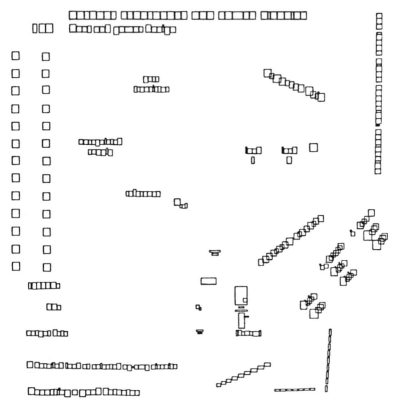

Figure 2.3 Component rectangles after area/ratio filtering.

height of the connected components. This resolution provides a threshold for noncollinearity of the components of text. Next, the Hough domain is scanned for extraction of the collinear components. The Hough domain is scanned first only for horizontal and vertical strings, then for all others. Further, for each of these scans, the Hough domain is searched to locate longer strings before locating shorter strings. This order of searching prevents incorrect segmentation of strings, as shown in Fig. 2.4.* When a potential text string is identified in the Hough domain, a cluster of cells centered around the primary cell is extracted. This is done to ensure that all

---

*The incorrect segmentation as illustrated in Fig. 2.4 occurs because components, once identified as belonging to a text string, are not considered to be a part of any other string. This is done to decrease the number of searches.

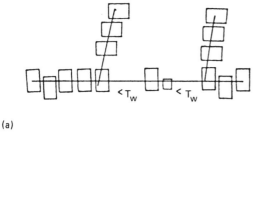

(a)

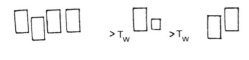

(b)

(c)

Figure 2.4 Effect of destructive line overlap: (a) a long text string connected to two shorter strings, (b) result of processing shorter strings first (word gap exceeds the threshold T_w), and (c) result after processing longer strings (phrases with less than three characters are not removed).

characters belonging to a text string are grouped together. The degree of clustering starts at 11 neighbors in ρ, but is modified subsequently to correspond to the average height of the extracted string (Fletcher and Kasturi/1988, Gattiker/1988).

To decrease the time spent scanning the Hough domain, a pyramidal reduction of the resolution is used. In this case, the reduction in resolution is 3 in ρ and 2 in θ, producing an array that is at each level one-sixth the size of the previous level. The method of reducing the resolution is a maximum operator, so that the maximum string length at the base is represented in the top level. There are five total levels in this implementation, reducing the resolution by a total factor of $6^4 = 1{,}296$. This means that the scanning of the Hough domain is reduced computationally by a factor of 1,296, but the individual string-extraction time will be increased. The worst case for descending the pyramid to the corresponding cell in the base Hough array is $4 \times 6 = 24$. This implementation of a pyramid system, therefore, presents considerable time savings as long as less than 1 out of 24 discrete cells requires descent. This condition is almost always satisfied in a typical document.

2.2.4   Separation of Components into Words and Phrases

After a collinear string is extracted from the Hough domain, it is checked by an area filter, which is similar to the one discussed earlier, so that the ratio of the largest to smallest component in the group is less than 5. This is necessary to prevent large components, which do not belong to the string under consideration (but having their centroids in line with the string), from biasing the thresholds such as intercharacter and interword gaps. The components are further analyzed to group them into words and phrases using the following rules:

1. The intercharacter gap $\leq A_h$
2. The interword gap $\leq 2.5 \times A_h = T_w$

Here, A_h is the local average height and is computed using the four components that are on either side of each gap.

This string is then refined according to some further heuristics. First, there must be no phrase containing fewer than three components. Also, there can be no more than two single-component words in a row. If the phrases do not conform to these rules, the string is reorganized and components are dropped until only valid text strings are left.

2.2.5   Postprocessing to Refine the Segmentation

The algorithm described above classifies broken lines (e.g., dashed lines) and repeated characters (e.g., a string of asterisks) as text strings, since they satisfy all the heuristics for text strings. However, it is desirable to label such components as graphics. On the other hand, text strings in which some of the components are connected to graphics are incorrectly segmented (e.g., underlined words in which characters with descenders touch the underline). Thus, the strings that are identified are further processed to refine the segmentation.

(A) Repeated-Character Filter

Strings made up of a character repeated many times are separated from other strings by checking for consistency in variance of the pixel density, the area, and the ratio of the length of the sides of the rectangle. These three variances are normalized by dividing by their respective thresholds. Whenever the product of the normalized variances is less than one, the string is classified as graphics. This method of combining the three features allows for some latitude in choosing the individual thresholds.

(B) Dashed-Line Filter

Although the above filter may work for certain dashed lines, it is easily seen that dashed lines are quite often irregularly broken, either for variety (e.g., lines made up of alternating short and long segments) or because they are hand-drawn. Since dashed lines are so common in graphics, a specialized algorithm for their recognition is required. The only reliable characteristic of dashed lines is that the variance in thickness is small. Since this information is not available in the connected-component data, it is obtained from the image. The thickness of the components at each point in a direction perpendicular to the direction of the string is calculated. Each thickness will contribute to the variance calculation. The variance in thickness is then compared to a single threshold. In these calculations, care is taken to include only those pixels that belong to the component under consideration, even when there are pixels belonging to other components within the rectangle.

(C) Text Connected to Graphics

A much more complex problem arises when text characters are connected to graphics. A special case of this problem occurs frequently when text strings with descenders are underlined. A typical case is shown in Fig. 2.5(a). Note that there are two components, y and g, embedded in the underlining. In this example, the underlines along with the characters y and g are classified as graphics and the string is classified as consisting of two words *an* and *ima e*. Note that the word *ima e* is not divided into two words simply because the gap resulting from a missing character will not always enforce a word gap ($\geq A_h$). Thus, wherever the gap between two characters is at least $0.5 \times A_h$, the string is checked for missing characters. It is assumed that the character is connected to the graphics only on one of the sides (not necessarily the bottom side). The basic approach to locate missing characters is to grow a three-sided box around the free sides (not connected to the graphics) of the component in question. The open side of the box corresponds to the side in which the character is connected to the graphics. The procedure is described below.

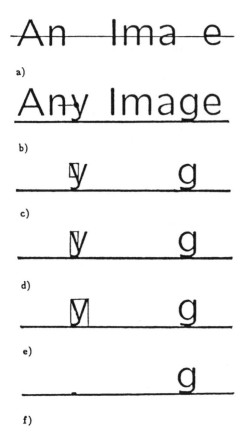

Figure 2.5   Algorithm for component separation: (a) input string with centerline illustrated, (b) contact with foreign component made, (c) three-sided box grown, (d) bottom of box extended, (e) box regrown, and (f) component severed, leaving graphic portion behind.

The centerline of the text string is found, and a line between the two components around the gap is checked for contact with connected components. If one is contacted that does not belong to the text string (Fig. 2.5[c]), it is investigated for removal. A three-sided box oriented at the angle of the text string is grown until it completely encloses the component on the three sides, as shown in Fig. 2.5(c). This example illustrates the case where the bottom is the open side, but, in fact, each side is checked in succession to determine the open side. While growing this box, if any dimension of the box exceeds two times the average character height, the process is aborted. If a valid box is grown, the open side is extended one row at a time outward until the height of the box reaches twice the average character size. While extending the box, the foreground pixel density along each new line is calculated. The line with the highest foreground pixel density is identified as the transition between the character and the graphic, as long as the density exceeds 75%. Note that when a character is connected to an underline, as in Fig. 2.5(d), the foreground pixel density reaches 100% along the line of contact. Finally, the box is extended along the other three sides, if necessary, as shown in Fig. 2.5(e). If these operations form a box that satisfies the size and positional constraints for a component in the text string, it is included as a part of the text string, and the graphics image is modified by deleting the pixels within the box (Fig. 2.5[f]).

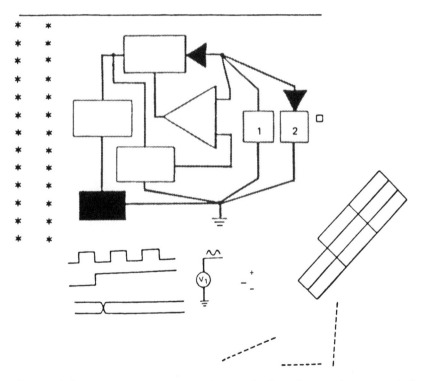

Figure 2.6   Test image in Fig. 2.1 after application of text-string-removal algorithm.

2.2.6   Experimental Results

The performance of the algorithm was evaluated on a number of test images. The output of the algorithm corresponding to the image of Fig. 2.1 is shown in Fig. 2.6. Note that all the text strings have been correctly identified and removed, while retaining all graphical components (including broken lines). The algorithm is implemented in Pascal on a VAX 11/785.

2.3   GENERATION OF LINE-DESCRIPTION FILE

The image output from the text-string-separation algorithm consists of graphics and isolated symbols. This image is processed further to generate a line-description file to represent thin entities (straight lines and curves) and boundaries of solid (completely filled) graphical components. Solid objects are located and separated from thin entities using erosion-dilation techniques. After skeletonization and boundary tracking, various types of lines (continuous, dashed, straight, curved, etc.) are identified and their attributes are computed by this algorithm. The algorithm generates a file describing the location and attributes of core lines (for thin entities) and boundaries (for solid objects). The algorithm consists of the following steps:

1. Separation of solid graphical components
2. Skeletonization and boundary tracking
3. Detection of straight-line segments
4. Detection of dashed lines

These steps are briefly described in the following sections. The detailed algorithm may be found in (Shah/1988).

2.3.1   Separation of Solid Graphical Components

The image is preprocessed to eliminate single-pixel-thick voids and protrusions. It is then eroded by N pixels, where N is a threshold that determines the maximum thickness of entities that are represented by their skeletons. Thus, all thin entities are completely eroded, leaving behind the eroded solid objects. The eroded image corresponding to Fig. 2.6 is shown in Fig. 2.7. This eroded image is dilated by N pixels to recover the pixels of solid objects that were removed during erosion. However, since dilation is not an exact inverse of erosion, additional processing steps are applied to completely recover the object. The *solid object image* is then subtracted from the original image to obtain the *thin entities image*.

Figure 2.7   Test image eroded by seven pixels.

2.3.2   Skeletonization and Boundary Tracking

The skeletonization algorithm described by Harris et al./1982 is applied to the thin entities image. Initially, all the graphics pixels are labeled as type 3, and the background pixels are labeled as type 0. The algorithm uses a set of 3×3 operators that generate a marked skeleton. After processing, each pixel is marked with one of the following four labels:

Type 0: pixel that is not a skeleton point
Type 1: skeleton end pixel
Type 2: skeleton internal pixel
Type 3: skeleton node pixel

An *order routine*, details of which can be found in (Shah/1988 and Alemany/1986), is then applied to the thinned image. This routine tracks all connected pixels in the skeleton image and generates an ordered list of pixels for each line segment in the skeleton. This routine also creates a file that contains the location of all junctions of core-line segments.

Depending on the orientation of the scanning, the skeletonization algorithm may result in the occurrence of spurious hairs. Such spurious hairs are mainly due to noise pixels or digitization artifacts. Thus, if a short segment is connected to at least two segments that are longer than twice its length, it is deleted and the line-segment description file is appropriately modified. Note that isolated short segments, which are possibly due to dashed lines, are not affected by this cleaning.

A *thickness routine* is then applied to compute the average thickness of line segments. The routine traverses the core line of each line segment and calculates the thickness at every point in the core line in four directions:

vertical, horizontal, and the two diagonal directions. To locate the junctions of lines of different thickness, it also calculates the running average of the line thickness. If the thickness at any three consecutive points exceeds 1.5 times the average thickness of the line traversed so far, the end of the line segment is marked and the beginning of a new line segment is noted.

An edge-enhancement operation is applied to the *solid object image* to locate boundary pixels. All pixels along the boundary are retained, and interior points are set to 0. The *order routine* described above is applied to this image to link all edge pixels and generate a boundary-description file. Figs. 2.8 and 2.9 show the *thin entities image* and its skeleton.

2.3.3 Detection of Straight-Line Segments

Note that when two straight lines of equal thickness meet at a point (e.g., corner of a polygon), the linked list generated by the *order routine* contains a single entry in its output that contains all the pixels in the two core lines.

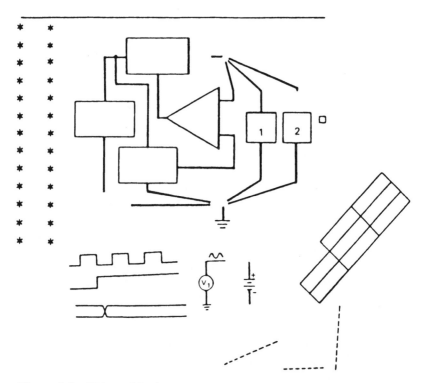

Figure 2.8 Thin entities image.

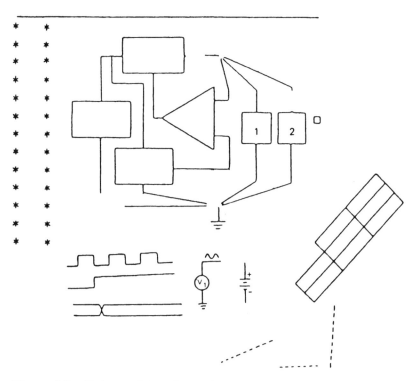

Figure 2.9   Skeleton of the thin entities image.

Similarly, straight lines and curves merging at a point do not trigger a new entry in the linked list if their thicknesses are the same. A separate entry is generated when three or more lines meet at a point. Thus, the linked list is processed further to split lines into straight-line segments and curves. The slope of a line formed by connected pixels is calculated and is split at points of significant change in slope. All straight-line segments are identified by their end points and thicknesses. Line segments with continuously varying slopes are classified as curves, and a file containing an ordered list of all core-line pixels in each such segment is generated.

2.3.4   Detection of Dashed Lines

The output of the line-segment-description algorithm generates a separate listing for each segment of a broken line. Thus, it is necessary to link these segments to generate a simple description for each dashed line. This algorithm, based on the technique described in (Alemany/1986), is designed to distinguish among dashed lines formed by different patterns. For example, in Fig. 2.10(a), the algorithm detects six lines: AA, BB, CC, DD, EE, and FF. In this example, line DD is distinguished from the others since it is continuous; line EE is distinguished from AA, BB, CC, and FF since it is made

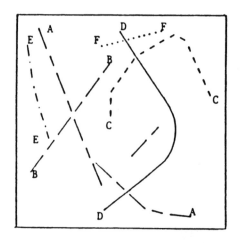

(a)

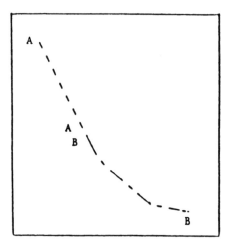

(b)

Figure 2.10 Types of continuous and broken lines: (a) lines formed by different patterns; (b) two lines having different patterns merging at a point.

of dashes and dots;* line FF is distinguished from AA, BB, CC, and EE since it is made up of dots; line CC is distinguished from AA, BB, EE, and FF since it is made of segments of equal length; and line AA is separated from line BB, even though they intersect each other, by grouping segments of similar slopes at the intersection.

For a line to be detected as a dashed line, it must contain at least four segments. The algorithm can also detect a change in the pattern of a line and properly identify the two parts (see Fig. 2.10[b]). A dashed/dotted line may be made up of segments of equal lengths or may consist of alternating longer and shorter segments. The algorithm identifies four segments having lengths l_1, l_2, l_3, l_4 and intersegment gaps $g_{1,2}$; $g_{2,3}$; $g_{3,4}$ as a dashed line, if the following conditions are satisfied:

$$l_i \leq T_{l2} \qquad i = 1, 2, 3, 4$$
$$T_{g1} \leq g_{i,i+1} \leq T_{g2} \qquad i = 1, 2, 3$$
$$l_i \simeq l_{i+2} \qquad i = 1, 2$$
$$g_{i,i+1} \simeq g_{i+1,i+2} \qquad i = 1, 2.$$

Here, Ts are length-threshold parameters.

The algorithm may be logically divided into the following three parts:

1. Segment- and gap-length computation
2. Neighboring-segment detection and sorting
3. Segment linking and dashed-line detection

(A) Segment- and Gap-Length Routine

The file containing a list of line segments is input to this routine. This routine examines each of the line segments and retains all the segments with length in the range of 1 to T_{l2}. The segments that are retained are given an identification number, L_i, $1 \leq L_i \leq K$, where K is the number of segments retained. The two ends of each segment are labeled as start (1) and end (2) to distinguish them from each other.

(B) Neighboring-Segment Detection and Sorting

The gap lengths between segment pairs are compared with the gap-length threshold to identify all segment pairs whose end points are within the range T_{g1} and T_{g2}. Several data arrays are generated in this step:

---

*In this discussion, a dot is a component that has up to four pixels in the thinned image. Note that a single-pixel component in the original image is lost in the preprocessing and thinning operations.

1. The identification numbers of all segments that are within the gap threshold from the start point of segment L_i are entered along the row L_i of the plane 1 of array L_{close}, and the corresponding segments at the end point are stored in plane 2.
2. The entries in the array L_{close} simply identify all the segments, L_j, where $L_j \neq L_i$, that are within the gap threshold from a particular end point of the line L_i. However, to pinpoint which one of the two end points of L_j is being referred to, another array P_{close} is generated. P_{close} and L_{close} have identical dimensions, and the values in P_{close} are either 1 or 2 to represent the start and end points, respectively, of the corresponding L_j in L_{close}.
3. The corresponding gap lengths are stored in an array S_{space}.
4. A two-dimensional array *count* contains the number of segments that are in the neighborhood of each L_i at each of its ends.

To properly track lines at their intersections, it is necessary to link segments of similar slopes. To accomplish this, the entries in both the planes of array L_{close} are rearranged by sorting each row according to the difference between the slopes of the reference segment and all the other segments in the row. If the difference in slope is small, the segment is placed in front of the list. If the length of the segments is small (less than 4 pixels), the slope of a line joining the first pixel of the small segment and the start or end pixel (as appropriate to the case) of the reference segment is calculated. If both the segment and the reference segment are small (less than 4 pixels), implying a dotted line, the segment is placed in front of the list. The difference between this slope and the slope of the reference segment is used as a criterion for sorting the segments.

(C) Segment Linking and Dashed-Line Detection

In this part of the algorithm, the segments are linked until all the dashed lines are detected. The algorithm extends the line by two segments (possibly alternating longer and shorter segments) at a time. The procedure may be summarized as follows:

1. Set $L_i = 0$.
2. Increment L_i by 1. If $L_i > K$, then stop; else if the segment L_i is already marked as a part of a line, then go to 1; else select L_i as the first segment of a new line. Select the start end of this line that is identified by setting $k_1 = 1$.
3. From the array L_{close}, locate the segment that is the best candidate to continue the line at the terminal k_1 of segment L_i. Let the segment be called M_1; compute l_{M1}; locate the other end, k_2 ($k_2 = 1$ or 2), of the segment M_1 using the information in P_{close}; go to step 4. If M_1 cannot be found or if M_1 is already identified as a part of

the line under consideration and if $k_1 = 1$, then set $k_1 = 2$; go to step 3. Else (note that this point is reached when the line cannot be extended further), go to step 2.

4. Find the best candidate to continue the line at the terminal k_2 of segment M_1. This yields M_2, l_{M2}, and k_3; go to step 5. If M_2 cannot be found, then go to step 3.

5. If $l_{Li} \simeq l_{M2}$ and $Space_{Li,M1} \simeq Space_{M1,M2}$, then go to step 6; else go to step 4.

6. Find the best candidate to extend the line at the terminal k_3 of segment M_2. This yields M_3, l_{M3}, and k_4; go to step 7. If M_3 cannot be found, then go to step 4.

7. If $l_{M1} \simeq l_{M3}$ and $Space_{Li,M1} \simeq Space_{M2,M3}$, then label M_1 and M_2 as extensions of the line; set $M_1 = M_3$, $k_2 = k_4$; go to step 4. Else go to step 6.

2.3.5 Output of the Line-Description Algorithm

A portion of the output line-description file is shown in Table 2.1. The line-description file generated by this stage is processed by the graphics recognition and description algorithm described in the following section.

Table 2.1 Portion of the Line-Description File for Test Image

			Line-Segment Description						
	Coordinates					Attributes			
Line No.	Start		End		Type 1[a]		Gap length	$L1,3$	$L2,4$
						Thick	between		
	x	y	x	y			segments		
1	1076	18	1068	30	Co St	8	—	—	—
2	829	20	836	32	Co St	8	—	—	—
3	843	20	836	32	Co Cu	8	—	—	—
4	500	200	610	120	So St	—	—	—	—
5	500	200	720	315	So St	—	—	—	—
6	720	315	610	120	So St	—	—	—	—
113	457	25	451	35	Co St	7	—	—	—
114	842	21	831	33	Co Cu	5	—	—	—
115	200	1508	360	1512	Br St	5	6	14	13

[a] Co St = Continuous Straight Line, Co Cu = Continuous Curved Line, So St = Straight-Line Segment of Solid Object Boundary, So Cu = Curved Line Segment of Solid Object Boundary, Br St = Broken Straight Line, and Br Cu = Broken Curved Line.

2.4 GRAPHICS RECOGNITION AND DESCRIPTION

This section describes algorithms for recognition and description of graphics. The objective is to obtain a succinct description of the contents of the
graphics. As an example, the algorithm is designed to describe the image
shown in Fig. 2.11 as consisting of several simple shapes interconnected by
lines. In other words, the algorithm locates loops of minimum redundancy
that are necessary and sufficient to describe the image. For the above example, the algorithm locates seven simple loops (e.g., loops that can be described as regular polygons) and two complex loops, as shown in Figs. 2.12
and 2.13. The simple loops are recognized as one of the known shapes and
are described by their location and orientation. The complex loops are then
analyzed to check whether they can be described as several overlapping simple loops. For example, the loop H in Fig. 2.11 can be described as a large
rectangle that is occluded by the polygons G, F, and E. Such a description is

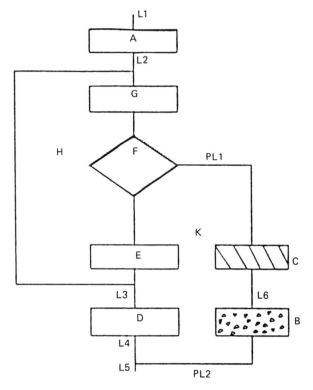

Figure 2.11   Input graphics image.

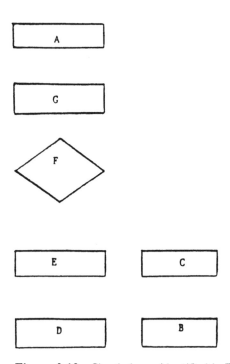

Figure 2.12   Simple loops identified in Fig. 2.11.

given only if it is possible to connect collinear segments to obtain a simple loop (i.e., the algorithm does not hypothesize the presence of corners that are not in the image). For example, loop K is not described as an occluded rectangle. For this loop, even if the collinear line segments are connected through the rectangles D, E, B, and C, it cannot be recognized as a regular polygon. Thus, it is divided into individual line segments.

Note that in this example of a flowchart, it may not be meaningful to describe the loop H as an occluded rectangle; the objective is to obtain a succinct description file that has adequate information to reconstruct the original image. The algorithm is a generic graphics-description system and has no knowledge about the context of data or the appropriateness of a particular description. The algorithm also identifies and describes hatching and filling patterns, as well as the spatial relationships among the various entities. The output corresponding to the graphics in Fig. 2.11 is given in Tables 2.2, 2.3, and 2.4. Table 2.2 contains all the primitive shapes and their attributes, such as their location, width, height, filling/hatching pattern, etc. Table 2.3 contains a listing of all interconnecting lines (lines that contain

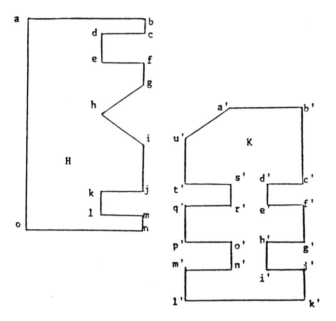

Figure 2.13   Complex loops identified in Fig. 2.11.

more than one straight-line segment are denoted as polylines in this table).
Table 2.4 gives the spatial relationships among the various entities.

The image reconstructed from the information contained in these tables
is shown in Fig. 2.14. Clearly, the reconstructed image is almost identical to
the original image (a routine to draw filling details is not yet implemented).
The algorithm consists of the following steps:

1. Generation of loops with minimum redundancy
2. Recognition of overlapping shapes
3. Detection of hatching and filling patterns

These steps are briefly described in the following sections. The algorithm
is an extension of the procedure described in (Bow and El-Masri/1987 and
Bow and Zhou/1986).

2.4.1   Generation of Loops with Minimum Redundancy

The simple graphic of Fig. 2.11 is made up of 57 line segments, excluding
the segments due to hatching lines and filling symbols. However, there are
74 loops that can be formed using these 57 segments. Thus, the problem
is the selection of the nine loops shown in Figs. 2.12 and 2.13. As a second

Table 2.2 Primitive Shapes in Fig. 2.11 and Their Attributes

Primitives	w	h	Center Point	Hatching	Theta
			Attributes[a]		
Rectangle A	120	30	(240, 465)		0.0
Rectangle B	100	35	(400, 102)	small shape fillings	0.0
Rectangle C	100	30	(400, 185)	sing. hatch a = 123.69 d = 18	0.0
Rectangle D	120	35	(240, 102)		0.0
Rectangle E	120	30	(240, 185)		0.0
Rectangle G	120	35	(240, 387)		0.0
Rectangle H	160	275	(160, 287)		0.0

	e	alpha	center point	hatching	theta
Rhombus F	72.11	33.69	(240, 300)		0.0

[a] w = width of rectangle, h = height of rectangle, theta = angle of a given shape with the x-axis, e = length of rhombus edge, alpha = half inner-angle at leftmost corner of rhombus, a = angle of hatching patterns with the x-axis, and d = spacing between hatching patterns (angles in degrees, lengths in number of pixels).

example, consider a large rectangular table with horizontal and vertical partitioning lines. Such a table contains a number of loops, each of which is a rectangle. In such a case, it is desirable to describe the graphics as a single rectangle and a few lines rather than as a collection of a large number of rectangles. The algorithm for obtaining such a description is described below.

A few notations, described next, are helpful to understand the algorithm. A line segment is designated as a terminal line segment, T, if it has no neighbors at one end. If a line segment has one and only one neighbor at both ends, it is designated as a path line segment, P. If a line segment has two or more neighbors at either of its ends, it is designated as a branch line segment, B. Clearly, T segments cannot be a part of a closed loop. Thus, all chain clusters that begin with T (see Fig. 2.15 for several examples of such clusters) are removed up to the branching point. In our example, only two line segments are removed during this step. After these segments are removed, the remaining segments are redesignated and the process is re-

Table 2.3   Interconnecting Lines and Their Attributes

			Connection	
Polyline		Segments	From	To
PL1	S1:	(300, 300) , (400, 300)	F	C
	S2:	(400, 300) , (400, 200)		
PL2	S1:	(400, 85) , (400, 50)	B	L4, L5
	S2:	(400, 50) , (240, 50)		

			Connection	
Line	Head	Tail	From	To
L1	(240, 500) ,	(240, 480)	Space	A
L2	(240, 450) ,	(240, 425)	A	H
L3	(240, 150) ,	(240, 120)	H	D
L4	(240, 85) ,	(240, 50)	D	PL2, L5
L5	(240, 50) ,	(240, 40)	PL2, L4	Space
L6	(400, 170) ,	(400, 120)	C	B

peated, if necessary. The segments that remain are examined and classified into three categories as described below:

1. *Self loops* are identified by chain clusters of the form PP, PPP, PPPP, ... or BPB, BPPB, BPPPB, etc. The segments in such loops cannot be a part of any other loop. Rectangle A in Fig. 2.11 is an example of a self loop. Such loops are removed and the designation of other line segments are upgraded and considered for removal, as necessary.

Table 2.4   Spatial Relationships Among Graphic Primitives

Primitives	Spatial Relations
E, F, G	Overlap H
A, G, F, E, D	Top $\longrightarrow$ Down
C, B	Top $\longrightarrow$ Down
D, B	Left $\longrightarrow$ Right
E, C	Left $\longrightarrow$ Right

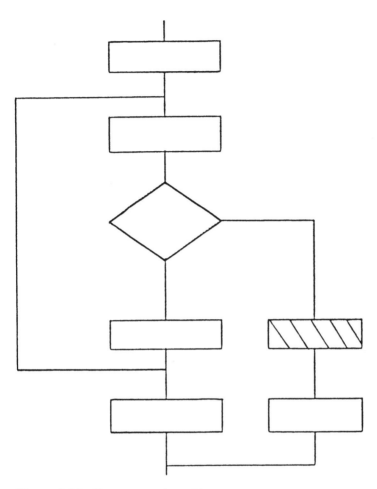

Figure 2.14   Reconstructed graphics.

2. All loops that can be formed by six or fewer sides and that can be recognized as a known shape (e.g., regular polygon, trapezoid, parallelogram, etc.) are located using a depth-first search technique. Such loops are called *simple loops*. At any branching point, if a collinear segment exists, it is used as the next element of the chain. Thus, a side of a simple loop may contain many collinear segments. This rule ensures that the largest possible number of simple loops are found before finding other loops. Extensive backtracking becomes necessary in this procedure. Simple loops are then analyzed

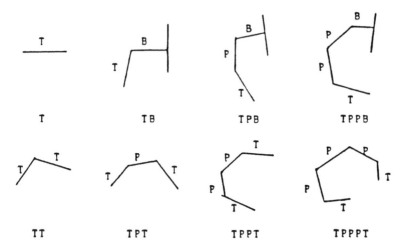

Figure 2.15   Examples of chain clusters that cannot be a part of a closed loop.

to identify their shape (regular polygon, parallelogram, trapezoid, etc.) The simple loops in our example are shown in Fig. 2.12.

3.   All other line segments that belong to loops and that cannot be included in a simple loop are grouped into *complex loops*. Note that such loops contain more than six sides and may contain segments that are also part of other loops. The complex loops in our example are shown in Fig. 2.13.

2.4.2   Recognition of Overlapping Shapes

Assuming that all shapes are convex polygons, an attempt is made to describe complex loops as overlapping simple loops. This is done by joining collinear but disconnected edges in the complex loops to obtain a convex polygon, if possible. In our example, the complex loop H consists of 15 edges and shares the segment sets [cd, de, ef], [gh, hi], and [jk, kl, lm] with the simple loops G, F, and E. Connecting vertices c–f, g–i, and j–m would result in a rectangle (a, b, o, n) with overlapping simple loops G, E, and F. Following the same reasoning, the complex loop K would yield a convex shape (a', b', k', l', u'), but it is not a recognizable shape. Thus, it is divided into individual lines. The description generated by the system is shown in Tables 2.2, 2.3, and 2.4.

2.4.3 Detection of Hatching and Filling Patterns

The algorithm detects the hatching and filling patterns in the graphics. This operation is essential to generate a meaningful description file. Note that the individual segments that form the hatching patterns are not grouped together to form simple loops or complex loops; otherwise, an enormous number of loops would be generated. Thus, all lines (if any) within each loop that have been located in the previous step are analyzed to determine their orientation and spacing. The algorithm detects single and double hatching patterns. Table 2.2 shows the description of the hatching generated by the system.

Some of the graphics are filled with small shape fillings, as shown in Fig. 2.11. Each of these shapes is recognized as a self loop. The algorithm also notes that many such loops are enclosed by a larger loop (loop B in the example). Whenever a large number of self loops are enclosed by another graphics primitive, the small shape descriptions are deleted from the output file and the attribute of the enclosing loop is modified, as shown in Table 2.2.

2.5 STRUCTURAL DESCRIPTION OF GRAPHICS MADE UP OF CURVED SEGMENTS

In the previous section, we described an algorithm that recognizes and describes graphical primitives made up of straight-line segments. In this section, we describe the procedure that generates a description of graphics consisting of circles. Curved segments that are not circles are approximated as concatenated circular arcs.

2.5.1 Detection of Circles

The circle detection algorithm consists of the following steps:

1. Estimation of radius and center point
2. Recognition of circles
3. Representation of curves by circular arcs

These steps are briefly described in the following sections.

(A) Estimation of Radius and Center Point

All curved segments from the output of the line-description algorithm are processed by this step. Three points on each curve are selected, one near each end and one at the midpoint. Using this information, the center and the radius of a circle passing through these three points are calculated. To detect whether the segment under consideration is circular or not, the segment is repeatedly divided into halves and the radius of each segment is calculated. If these values fit within a tolerance, it is classified as a circular segment.

(B) Recognition of Circles

The algorithm for finding polygonal loops of minimum redundancy was described in Sec. 2.4. A similar technique is also applied to find curved loops. All curved segments are sorted according to their lengths. To start with, the largest circular segment is chosen, since the error in its estimated radius is expected to be small compared to that of shorter segments. Each neighbor of the current segment is considered, one at a time, as a potential extension of the current segment, and the neighbor that is the best fit is selected. Here, the best fit is defined as a segment that, when used as an extension to the current segment, results in a minimum change in radius. If no such neighbor is present, the system backtracks to the recently chosen neighbor

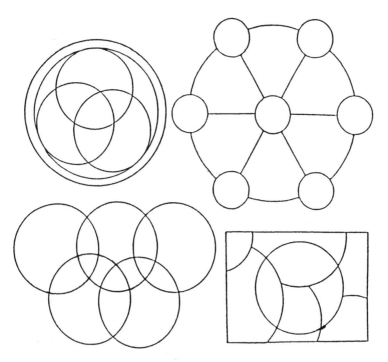

Figure 2.16 Test images of circles.

and selects the second-best neighbor at that junction. After all complete circles are detected, the algorithm evaluates all other circular segments to identify circles that are potentially overlapped by other objects, resulting in incomplete circles.

The output generated by the system corresponding to the input shown in Fig. 2.16 is shown in Table 2.5. Note that all the circles have been identified, including the large circle on the top right (circle 19 in the table), which is overlapped by six smaller circles.

2.5.2   Representation of Curves by Circular Arcs

All curved lines that do not belong to a detected circle are approximated as a concatenation of circular arcs using the algorithm described in (Honnenahalli/1987). The algorithm begins processing the line from one end and computes the radius and center by forming two segments (see Fig. 2.17[a]). The lengths of the segments are doubled, and the center and radius are computed again (Fig. 2.17[b]). This process is repeated until the radius or the center location changes by a significant amount compared to the pre-

Table 2.5   Circles Detected in Fig. 2.16

Circle	Radius	Center X	Center Y
1	381	493	469
2	228	1193	890
3	230	1195	257
4	338	495	473
5	221	1472	687
6	233	1455	415
7	236	1418	1548
8	233	1180	591
9	95	917	1647
10	210	596	554
11	209	368	463
12	98	129	1654
13	97	529	1864
14	96	906	1190
15	92	125	1207
16	96	509	980
17	210	548	359
18	97	519	1422
19	457	509	1422

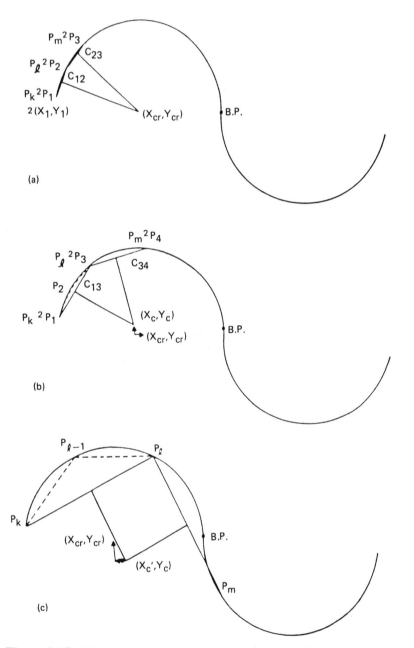

Figure 2.17 Representation of curves by circular arcs: (a) initial estimation of radius and center, (b) radius and center within tolerance, and (c) radius and center outside tolerance.

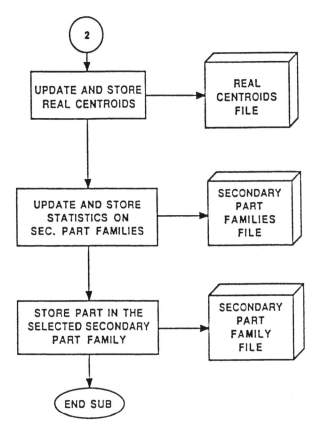

Figure 2.18   A simple test image.

vious values (Fig. 2.17[c]). When this happens, the position of the break point is located by backtracking. Note that a segment of a circle is represented by a single circular arc, whereas other curves are approximated by a concatenation of several arc segments.

The final output of this algorithm is a file that contains the location and size of all circles in the image. All other segments that do not belong to any circles are described by their location and attributes.

2.6   EXPERIMENTAL RESULTS

The performance of the graphics-recognition system has been evaluated using a number of test images. In this section, the performance of the algorithm on a simple test image, shown in Fig. 2.18, is presented. The output of the text-string-separation algorithm is shown in Fig. 2.19. Note that the number 2 is not removed, since it is not a part of any string. The simple polygonal loops in this image are shown in Fig. 2.20. These are identified as triangles, rectangles, and parallelograms by the recognition system. Note that although the boxes are not exact rectangles, they are recognized by the system as rectangles. An image reconstructed using the final description generated by the system is shown in Fig. 2.21. The circle and ellipse are not redrawn in this image.

Note that the polygons reconstructed are regular, although in the original data they may have irregularities within a specified tolerance. However, this exact reconstruction has its own disadvantages. For example, the rectangles and parallelograms do not have common edges. This can be corrected

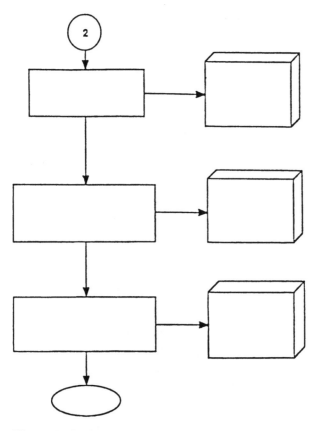

Figure 2.19   Image after text separation.

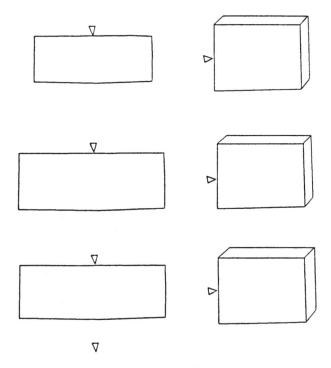

Figure 2.20   Polygonal loops in Fig. 2.19.

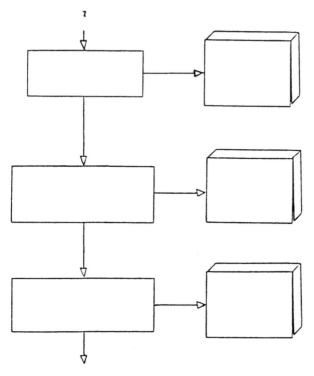

Figure 2.21   Image reconstructed from output.

using the additional information that is available in the output tables and additional heuristics.

2.7 SUMMARY AND CONCLUSIONS

A graphics-recognition and interpretation system has been presented. The system consists of image-processing and pattern-recognition algorithms implemented in software. The high-resolution binary input image is segmented into a text-string image and a graphics image. Lines of various types are identified in the segmented graphics image. Line segments are analyzed to find closed loops that form graphical primitives of known shapes. Curved lines are approximated by concatenated circular arcs. Spatial relationships among various primitives are described. Hatching and filling patterns are identified. The system generates an output description file that has adequate information to faithfully reconstruct the original. The algorithms are computationally intensive and require CPU times on the order of minutes on a VAX 11/785.

ACKNOWLEDGMENTS

We wish to acknowledge the assistance of Lloyd A. Fletcher, Juan Alemany, Jayesh Shah, Wassim El-Masri, Umesh Mokate, James Gattiker, Chingchuan Shih, Bin Zhou, and Suresh Honnenahalli in designing, implementing, and testing of the system described in this chapter. We would also like to thank Greg Wallace and Babu Obilichetti for many helpful suggestions and stimulating discussions.

REFERENCES

Alemany, J. (1986). *An Image Processing System for Interpretation of Paper-based Maps*, M.S. Thesis, Dept. of Electrical Engineering, Pennsylvania State University, University Park, Pennsylvania.

Bley, H. (1984). Segmentation and preprocessing of electrical schematics using picture graphs, *Computer Vision, Graphics, and Image Processing, 28*: 271–288.

Bow, S. T. and Zhou, B. (1986). Automated generation of the structural description for graphics by computer, *Proceedings of the International Symposium on Mini and Microcomputers and Their Applications*, Austin, Texas, ACTA Press, pp. 58–62.

Bow, S. T. and El-Masri, W. (1987). Knowledge-based graphics understanding and description for document archival and retrieval, *Proc. SPIE Conf. Advances in Intelligent Robotic Systems, 848*: 640–647.

Bunke, H. (1982). Automatic interpretation of lines and text in circuit diagrams, *Pattern Recognition Theory and Applications* (J. Kittler, K. S. Fu, and L. F. Pau, eds.), D. Reidel, Boston, pp. 297–310.

Chen, W. H., Pratt, W. K., Hamilton, E. R., Wallis, R. H., and Capitant, P. J. (1980). Combined symbol matching facsimile data compression system, *Proc. IEEE, 68*: 786–796.

Ejiri, M., Kakumoto, S., Miyatake, T., Shimada, S., Matsushima, H. (1984). "Automatic Recognition of Design Drawings and Maps," Proceedings of 7th International Conference on Pattern Recognition, Montreal, pp. 1296–1305.

Fletcher, L. A. and Kasturi, R. (1988). A robust algorithm for text string separation from mixed text/graphics images, *IEEE Trans. PAMI, 10*: 910–918.

Gattiker, J. R. (1988). *An Improved Algorithm for Text String Separation from Mixed Text/Graphics Images*, M.S. Thesis, Dept. of Electrical Engineering, Pennsylvania State University, University Park, Pennsylvania.

Harris, J. F., Kittler, J., Llewellyn, B., and Preston, G. (1982). A modular system for interpreting binary pixel representation of line-structured data, *Pattern Recognition: Theory and Applications*, D. Reidel, Boston, pp. 311–351.

Honnenahalli, S. (1987). *Piecewise Representation of Curves by Circular Arcs Through Detection of Break Points*, M.S. Thesis, Dept. of Electrical Engineering, Pennsylvania State University, University Park, Pennsylvania.

Huang, C. L. and Tou, J. T. (1986). Knowledge based functional symbol understanding in electronic circuit diagram interpretation, *Applications of Artificial Intelligence III, Proc. SPIE, 635*: 288–299.

Karima, M., Sadhal, K. S., and McNeil, T. O. (1985). From paper drawings to computer aided design, *IEEE Computer Graphics and Applications*: 24–39.

Shah, J. D. (1988). *Vector Representation of Raster Scanned Images*, M.S. Thesis, Dept. of Electrical Engineering, Pennsylvania State University, University Park, Pennsylvania.

Wahl, F. M., Wong, M. K. Y., and Casey, R. G. (1982). Block segmentation and text extraction in mixed text/image documents, *Computer Vision, Graphics, and Image Processing, 20*: 375–390.

3

Automatic Recognition of Engineering Drawings and Maps

Masakazu Ejiri, Shigeru Kakumoto,
Takafumi Miyatake, Shigeru Shimada,
and Kazuaki Iwamura

Hitachi Ltd.
Kokubunji, Tokyo, Japan

3.1 INTRODUCTION

For the past twenty years, the ability to automatically input drawings directly into a computer system has remained an elusive goal in the field of information processing. Perhaps the main reason for this lack of success is that the algorithms needed to carry out automatic recognition of complex drawings tend to be unwieldy. Furthermore, the processor technology needed to support an analysis of actual working-sized drawings within a reasonable amount of time has become attainable only recently.

Numerous needs for realizing automatic drawing recognition have been identified so far. These can be classified broadly into the following three categories:

1. Need for an efficient way to input design drawings to CAD systems

Based on the paper "Automatic Recognition of Design Drawings and Maps" by Ejiri et al. appearing in *Seventh International Conference on Pattern Recognition*, Montreal, Canada, July 30–August 2, 1984, pp. 1296–1305. © 1984 IEEE.

2. Need for an effective way to manage the multitude of existing drawings using a computer
3. Need for an efficient means of executing planning and maintenance tasks based on displayed digital drawings

It is indeed very difficult to realize a technology for inputting drawings that can satisfy all of these demands. The main reason is that various types of complex drawings exist, each drafted according to very different rules. Furthermore, explicit expressions are apt to be omitted in the drawings, thus making automatic input even more difficult.

With regard to the first category, there are basically two methods for inputting drawing data. One is an interactive method in which the drawing data are completed with the aid of a workstation and sophisticated software for graphics editing and CAD. Although this is often considered a smarter, more modern method for design automation, it usually forces the designer into a lengthy and unnatural posture for designing. Furthermore, the resolution of most CRTs (cathode ray tubes) does not allow the entire drawing to be displayed at one time. Partial display of a drawing sometimes interferes with the designer's thought process and is one of the shortcomings of this method, especially when a large-scale drawing must be designed. In the other method, the coordinates of lines in a design drawing can be indicated manually using a so-called manual digitizer. This method does not usually impose a limitation on the designer, since the traditional design method can be used, allowing an unobstructed view of the complete design. The operator of the digitizer, however, must endure a lengthy and monotonous point-and-pick type of inputting task, in which the coordinates of each line in the drawing are specified sequentially.

With regard to the second category, the objective of drawing recognition is to eliminate the need to store the voluminous amounts of paper that have traditionally been handled in the machinery, aerospace, ship-building, automobile, and construction industries. Storing such drawings in a computer database would allow for quick reuse of data when modifying or improving previous designs.

The third category, automatic recognition of maps, is the most rewarding, especially for urban planning and for gas, water, electric, and telephone utility management. Until recently, the only method for inputting drawings in the second and third categories was also by use of a manual digitizer.

Some researchers, foreseeing the needs mentioned above, conducted studies in the 1970s to make drawing recognition a reality (Ejiri et al./ 1972, Kakumoto et al./1978). Most of these were experimental feasibility studies to investigate basic approaches to drawing recognition. In the 1980s, as a result of these studies, several machines capable of automating the digitizing task were constructed (Fulford/1981, Kakumoto et al./1983, Ejiri et al./1984).

The objective of drawing recognition is to represent the two-dimensional image obtained by an imaging device as a set of codes corresponding to each line, symbol, and character in the drawing. Coding a line means tracking the set of pixels that form the line and converting them into vector data consisting of a pair of point coordinates for the ends of each line segment. The coding of symbols and characters means finding the portions of the image where these are drawn, recognizing what they are, and converting them into corresponding codes defined in advance. Using such a coding technique, the image of a drawing can be compressed to a quantity of data that can be handled easily by a conventional computer system. Needless to say, the coding method, as well as the resulting set of coded data, should match subsequent data-handling paradigms in CAD and in other application systems. The methods pursued so far are not yet capable of fully recognizing drawings containing various symbols and characters. Nevertheless, several applications of drawing recognition have been put into operation successfully for certain types of engineering drawings and maps where the symbols and characters are simple and relatively few.

This chapter discusses various techniques related to the recognition of engineering drawings and maps. Automatic digitizers based on these techniques are also presented, together with the state of the art in the application of drawing-recognition techniques. An intelligent map-based information-processing system is also introduced as a typical example that uses digitized data.

3.2   DRAWING-RECOGNITION TECHNIQUES

There are many basic procedures for recognizing drawings. Some important processing procedures are explained in this section. Since images in drawings usually involve an enormous amount of pixel data, efficient and fast processing procedures are keys to drawing recognition.

3.2.1   Input Methods

The first step in automatic drawing recognition is to obtain a stable and precise image of the drawing using a scanner. There are basically two types of scanners: a drum type and a flatbed type. The drum scanner uses a rotating drum and a point-sensor to traverse the drawing in the direction of the drum axis and acquire two-dimensional information from the drawing. The flatbed-type scanner uses a linear sensor mounted on a servo-controlled carrier. The carrier is also mounted on a servo-controlled arm, thus providing X and Y motion in the plane of the flat table on which the drawing is set. The carrier is usually controlled to move at a constant speed to obtain two-dimensional information from a narrow area on the drawing, whose width corresponds to the length of the linear sensor. Then, the arm is moved stepwise to another area after the carrier finishes one full stroke of motion.

There is another type of scanner, called a continuous-roll-feed type, that functions somewhat as a combination of the two types discussed above, where a drawing is fed into the scanner at a constant speed. A linear sensor scans the entire width of the drawing while it is being fed under the sensor. The difficulty, however, arises in obtaining a high-resolution sensor capable of long width coverage and in ensuring the uniformity of the paper-feeding velocity for the entire coverage. Thus, its use is limited to relatively small drawings.

In all of these scanners, scanning pitch, which corresponds to pixel size in the obtained image, is perhaps the most important parameter. As the pith becomes finer, the image is reproduced with greater fidelity in the scanner memory; however, memory size becomes concomitantly larger. Also, the time required for image processing tends to increase beyond all practical limits. Given that most drawings have line widths of 0.2 to 0.5 mm, 0.1 mm is usually adopted as the standard scanning pitch. However, such standard pitches as 200 dots/inch (8 lines/mm, 0.127 mm pitch), 300 dots/inch (12 lines/mm, 0.085-mm pitch) and 400 dots/inch (16 lines/mm, 0.064-mm pitch) can also be used, depending on the application.

Using a 0.1-mm resolution, the maximum allowable paper size for a design drawing (for example, ISO size A0—1,189 × 841 mm) corresponds to an image with 10^8 (11,890 × 8,410) pixels. Assuming one byte is assigned to one pixel for ease of storing color codes in color drawings and also various flags for later analysis, this would require a system with 100 Mbytes of image memory—still far too large to be implemented in a scanner. Therefore, several smaller buffer memories, typically four, are usually implemented. In this approach, image partitioning should be done so that neighboring image portions overlap each other. This prevent loss of information at the partition boundary and makes it possible to combine the coded information of the partitioned areas later. Data scanned for a partitioned area of the drawing are stored in one of the buffers. Then, an image processor is assigned to process the data in that buffer, while another area is being scanned at the same time and the results stored in another buffer. This process can be executed continuously by using multiple processors and buffers in parallel, thus eliminating the overhead of scanning time. Examples of image partitioning are shown in Fig. 3.1. A

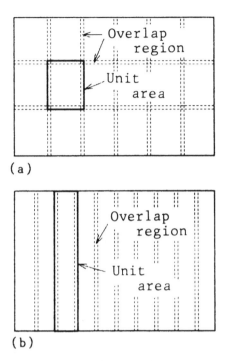

(a)

(b)

Figure 3.1 Examples of image partitioning: (a) two-dimensional partitioning; (b) one-dimensional partitioning.

typical example of a partitioned unit area contains 512 × 2,048 pixels and corresponds to 1 Mbyte.

3.2.2 Detection of Colors

In monochrome drawings, the sensor output at each pixel position is thresholded into a binary value having logical one on black lines and logical 0 on white background. If the drawing to be input is relatively old, or if the drawing is a copy, a shading effect is sometimes encountered. In such cases, simple thresholding using a fixed threshold value usually does not yield a good-quality binary image. A sophisticated thresholding algorithm is generally needed to obtain a system capable of adapting itself to the quality of the drawings. A histogram of gray-level pixels in a local area can serve as a cue for deriving an optimum threshold value in each local area.

In certain drawings, such as diagrams of LSI (large-scale integrated) cells, color pencils are sometimes used. To detect these colored lines, the color-brightness of each pixel is detected, converted into a color code, and stored in the image buffer. This is usually done in the real-time mode as sensor scanning proceeds.

Since the scanning-spot size is usually set as shown in Fig. 3.2, both the line color and background color (usually white) are observed in mixed form at the boundary of a line; therefore, it is sometimes difficult to obtain a stable and precise color code at the boundary. Furthermore, the point where two different-colored lines cross yields an even more complex mixture of colors. Consequently, it is preferable to assign an uncertainty code rather than to try to specify the color at the boundary and the cross points.

For real-time color detection, multiple color filters can be installed in the scanner in combination with an equal number of photometric sensors. The output of each sensor is digitized and fed into a code-conversion circuit that calculates the equation:

$$\left| \frac{A(i)}{\sum_i A(i)} - S(i,j) \right| < T(i,j) \qquad \text{where } i = 1, 2, \dots, n \tag{3.1}$$

Here, n is the number of sensors, each coupled with a different color filter, $A(i)$ is the absorption rate detected by the ith sensor, $S(i,j)$ is a standard value for the jth color, and $T(i,j)$ is the prescribed tolerance. Satisfying Eq. (3.1) for all i values results in color code j. This color conversion can be executed at high speed using a table-lookup method, as will be shown later. It has been experimentally confirmed that at least six colors (typically, magenta, green, cyan, yellow, purple, and black) can be detected reliably using a white light source and four filtered photomultipliers, each having a

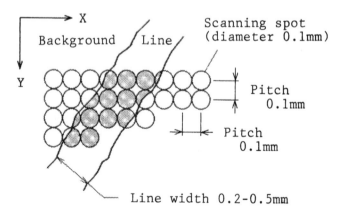

Figure 3.2   Geometry of a line and scanning spots.

maximum sensitivity at red, green, blue, and infrared. The detected color codes are stored in an image buffer, a small portion of which is shown in Fig. 3.3.

3.2.3 Detection of Lines

The line-detection process is then initiated to manipulate binary pixel data in the case of monochrome drawings or the color-coded pixel data in the case of color drawings. Various methods of detecting line segments have been developed. The most common is to apply a thinning algorithm (Hilditch/1969) to obtain medial line images and then to track the medial lines by using eight connected neighbors within a 3 × 3 window. This method is illustrated schematically in Fig. 3.4. One of the shortcomings of

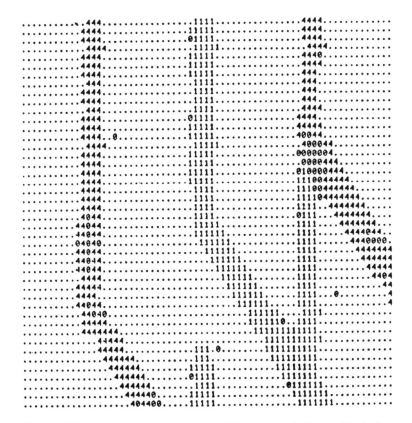

Figure 3.3 Detected color codes stored in an image buffer: 1: black, 0: uncertain, 4: red, ·: white (background).

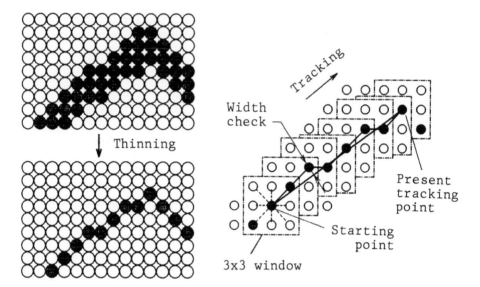

Figure 3.4 Medial line-tracking method.

the thinning method is that the entire image must be scanned a number of times, a time-consuming process. In addition, this method usually causes unnecessary twigs in the data, which sometimes make subsequent processing more difficult.

Another method is to search the logical 1 pixels or the same color pixels in an image buffer, without thinning, and find the longest vector drawable on the line segment. This process is shown in Fig. 3.5. Even if the pixels at a cross point are uncertain, as shown already in Fig. 3.3, the algorithm searches beyond the cross point to see if the same color line continues. When the longest vector is found, line tracking is suspended to store the coordinates of both ends of the vector. Then, the pixels that make up the line are flagged to show that the search has been completed. New tracking starts from the last point just stored. Thus, line images can be represented as a set of vector data, with each vectored line consisting of a pair of point coordinates. This method is reasonably fast when a drawing consists mainly of long straight lines; however, some kind of postprocessing is usually necessary to rectify the obtained data in order to increase the coding precision. Grid locking is one such postprocessing technique and will be discussed later.

Using these line-detection methods, a straight line is usually represented by a long vector, while a curved line can be approximated by a set of short

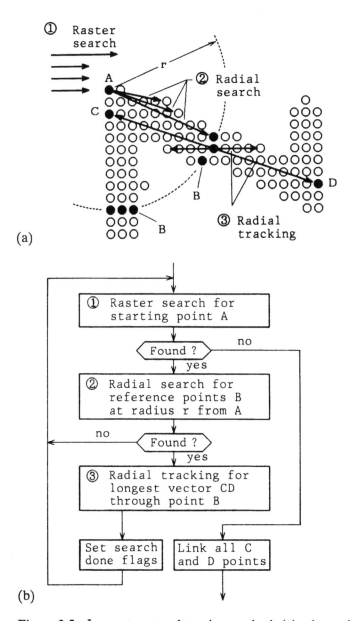

(a)

(b)

Figure 3.5 Longest vector-detection method: (a) schematic of line detection; (b) flow of line-detection algorithm.

vectors, each having gradually changing directions. One example of a data structure for vectors obtained with this method is shown in Fig. 3.6.

3.2.4 Recognition of Line Types

The set of line segments (vectors) obtained using the above methods can be processed to find broken lines by extracting subsets of short line segments. Figure 3.7(a) shows an example of a freehand drawing. The main objective in line-type recognition is to find the route for each line type, as shown in Fig. 3.7(c), by analyzing the set of short line segments shown in Fig. 3.7(b). For this purpose, a two-level recognition algorithm is provided: one level is for local recognition, the other is for global recognition. In local recognition, one line segment—for example, P_1P_2 in Fig. 3.7(b)—is chosen as the starting segment. At the end of the segment, a small search area is generated around P_2, and point P_3 is found within this area as a possible connecting point. The direction of segment P_3P_4 is checked to see if it lies in almost the same direction as the preceding segment P_1P_2. If so, the length of the segment is checked and is accumulated for statistical analysis to determine the line type. In this case, the analysis indicates a simple dashed line as the most probable line type.

This process proceeds and pauses at point P_8, because the length of segment $P_9P_{10}P_{11}$ does not match the lengths of the segments that have

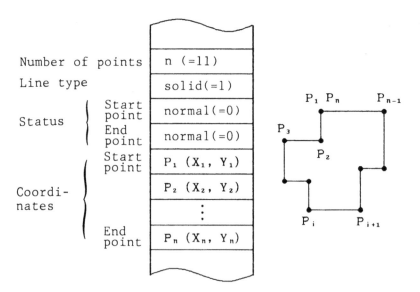

Figure 3.6   Data structure for lines.

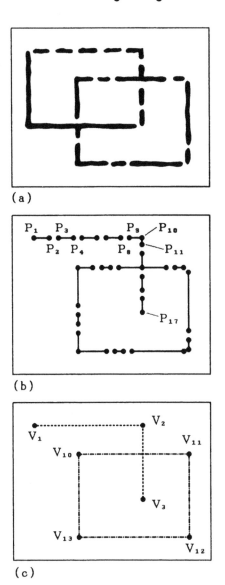

(a)

(b)

(c)

Figure 3.7 Recognition of broken lines: (a) image of freehand drawing, (b) vectorized short-line segments, and (c) recognized results.

been tracked until then. Therefore, P_8 is flagged as representing a backtrack point, and a probable line type is stored. The process is begun again at P_{11} until P_{17} is reached. In this case, since no mismatch occurs in line length between P_{11} and P_{17}, backtracking to P_8 is not initiated. If a mismatch were to occur, the process would restore the line type up to P_8, and new tracking would begin at P_8.

All broken lines can be found sequentially by this method; however, several obscure points that were flagged during the local recognition process still remain pending. At this point, the global recognition process is initiated to check for conflicts in the results. In this case, syntax rules are used, most of which are peculiar to the drawings in each application field. Examples of such rules are shown in Table 3.1. If the global processing finds a conflict, and the connection or disconnection of points is required, the local recognition procedure is again applied to these points, until all conflicts are resolved. In this example, detected line segments P_1P_8 and $P_{11}P_{17}$ are again combined with the segment $P_9P_{10}P_{11}$ to find P_{10} as a corner point V_2. Fig. 3.8 shows another example of the results of broken-line recognition (Shimada et al./1986).

Table 3.1   Examples of Syntax Rules

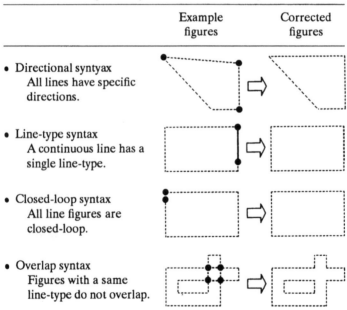

	Example figures	Corrected figures

• Directional syntyax
   All lines have specific
   directions.		

• Line-type syntax
   A continuous line has a
   single line-type.		

• Closed-loop syntax
   All line figures are
   closed-loop.		

• Overlap syntax
   Figures with a same
   line-type do not overlap.		

Note: Solid circles show conflicting points.

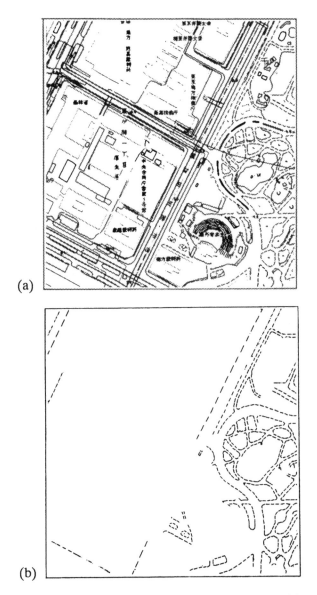

(a)

(b)

Figure 3.8 An example of broken-line recognition: (a) original drawing (district map); (b) recognized results.

3.2.5   Recognition of Characters and Symbols

The method for recognizing characters and symbols in a drawing usually consists of the following two steps: (1) detecting areas within the drawing that contain characters or symbols, and (2) determining character or symbol types in each detected area. Unlike ordinary OCRs, this area-detection function is absolutely essential in the case of drawing recognition, since the characters and symbols routinely coexist with other complicated lines and figures.

Area detection can be accomplished by one of the following four methods. The first uses a labeling technique by which concatenated pixels are grouped. An isolated and relatively small group is regarded as a candidate for a character or a symbol. The second method uses loop information, since loops are often a part of characters and symbols. The third method is based on the grouping of short line segments, and the fourth is based on the density of figure data. The first and second methods can be applied reliably to relatively simple drawings that have been carefully drafted. The third and fourth methods are applicable to more complex drawings; however, the rate of correct detection is not always satisfactory.

As for recognition of the character or symbol types, there are basically two methods: one is a pattern-matching method; the other is a structure-analysis method. In the pattern-matching method, a candidate area is matched with standard patterns of characters and/or symbols. This method, however, does not always provide adequate results, since characters and symbols in drawings are often hand-drawn with various sizes and orientations. In addition, some symbol types are very similar, making recognition difficult. Hence, some method capable of discriminating slight differences in shape would be most useful in the drawing-recognition task. This capability can be provided by the structure-analysis method.

In the structure-analysis method, features of a pattern within the candidate area are calculated and compared with features of the characters and/ or symbols to be found. A typical structure-analysis method uses concavity/convexity information in the contour of the pattern in a candidate area and decomposes the contour into simple patterns. This shape information is then converted into a quantitative feature vector (Yamamoto et al./1976). Another typical structure-analysis method uses Fourier analysis, and the resulting coefficients are used as a feature vector to be compared (Granlund/ 1972).

3.2.6   Coding of Solid Figures

Drawings consisting mainly of lines can be coded by combining the methods described above. For more complex drawings where complicated solid figures are drawn, additional methods are required. Liquid-crystal patterns are typical examples of solid figures. To input these drawings, the contour of each solid figure must be vectorized automatically. A technique based on chain coding in eight directions can be adopted, as shown in Fig. 3.9. In most cases, it is desirable to convert the contour into vectors having suitable lengths: longer vectors for straight portions, and shorter ones for curved portions with larger curvature. By evaluating the deviation of the chained pixels from a straight line that connects the starting point and the present tracking point, a decision can be made automatically concerning whether the present tracking point should be the end point of the vector tracked until then. This method yields adaptive vector lengths depending on the contour curvature and effectively minimizes the volume of coded data, even for drawings with very complicated, curved patterns. In this case, a contour is always tracked by examining the filled area to the right of the contour, so that white blanks in a black figure can be distinguished from black figures.

One convenient feature of contour tracking compared to line tracking is that it intrinsically yields sets of vectors that make up a closed loop. This simplifies checking for conflicts that might be contained in the final result. If the chain codes, rather than their vector approximates, are stored as a final result, the original shapes in the drawing can be completely preserved. Nevertheless, this chain-coded contour representation retains less data than is contained in the original image.

In addition to coding solid figures, the contour-coding method can also be used to code line drawings. In this case, a line is represented as a closed

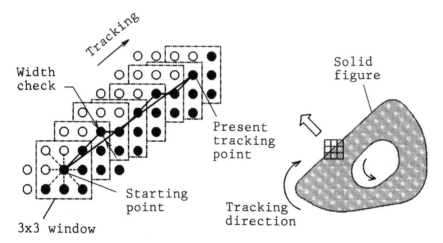

Figure 3.9   Contour-tracking method for solid figures.

loop contour. In each loop, narrowly separated parallel contour vectors having opposite directions correspond to two sides of a single line. From these pairs of contour vectors, medial lines can be calculated easily, thus providing another method of detecting lines.

3.2.7 Run-Length Coding

Since a drawing consists of a huge amount of pixel data, drawing recognition usually necessitates a fairly large working memory area and lengthy processing time. Thus, reducing the number of processing operations becomes very important. One possible solution is to use direct processing of a compressed image made by the run-length coding method. The number of runs in a drawing is approximately 1/80th the number of pixels [average based on eight standard facsimile images of CCITT (International Telegraph and Telephone Consultative Committee)]. Since less data is processed, there is a possibility of reducing the total number of processing operations.

Image processing using run-length-coded images is accomplished by performing a scalar operation on the position coordinates of the starting and ending points of a run, as well as on the length of the run. For example, a figure can be contracted in the X direction (run direction) by adding $n/2$ to the coordinate of the starting point of a black run and by subtracting $n/2$ from the coordinate of the ending point, where n is the size to be contracted. Contraction in the Y direction (perpendicular to the run direction) is also performed by checking vertically for overlapping runs. However, conventional image processing using binary pixel data necessitates logical multiplication of $n \times n$ neighbors, thus resulting in lengthy execution time. More complex operations, such as contour tracking (Capson/1984) and thinning (Pavlidis/1986), can also be realized for run-length-coded images.

More and more large-scale drawings and documents are now being stored in computer systems in compressed digital form. Therefore, algorithms capable of directly processing compressed images are becoming increasingly important for efficient drawing recognition and document analysis.

3.3 AUTOMATIC DIGITIZERS

Based on the techniques described above, several types of automatic digitizers have been developed. One typical example is a drum-type digitizer with a color sensor for analyzing color drawings; another is a flatbed-type digitizer with a CCD (charge-coupled device) sensor for analyzing monochrome drawings.

3.3.1 Hardware Configuration

Typical examples of the two automatic digitizers (ADGs) are shown in Fig. 3.10 (Kakumoto et al./1983, 1985, Ejiri et al./1984). Both ADGs can accommodate up to A0 size (1,189 mm × 841 mm) drawings, which is the maximum paper size commercially available. The configuration of the color ADG is shown in Fig. 3.11. The monochrome ADG has a similar configuration, except for its scanning mechanism, scanning head, and color-coding circuit.

(a)

(b)

Figure 3.10 Automatic digitizers: (a) drum-type for color drawings; (b) flatbed-type for monochrome drawings.

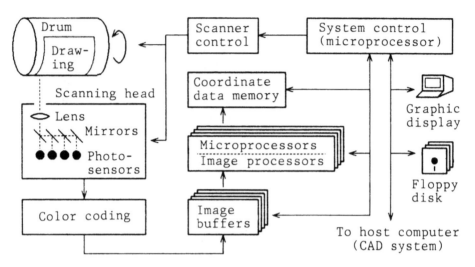

Figure 3.11 Configuration of automatic digitizer for color drawings.

In the color ADG, image signals from four photomultipliers are fed into a color-detection circuit, from which a color code is output automatically in real time, synchronized with drum rotation, to one of the four image buffers. Sixteen-bit microprocessors are used in conjunction with special microprogram-controlled image processors, for line detection and contour detection, to process the input image portions in the buffer memories.

One microprocessor is dedicated for controlling the operation of the entire system including image input, man-machine interaction, and filing, as well as for communication with the host computer. The scanning head of the monochrome ADG consists of a CCD-type linear sensor mounted on a servo-controlled carrier, which is itself mounted on a servo-controlled arm, thus realizing *XY* motion on the flatbed plane. The color-coding circuit is replaced with a thresholding circuit in the monochrome ADG to obtain binary image data from the CCD output.

3.3.2 Color-Detection Circuit

To calculate the color code in real time, a color-detection circuit is used, based on the principle mentioned earlier. The color-detection circuit is shown in Fig 3.12. Here, a combination of dichroic mirrors and trimming filters having a peak transmittance at 455 nm, 515 nm, 675 nm, and 765 nm, each with 25 nm full-width-half-maximal (FWHM), is adopted to cover the spectra of the six color pencils used. Spectral absorbances of the six color pencils are shown in Fig. 3.13. Analog intensity signals for each spectral

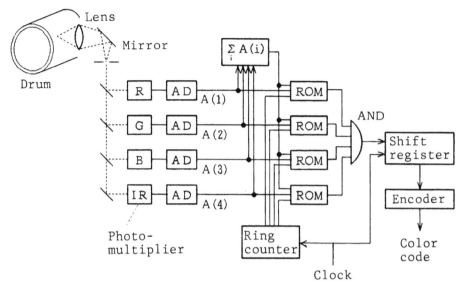

Figure 3.12   Color-detection circuit.

band from the four photomultipliers are converted to digital values by AD (analog-to-digital) converters and are then summed up by an adder circuit. The resulting sum and each digital value are fed into ROMs (read only memories), which contain predetermined tolerance values $T(i,j)$. These tolerance values are selected by a ring counter for each candidate color sequentially, and the color that outputs logical 1s from all ROMs is selected as the resulting color.   A shift register enables the storing of position information of all logical 1s, and an encoder outputs a color code. When two more photomultipliers are used, and when the sensitivity ranges of the band-pass filters are adequately chosen to cover the visible and infrared regions, a total of 12 commonly available colors can be discriminated reasonably well by this method.

3.3.3   Setting of the Scanning Area

The are to be scanned in a drawing is specified by manually positioning the ADG scanner at the upper left and lower right corners of the area. Coordinates for these points are read automatically from encoders attached to the scanner servomechanisms. The scanning area can also be specified by setting the scanner at the upper left corner of the intended scanning area, followed by keyboard input of numerical X and Y width data. After the scanning area is set, a starting switch initiates automatic scanning of the area

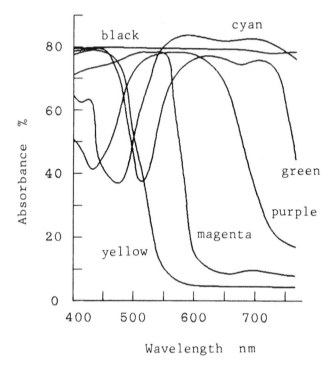

Figure 3.13 Spectral characteristics of color pencils.

at a constant pitch of 0.1 mm. This simple method is especially well suited for inputting roughly sketched drawings in which accuracy is not critical.

When fairly good accuracy is required, a minimum of three cross marks in the shape of a plus sign (+) can be placed at the corners of the drawing: upper left, upper right, and lower left. Once rough positions are input into the ADG, local images of the marked areas are picked up, and the precise locations of the marks are automatically detected by analyzing a histogram of the pixels projected on the X and Y in each area.

3.3.4   Grid Locking of Data

Sometimes, a sheet with 1-mm-interval grids is used in certain design fields where it is necessary for designers to draw lines on a grid to obtain the required design accuracy. This is not always easy to do, and there is no guarantee that every line will lie precisely on the grid. This situation is shown in Fig. 3.14. In such cases, all extracted coordinates should be pulled

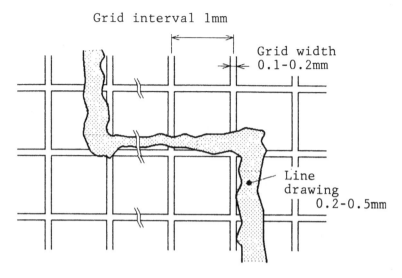

Figure 3.14 Position relation between grids and actual lines.

in and locked on the grid as if they were drawn exactly on the grid lines in the original drawing.

Since the detection of each grid line entails a lengthy tracking process and interferes with the detection of lines in the drawing, the grids are optically filtered out during the color-detection process. Instead, a grid-generating function can be added to the ADG. This grid generation is based on a parallelogram-type coordinate transformation, using the machine-read coordinates of the three cross marks mentioned above and the actual number of grid lines between the marks. This transformation controls the image-input timing signal originating from the encoders with a resolution of 0.01 mm, thus eliminating the effect of uniform sheet expansion. This function also allows for nonprecise positioning of the drawing on the ADG. All data are extracted on the basis of this generated grid, and detected coordinates are replaced by the nearest mesh points in the generated grid. Using this locking function, a drawing can be digitized into integer-valued data with a resolution of 1 mm. This corresponds to 1 micrometer in an actual design when the drawing is made with a magnification of 1,000, as in the case of LSI cell design. This function also allows designers to draw lines between grid lines. In this case, the ADG can also generate a hypothetical 0.5-mm-wide grid and pull the measured data onto those grid points. Thus, the grid-locking function provides a scheme for automatically rectifying or reforming the obtained data.

3.3.5 Other Functions

Perfect recognition is not always easy to attain. This is especially so when scratched or deteriorated drawings must be read. Therefore, an interactive data-modification function is a requisite for the ADG. In addition, coded data from old drawings are often displayed and modified interactively to yield new design versions. This is usually faster than modifying the original drawing and again inputting it with the ADG. For such situations, an interactive modification and editing function is implemented in the ADG and is executed in parallel while another drawing is being read. This function includes enlarging an arbitrary portion of the drawing; moving the cursor to an arbitrary point; connecting and disconnecting lines; and inserting and deleting lines, figures, characters, and symbols. Each task can be initiated either by keyed-in commands or by pointing to displayed function blocks. Final data processed by the ADG can be transferred to any commercially available CAD system using floppy disks or magnetic tapes. These data can also be sent directly to the CAD system or a host computer for further processing.

3.4 RECOGNITION OF ENGINEERING DRAWINGS

Automatic digitizers can be applied effectively to certain types of engineering drawings, mainly as a means of initial data input for subsequent CAD processes. The state of the art of applications is discussed in this section, together with some examples.

3.4.1 LSI Cell Diagrams

Perhaps the most fundamental process in LSI design is designing the cells used in LSIs. A cell diagram is often drawn on a single sheet, using color pencils in such a way that the cell's multilayered structure can be seen easily, thus allowing the designers to check for inclusions and overlapping areas between figures in different layers. This application uses the color ADG mentioned above. Here, solid lines and two types of broken lines can be recognized. The drawing rules for such broken lines are shown in Fig. 3.15. A total of 18 different lines, consisting of the 3 line types in combination with the 6 colors, can be used in one drawing. Since the extracted line data are automatically assigned one of 18 different line-type codes, it is easy to display an arbitrary layer either separately or in combination with other layers.

Normally, when colored pencils are used, line segments in one layer that overlay or underlay lines in another layer and that share the same XY coordinates result in confusion, not only for the machine but also for the

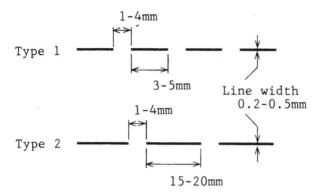

Figure 3.15 Drawing rules for broken lines used in LSI cell diagrams.

designer. Since overlaying a solid line on a broken line does not make sense when the goal is to represent distinguishable information, designers usually draw only one of the overlaying lines and omit the rest. A mark is then put on the line at an appropriate point to show where the overlay exists. In this case, a short perpendicular line segment, as shown in Fig. 3.16, is used to represent the existence of an overlapping, but omitted, line. The mark is the same color as the omitted line. These overlap symbols are considered to be no more than line segments just after line detection. However, within the set of recognized line segments, it is possible to find short, unconnected segments that can be considered as candidates for overlap symbols.

Since a line route with a perpendicular symbol on it indicates the existence of another line, other lines connected to the same route and having the same color as the symbol are searched to eventually find the closed loop. This loop finding is based on the fact that the lines drawn

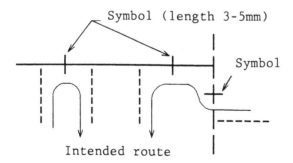

Figure 3.16 Overlap symbol.

in LSI cell diagrams essentially represent the contours of two-dimensional regions. Using this method, the drawing in Fig. 3.17(a) can be decomposed into lines, as shown in Fig. 3.17(b), and then interpreted as the separate figures shown in Fig. 3.17(c). This function is implemented as an application program for the ADG, and is especially useful in recognizing LSI cell diagrams (Shimada et al./1985).

In addition, an interpretation function for wires is added. In this function, open-ended line segments are regarded as the center lines of connecting wires in the LSI. These segments are automatically converted into a figure with some prescribed width, as shown in Fig. 3.18. By using the CAD software in the host computer, these wires are finally transformed into a combination of rectangles that will be matched in the later mask-fabrication process.

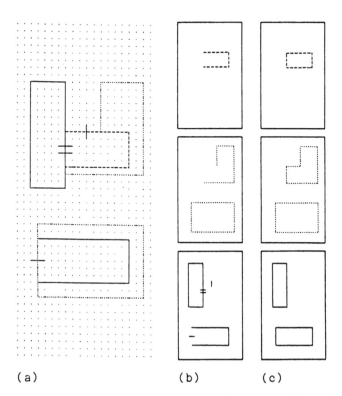

(a)   (b)   (c)

Figure 3.17   Recognition of overlapped lines: (a) original shapes (after line-type recognition); (b) detected and decomposed lines, and (c) interpreted shapes.

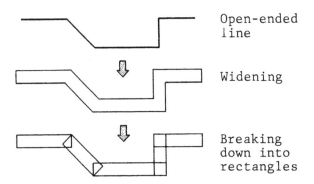

Figure 3.18   Interpretation of wires.

A typical LSI cell diagram is shown in Fig. 3.19(a). The original is drawn on a polyester sheet with mechanical color pencils. The result of coding using a color ADG is shown in Fig. 3.19(b). This is a printed image of the CRT on which the diagram is reproduced and displayed from the vector data. Further processing by CAD software in the host computer yields the result shown in Fig. 3.19(c), which is reproduced by an *XY* plotter. The results of the overlap symbol interpretation and the figure generation for wires are shown in this figure. Processing time for the ADG with a standard A3-size (420- × 297-mm) cell diagram is approximately 5 to 8 minutes depending on diagram complexity. This is one half the time required for conventional manual digitizing by a skilled person. When compared with the time required by an unskilled operator, this ratio can be reduced to less than 1/10th. Thus, shorter turn-around time can be achieved in the design process. This contributes to faster response in designing new products to meet market needs.

3.4.2   Liquid-Crystal Patterns

Liquid-crystal patterns usually consist only of solid figures, which are drawn on a sheet as black-and-white patterns. Therefore, the contour-tracking mentioned earlier is commonly used to code the patterns.

An example of an original drawing designed by an artist is shown in Fig. 3.20(a). This cartoon pattern is digitized by a monochrome ADG and is converted into a set of contour vectors, as shown in Fig. 3.20(b). A subsequent CAD process adds the wiring and electrode information, as shown in Fig. 3.20(c). The final output is then used to produce masks for the liquid-crystal panels. This example is a pattern for a game-watch that once gained wide popularity. Nowadays, liquid-crystal boards for almost

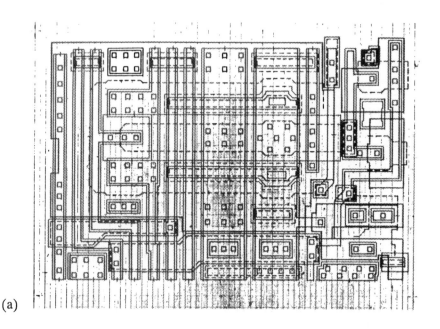

(a)

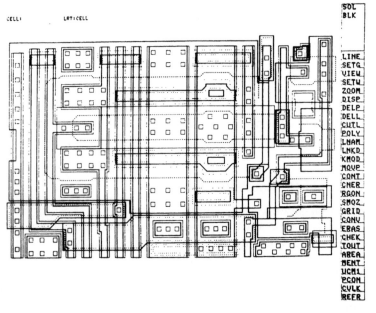

(b)

Figure 3.19 Recognition of LSI cell diagram: (a) LSI cell diagram (original: color), (b) result of coding (printed CRT-image), and (c) final result after CAD processing (*XY*-plotter output).

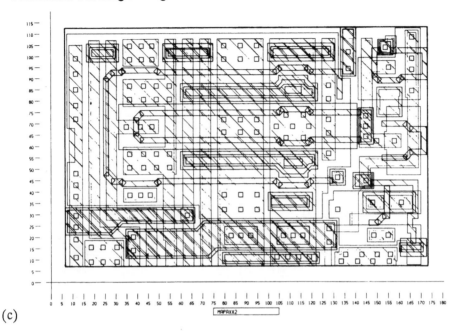

(c)

any kind of application can be produced using this method. In most cases, the input time is reduced from 1/5th to 1/20th depending on the complexity of the drawings. One direct benefit of the ADG in this application is that it makes it possible to input drawings that are too complicated to cope with using the conventional manual method.

3.4.3 Printed-Circuit Patterns

In recent years, diversified customer needs for electronic products have served as the driving force behind the development of smaller, more functional products with a shorter product life cycle. These needs can be met by a variety of high-precision and high-density PCBs (printed circuit boards) designed using CAD systems. Consequently, an ADG capable of inputting original design drawings to the CAD system at high speed is becoming increasingly necessary.

To simplify the inputting task, drawings are sometimes drafted with certain restrictions. In one such design application, the circular terminal symbols and connecting lines are drawn on the grid of a piece of specially designed graph paper (Masui et al./1980). In this application, parallel processing hardware is developed that is capable of recognizing symbols, as well as lines with their width information.

(a)

(b)

Figure 3.20 Recognition of liquid-crystal pattern: (a) original design, (b) result of contour detection (printed CRT-image), and (c) final result after CAD processing (*XY*-plotter output, in part).

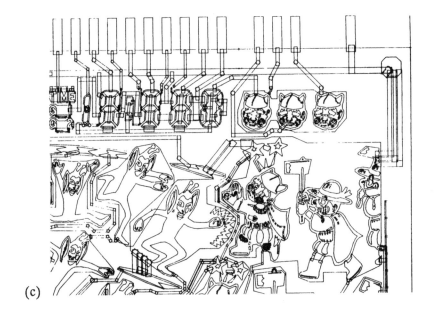

(c)

Greater flexibility is required if PCB patterns are to be drawn freely on unlined paper. In one such design application for hybrid (i.e., analog and digital) PCBs, patterns are drafted using contour lines (Takagi et al./ 1982). In this case, the contour lines in an image are tracked using a 2-D mask operator. Another application uses a unique method for symbol recognition, where an area containing a symbol is first extracted and the outside portion of the symbol in the detected area is then filled with logical 1s. The inner contour is then tracked. Again, the inner contour of the remaining area inside the symbol is tracked, until all inner portions are filled. This repeated filling and tracking and sequential finding of the inner pattern contour facilitates the recognition of symbols (Kikkawa et al./1984).

The approaches just discussed use extensive image processing in which pixel data are accessed directly. When applied to a large drawing, a large-scale memory area is needed. In contrast are approaches in which the drawing data are coded in an early processing stage, and the resulting vector data are used extensively for subsequent processing. Figure 3.21 shows an example of the wire-connection pattern drafted as a line drawing. This pattern is then vectorized, and circular symbols are automatically generated at each end of the wires. Figure 3.22 shows an example of a hybrid PCB, drafted as a silhouette pattern. Contour lines are first tracked and coded to vector data. These vector data are then contracted in size to find circular symbol areas, and the extracted areas are used to measure the original

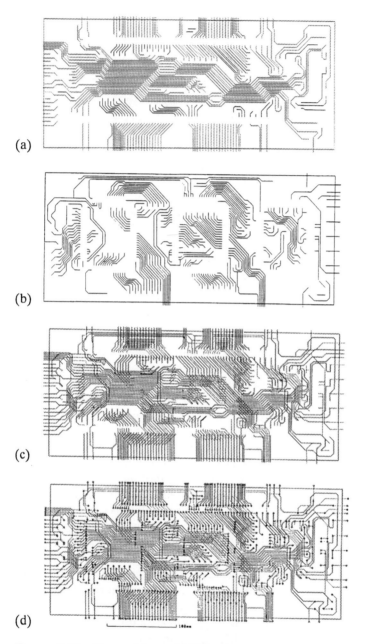

(a)

(b)

(c)

(d)

Figure 3.21 Recognition of PCB wire pattern: (a) input wire pattern (front side), (b) input wire pattern (back side), (c) registration of two input patterns, and (d) through-hole circle generation.

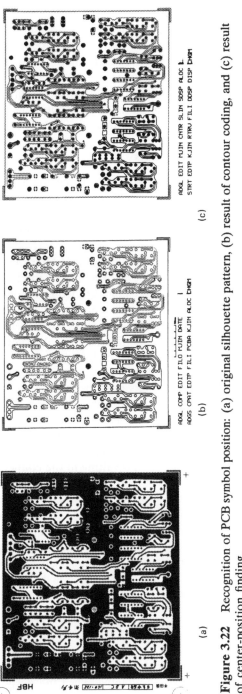

Figure 3.22 Recognition of PCB symbol position: (a) original silhouette pattern, (b) result of contour coding, and (c) result of center-position finding.

symbol sizes and their center positions. Figure 3.23 shows another example of a PCB, where contour lines and medial lines are vectorized. Based on these results, a network can be generated by finding the connection between circular through-hole symbols.

There is still no single standard method for designing PCBs; therefore, various approaches for cost-minimum designing have been tried. An ADG is expected to be one of the effective devices for this purpose.

3.4.4 Logic-Circuit Diagrams

Logic-circuit diagrams usually consist of three main elements: symbols representing logic-circuit components, lines representing connecting wires, and characters and numbers representing component names and relational attributes. An example of a logic-circuit diagram is shown in Fig. 3.24(a). This is a reduced image of an A4-sized portion, derived from the A2-sized original diagram digitized with a resolution of 16 lines/mm. The recognition of such logic-circuit diagrams can be executed by the following sequence. First, characters and numbers are separated from component symbols and wire lines, based on a size-and-relation feature that assumes that characters and numbers are small, isolated figures. Next, for the extracted character-and-number image, contour tracking is executed. A structure analysis of the resulting set of points coordinates that make up the contour is then executed to recognize the characters and numbers.

The part of the image containing the remaining component symbols and wire lines is thinned, and the resulting medial lines are converted to a set of point coordinates. Figure 3.24(b) shows the results of the thinning operation applied to the remaining symbol-and-line image. From this, feature points such as end points, branch points, cross points, and corner points are extracted and analyzed, resulting in the connection information for the wire lines. Loops also can be extracted from the medial line data. By analyzing the shape of the loops, component types are recognized in one of the four orientations in the horizontal and vertical directions.

Finally, recognized components and characters are replaced with pre-scribed symbols and fonts with standard shapes, and recognized wire lines are straightened to be horizontal and vertical, as shown in Fig. 3.24(c). Other final outputs in the drawing-recognition task include a table of components and a table of wiring information between each component.

3.4.5 Chemical-Plant Diagrams

In the petroleum, chemical, and nuclear power industries, piping and instrument drawings are used widely in the planning, design, construction, and management of plants. To effectively use the design data contained

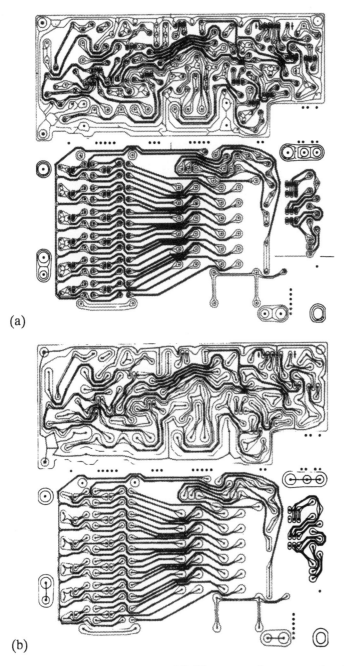

(a)

(b)

Figure 3.23 Recognition of PCB connection network: (a) medial-line detection; (b) connection-network finding.

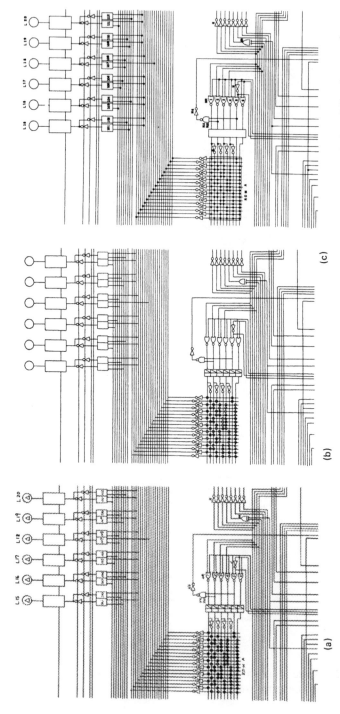

Figure 3.24 Recognition of logic-circuit diagram: (a) digitized original image, (b) thinning for symbol and line image after removing characters and numbers, and (c) final result after straightening and symbol and font replacement.

in such drawings, automatic recognition of hand-drawn plant diagrams is considered as being an initial data input for a CAD system (Furuta et al./ 1984). Figure 3.25(a) shows a part of one such diagram, in which an attempt has been made to detect characters, lines, and symbols.

Here, characters are usually extracted by checking the size of a rectangular area generated around a set of connected vectors. Symbols with smaller, clearly defined forms can be recognized using a combination of feature-matching methods based on the structure-analysis and pattern-matching methods. First, the features of a symbol, such as the existence of any solid areas, the number of loops, the ratio of symbol area to surrounding rectangular area, the peak position of the projected image, the shapes of the inner and outer contours, etc., are extracted. Then, these features are evaluated at each branch point of a predetermined decision tree, until the symbol is classified into one of a number of symbol groups, each consisting of similar symbols. Finally, pattern matching is executed with standard symbol patterns within the detected group.

Larger symbols with previously undefined forms, such as the chemical vessel at the left in Fig. 3.25(a), can also be analyzed and extracted by the repetitive execution of complicated line-and-arc fitting, interpretation, and conflict-finding techniques. The final result obtained is shown in Fig. 3.25(b), which is edited to some extent by manual operations. For complete automation of the process, however, this approach is far from perfect. A more intensive study is needed before this approach finds wider usage in actual applications.

3.4.6   Mechanical-Engineering Drawings

Automatic recognition of mechanical drawings offers tremendous application potential in the machinery, aerospace, ship-building, and automobile industries; in most of these companies, there are huge piles of drawings, many of which are still active. The volume of such drawings continues to grow, since CAD systems have not yet made many inroads into the mechanical design field.

In the case of periodic maintenance of machinery, drawings are usually distributed to the maintenance people to facilitate maintenance and repair. Thus, some kind of computer system is needed for storing and managing drawings. This system should allow quick referencing and on-demand reproduction. In this case, the drawings can simply be converted to compressed digital form without any further processing. Of course, the resulting coded data should, at least, be adequate for re-

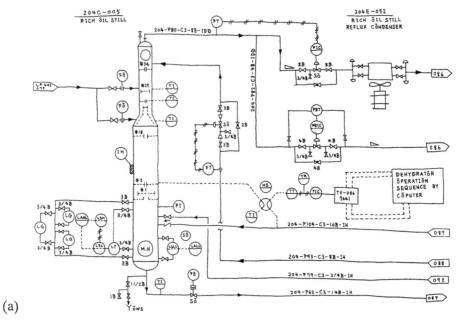

(a)

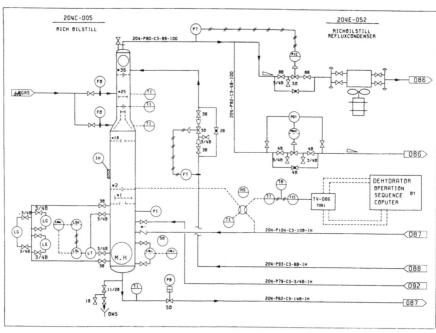

(b)

Figure 3.25 Recognition of chemical-plant diagram: (a) original image; (b) detected and rectified result. Source: [M. Furuta, Information Technology Promotion Agency, Tokyo, Japan]

producing drawings that approximate the original drawings as closely as possible.

On the other hand, when more flexible design and fabrication automation is the goal, intensive recognition of drawings is a requisite. Recognition of mechanical-engineering drawings poses somewhat different problems from recognition of other kinds of drawings. This is because the object expressed in a drawing is usually complex and intrinsically three-dimensional. Therefore, the resulting output of the drawing-recognition task must be a three-dimensional model that can be handled easily by CAD systems and NC (numerically controlled) machines (Hofer-Alfeis/1986). If this type of high-performance recognition can be obtained, it will become possible to build a CAD database of standard mechanical parts automatically. This method would allow for the repetitive use of data for variational designs and could be used to generate fabrication data for NC machines. Unfortunately, the current state of the art of drawing recognition does not yet allow for this level of application.

In the recognition of mechanical-engineering drawings, one problem arises in the binarization process of existing drawings, where degradation problems are apt to be encountered. This necessitates floating-type binarization using a locally adapted threshold rather than a global one, in order to always obtain good and reliable images.

After vectorization, symbols must be discriminated from each other. These symbols include dimensioning symbols (witness lines, leader lines, dimension numbers, and tolerances), symmetry symbols (dash-dotted center lines), sectioning symbols, and production symbols (surface descriptions). Some of the symbols in mechanical drawings have no specific meaning by themselves; however, they take on meaning when combined with other lines and symbols in the drawing and, thus, provide very important information. Consequently, these symbols must be analyzed, integrated, and interpreted in connection with the lines representing the object.

It is then necessary to reconstruct the two-dimensional geometry by rectifying the vector data according to the dimension information derived in the above process. Then, by combining views, sections, and details represented in the drawing, a three-dimensional model can be reconstructed. There are already several approaches to this 3-D reconstruction problem; however, these are still basic (Aldefeld/1983, Sakurai and Gossard/1983, Aldefeld and Richter/1984).

The three processes described above (i.e., integrated symbol recognition, 2-D reconstruction, and 3-D reconstruction) pose difficult problems and need to be studied more intensively before an efficient drawing-recognition method can be established. A knowledge-based method with a scheme that controls top-down and bottom-up analysis might be one of the approaches toward obtaining an eventual solution.

3.4.7 Other Applications

Drawing-recognition techniques are applicable to many other fields, such as recognizing the texture patterns of fabrics in the textile industry, the paper patterns of dresses in the fashion industry, and the architectural plans of room layouts in the construction industry. Recognition of weather charts is also being studied, aiming at realizing a long-term climate database. Technical graphics, such as those used in repair manuals for automobiles and office-automation equipment, can also be considered to fall within the application field. One extraordinary application of an ADG is for determining the DNA (deoxyribonucleic acid) sequence by analyzing the autoradiograph obtained in electrophoresis. In this case, films to be read are not binary-valued, and an intensive peak analysis of stripes in a poorly contrasted gray-level image is necessary to determine the sequence of the four types of nucleic acids. In the future, the fields of application can be expected to increase even more as the technology improves.

3.5 RECOGNITION OF MAPS

In local governments and public utility companies, information on facilities—such as the location of electric cables, conduit pipes, telephone cables, and various other properties—is managed with the aid of district maps. In urban planning and national land planning, more complex topological maps are used extensively. To obtain a computer-based management and planning system, it is necessary to be able to enter commercially available maps directly into the computer, thereby creating a useful map-based information-processing system of appropriate scale for each objective.

3.5.1 Inputting Maps

Maps are one of the most complicated types of drawings. Unfortunately, drawing-recognition technology is not yet sophisticated enough for automated map reading. One simple yet reliable method for inputting maps is to trace each layer of information contained on the maps separately onto different sheets of paper. Each sheet is then read by a monochrome ADG and is assigned to one of the multiple layers. The computer registers all layers by checking the position of standard marks put on each layer. This method is also used when district maps are first produced. Here, each map layer is drafted based on firsthand survey data and is input by an ADG.

Figure 3.26 shows an example of a house-frame layer that has been first input and then output by the plotter. This layer is then combined with a road-boundary layer and a typeset character layer for printing. Recently, these layer data are becoming available on the market as digital maps in CD-ROMs (optical disks).

Another simple yet reliable method is, again, to trace maps with colored pencils and to use the color ADG described earlier. Using this method, it is possible to construct a rather small-scale map-based information-processing system. Each color can be assigned to one of the hierarchical levels. Thus, the map data are expressed internally as a multilayered structure.

In general, tracing maps is a costly process, and human errors and degradation of precision are apt to be encountered. Therefore, a more direct input method is preferable. For example, all information on a map is first vectorized automatically using an ADG. Then, an operator points to

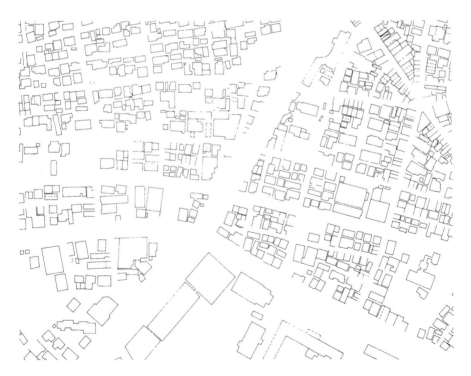

Figure 3.26 Layer of house frame for district-map production (*XY*-plotter output, in part).

specific vectors on a display, and these vectors are discriminated from the others and memorized into a specific layer. To achieve high efficiency, it is desirable that the largest number of line vectors be handled in the least number of operations. For example, when the operator assigns altitude data to a displayed contour vector, the same altitude data are automatically added to all line vectors connected to this vector to make up an equi-altitude contour. When the operator adds the altitude interval data and designates the descending or ascending direction among contour lines, an automatic assignment of altitude values begins sequentially from the nearest contour lines inward and outward, until a conflict is found, at which point the computer asks the operator to make a judgment.

For such interactive layering, it is sometimes helpful to rotate the map data on a screen for easy pointing. Since the map data are already vectorized, the rotation can be easily and quickly executed, as shown in Fig. 3.27. In this case, buildings are designated as one of the layers.

By using the above-mentioned techniques together with the conventional figure-editing techniques, an efficient man-machine interactive layering can be achieved, combining higher level human judgment and lower level, yet efficient, computer decisions. Data remaining untouched by this interactive method can be stored as another layer, representing background figures. This method facilitates adequate layering of maps for each application, yet minimizes the amount of data storage.

3.5.2 Extraction of Fundamental Figures

To automatically derive each layer (or subject map) directly from more complicated topographical maps, sophisticated image-processing technology must be applied. The map shown in Fig. 3.28(a) depicts a portion of the greater Tokyo area, digitized with a resolution of 10 lines/mm. This digitized image has $1,536 \times 1,536$ pixels and corresponds to an area of 4×4 km. The original map has a scale of 1/25,000 and was issued by the Geographical Survey Institute of Japan. To extract a road map from this image, a very complicated algorithm is needed, since the image generally involves such complicated patterns as symbols, characters, residential sections, rivers, railroads, etc. A recently developed algorithm can extract two groups of roads, each expressed as parallel lines, but having different widths in the map (Miyatake et al./1985). The algorithm is summarized as follows (refer to Fig. 3.29):

Let $P(i,j)$ be a binary image after preprocessing, by which isolated small figures are removed from an originally input binary image, and $\overline{P}(i,j)$ an inverse image of P. Then, a new image $Q(i,j)$ is obtained from

$$Q(i,j) = T(L(T(W(\overline{P}(i,j));a,b));c,\infty) \tag{3.2}$$

(a)

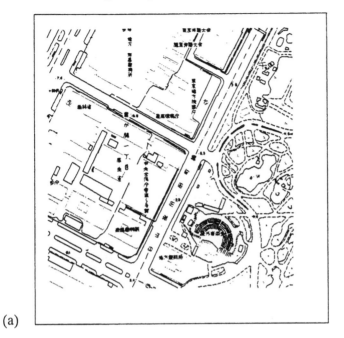

(b)

Figure 3.27 Map rotation for interactive layering: (a) vectorized original data; (b) rotated image.

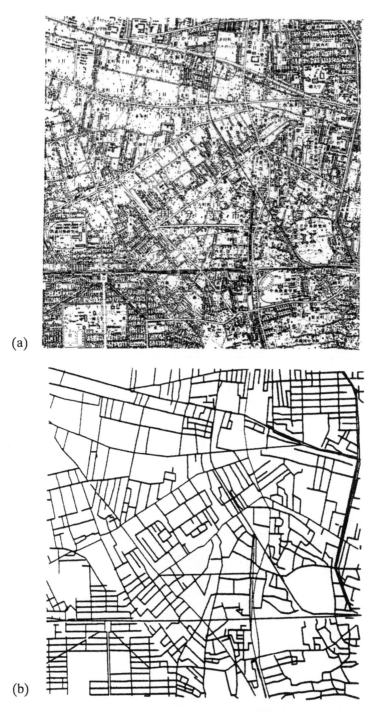

(a)

(b)

Figure 3.28 Extraction of roads from topographical map: (a) original digitized image; (b) final result after modification.

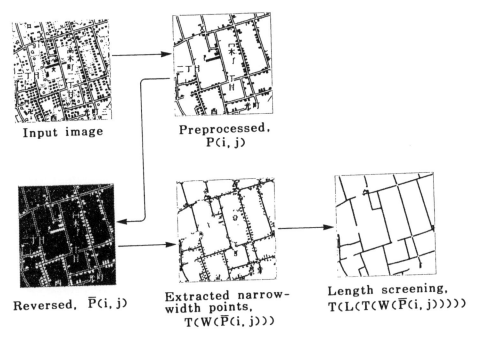

Figure 3.29 Road-extraction algorithm.

Here, W and L represent the width and length of a figure at point (i,j), as shown in Fig. 3.30. T is a thresholding function that extracts pixels with the value of $[a,b]$. Thus, the interior portion of the parallel lines having a width of a to b and length of more than c is extracted. This is followed by a concatenation process of the extracted pixels based on the expansion-and-contraction method, and the obtained image is tracked by a line-following technique using eight connected neighbors. The extracted data for both groups of roads are then combined, and a few modifications, mainly supplementing lines that were omitted in the original map because of character overlays, are added interactively to reproduce the final road map data, as shown in Fig. 3.28(b).

Residential areas with heavier populations are sometimes shown as hatched areas on the map. An example of hatched-area extraction, based on a pattern-matching technique, is shown in Fig. 3.31. Here, a hatched pattern with a 5×5 matrix is used as a reference pattern. When used in conjunction with the above road-extraction algorithm, this algorithm

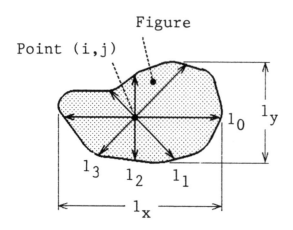

$$W(*)=Min(l_0, l_1, l_2, l_3)$$

$$L(*)=Max(l_x, l_y)$$

Figure 3.30 Definition of width and length functions.

makes it possible to extract even finer road information because residential districts usually have common boundaries with roads. The image processing realized by these algorithms, however, is still too lengthy and costly to be implemented with ADGs. Consequently, a more powerful, more carefully designed parallel image processor will be needed in the future.

3.5.3 Inputting Characters and Symbols

As mentioned earlier, recognition of characters and symbols in a drawing poses a problem different from that usually encountered in OCRs. In a map, characters are usually mixed with other figures and lines. Even characters in a single word are sometimes distributed and arranged separately from each other. Oblique arrangements are also common. To make this situation even more complicated, more than 4,000 different Kanji characters can be used in Japanese maps. Also, a character may be printed in various sizes and fonts, each of which is assigned to a different usage. Thus, the problem of automatic recognition of characters in topographical maps will not be solved for some time. However, automatic extraction of candidate areas where characters may exist is possible by finding a set of strokes

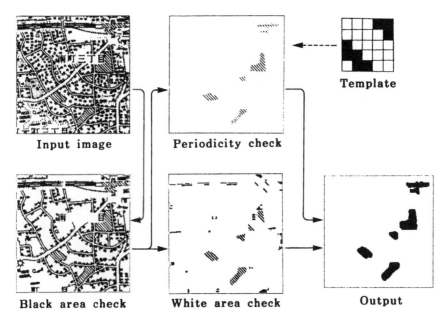

Input image   Periodicity check   Template

Black area check   White area check   Output

Figure 3.31   Extraction of hatched residential areas.

with appropriate vector lengths, as shown in Fig. 3.32. These areas can be converted easily to character codes with human assistance. To do so, a conventional method used in Japanese word processors can be applied effectively (Fujikata et al./1981). In this method, the phonetic reading of an automatically cursored Kanji character is keyed in, and the corresponding character candidates are displayed on the screen in an arrangement that duplicates the keypad on the keyboard. Pressing one of the keys inputs the desired character code. A character pattern generated from the code is then substituted for the image portion on the display, and the location of the character and its code are stored in memory.

Symbols representing post offices, schools, hospitals, or other landmarks can also be input by the same interactive method. They can also be input automatically by a pattern-matching technique, if the corresponding image portion is guaranteed to hold a symbol. This is because the number of symbols is rather limited compared to the number of Kanji characters. However, finding candidate symbol areas is generally more difficult, and an effective algorithm for this purpose has not yet been developed. When the extraction of roads and residential sections proceeds, the number of remaining uncoded images decreases, which facilitates the finding of symbol

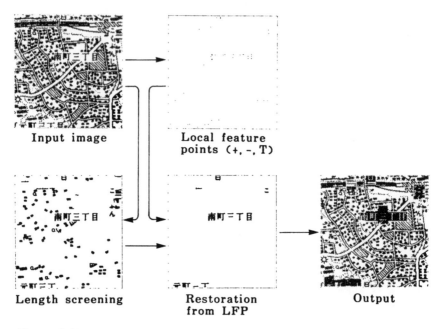

Figure 3.32   Finding candidate character areas.

areas. By accumulating such algorithms, automatic input of topographical maps may eventually be possible. These efforts will play an important role in the future as a key technology for fully automatic layering of more complicated maps.

3.5.4   Map-Based Information-Processing System

As it becomes easier to input maps to a computer system, the possibility of realizing a map-based information-processing system increases. There already exist several system products that meet the requirements for water, sewage, gas, and electric power utilities. Figure 3.33(a) shows such a system, called AQUAMAP, developed especially for managing water facilities for local governments. In this system, conduit pipes and their specifications and histories, such as diameters, materials, and installation date, are stored together with map data in a database (Tsutsui et al./1986). Figure 3.33(b) shows an example where conduit pipes are displayed with their attributes. When replacement or installation of new pipes is needed, the display quickly shows what valves should be closed and what pipelines will be

(a)

(b)

Figure 3.33   A map-based water-facility management system, AQUAMAP: (a) hardware configuration; (b) example of display.

affected when the valves are shut. Pressure distribution and water flow also can be simulated using this system.

The application of such systems to actual operation is just beginning, and more sophisticated systems will be needed in the future. As part of the basic research for building an intelligent map-based information-processing system for the future, a *gen*eralized *t*opographical *l*and-use *e*xpert (GENTLE) system has been prototyped on a mainframe computer using LISP language (Shimada and Ejiri/1986). Through the time-sharing mode of the computer, the system responds to menu-driven commands and also to natural-language commands given by an operator, infers its knowledge base and database, and outputs various data. The configuration of the database is shown in Fig. 3.34. In this system, *figure* data are expressed as a combination of structured subject maps (layers) and a nonstructured background map. The background map is represented as a set of vector data, having coordinate data as its component. In the structured subject maps, figure data are described as geometric relations between point data, line data,

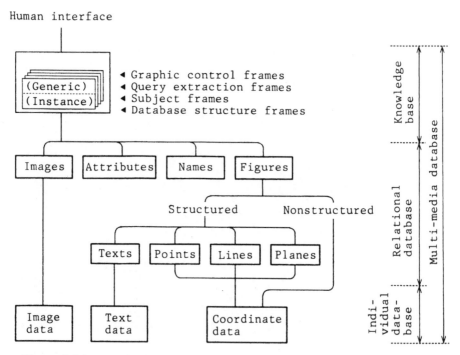

Figure 3.34 Configuration of multimedia database.

and plane data. These figure data are again combined with other data such as names, attributes, and images to form a relational database. For example, the name of a city is related to plane data representing the city area. These plane data are also related to the line data representing the city boundary, which in turn are related to the point data that represent the coordinates of the edge points of each vector constituting the boundary. Therefore, by inputting the city name, the system can easily display the city with a highlight. Furthermore, a knowledge base is used as an interface between the user and the underlying database. A summary of the database structure is also included within the knowledge base, thus resulting in an intelligent multimedia database system with efficient information-retrieval capabilities.

Once the map information is expressed in the combined form of knowledge base and relational database, various inference retrievals of information are possible among names, attributes, and figure components. Typical examples are as follows:

> Name-to-Figure. A keyed-in character set of an arbitrary length can be matched to a registered name, and corresponding figures are highlighted on the display. Figure 3.35 shows the Kasumigaseki area, the government center in Tokyo, where several buildings matched to the input "SHOU" are displayed with hatching, and a building matched with "GAIMU" is also displayed as a solid figure. Here, "SHOU" is a suffix of government office names, and "GAIMU" is a partial input of GAIMUSHOU, an official name of the Japanese Ministry of Foreign Affairs.
>
> Figure-to-Attribute. By moving the cursor to a figure on the screen, a table of various attributes related to the figure is displayed, as shown in Fig. 3.35. This example shows that the building in the lower left of the display is Government Office Building #4, shared by various offices appearing in the first column. Floor numbers and telephone numbers of these offices are also displayed in the table.

In addition to these references, various other references are possible, such as attribute-to-figure, figure-to-name, and figure-to-image. When a scene is stored in advance as an *image* in relation to a triangular *figure*, pointing to the figure causes the system to display the previously stored image of the city as seen from the baseline of the triangle in the direction of the apex, as shown in Fig. 3.35. When degrees of traffic congestion and/or maximum speed limits are added to roads as their attributes, a time-minimum route as well as a distance-minimum route can be found, as shown in Fig. 3.36, by a route-finding algorithm (Hart et al./1968).

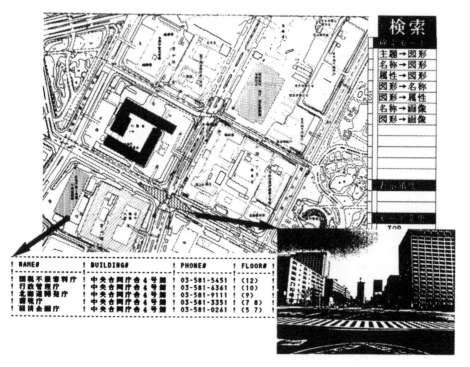

Figure 3.35   Examples of information retrieval in GENTLE.

In addition to the usual method of menu selection to indicate information retrieval, a natural-language interface is provided in GENTLE. Figure 3.37 shows the man-machine interface portion of GENTLE. Here, an input sentence (in this case, phonetic input of the Japanese language) is analyzed by using the knowledge base to eventually yield a query language for the database. One typical example of a sentence (in Japanese) has the form:

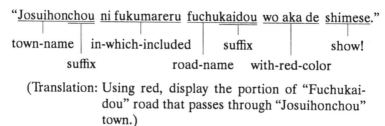

"Josuihonchou ni fukumareru fuchukaidou wo aka de shimese."

town-name | in-which-included | suffix | show!

suffix | road-name | with-red-color

(Translation: Using red, display the portion of "Fuchukaidou" road that passes through "Josuihonchou" town.)

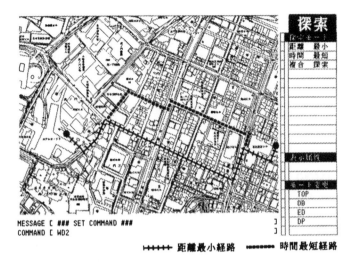

MESSAGE [### SET COMMAND ###]
COMMAND [WD2   ]

├┼┼┼┼┤ 距離最小経路   ━━━━━━ 時間最短経路

Figure 3.36 Route-finding in GENTLE (chained line shows distance-minimum route; dotted line shows time-minimum route).

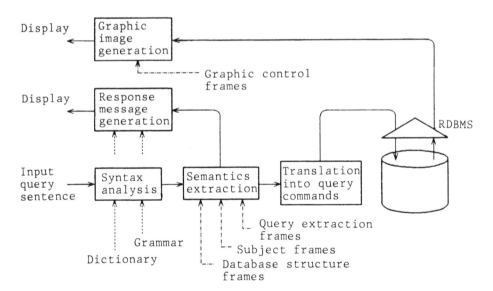

Figure 3.37 Natural-language interface of GENTLE. (Note: RDBMS stands for relational database management system.)

This language command involves a logical operation of figures between line data (e.g., road) and plane data (e.g., town). Though there is no such information stored directly in the database, the system infers the knowledge base, finds the strategy needed to obtain the portion of the road, and displays it using the designated color.

GENTLE is just one example of a future-oriented, map-based information-processing system and is typical of the many useful functions that can be installed in more powerful systems.

3.6 CONCLUSIONS

Drawing recognition is a technology that converts two-dimensional image data, such as lines, symbols, and characters contained in drawings, into a set of vector data, symbol codes, and character codes, respectively. The relationships among these are then analyzed, resulting in an understanding of the syntax and semantics contained in the drawings. This technology is useful not only as a means of initial data input for subsequent CAD processes, but also as a means of storing and accumulating drawing data in a CAD database for efficient management and usage.

From the viewpoint of image processing, drawing recognition can be characterized by the fact that it usually handles intrinsically binary-valued, albeit relatively large-scale, image data. Various types of drawings are drafted by different rules, and the variety and complexity of these drawings make it almost impossible to obtain a single method applicable to all types of drawings.

Although the type of recognition method varies depending on each application, there are common methods underlying each application. These basic methods and algorithms for drawing recognition have been explained here. Some of the more important algorithms among these have been implemented in drawing-recognition machines, called automatic digitizers. These are now finding wider use in many design and maintenance applications, especially in the electronic-engineering field and possibly in the mechanical-engineering field of the future. Application examples were also presented for LSIs, liquid crystals, printed circuit boards, logic-circuit diagrams, etc.

As an example of a possible future application, the state of the art of map recognition was also explained. As was shown, map recognition poses many different problems, although widespread use of map-based information-processing systems can be expected in the not-too-distant future. To accelerate this application, more extensive research must be conducted on image-processing and image-understanding technology, and more effective algorithms must be accumulated in addition to the ones mentioned in this chapter.

ACKNOWLEDGMENTS

The authors would like to express thanks to their colleagues, Mr. Hitoshi Matsushima and Mrs. Shigeko Otani, for their contributions to the studies described in this chapter.

REFERENCES

Aldefeld, B. (1983). On automatic recognition of 3D structures from 2D representations, *Computer Aided Design, 15*: 2.

Aldefeld, B. and Richter, H. (1984). Semiautomatic three-dimensional interpretation of line drawings, *Computer & Graphics, 8*: 4.

Capson, D. W. (1984). An improved algorithm for the sequential extraction of boundaries from a raster scan, *Computer Vision, Graphics, and Image Processing, 28*: 109.

Ejiri, M., Uno, T., Yoda, H., Goto, T., and Takeyasu, K. (1972). A prototype intelligent robot that assembles objects from plan drawings, *Trans. IEEE Computer, C-21*: 2.

Ejiri, M., Kakumoto, S., Miyatake, T., Shimada, S., and Matsushima, H. (1984). "Automatic Recognition of Design Drawings and Maps," Proceedings of 7th International Conference on Pattern Recognition, Montreal, pp. 1296–1305.

Fujikata, K., Ueda, H., and Ejiri, M. (1981). A dynamic Kanji allocation and selection technique for a Japanese word processor (in Japanese), *Trans. IEE of Japan, 101-C*: 1.

Fulford, M. C. (1981). The FASTRAK automatic digitizing system, *Pattern Recognition, 14*: 1–6.

Furuta, M., Kase, N., and Emori, S. (1984). "Segmentation and Recognition of Symbols for Handwritten Piping & Instrument Diagram," Proceedings of 7th International Conference on Pattern Recognition, Montreal, pp. 626–629.

Granlund, G. H. (1972). Fourier preprocessing for hand print character recognition, *IEEE Trans. Computer, C-21*: 2.

Hart, P. E., Nilsson, N. J., and Raphael, B. (1968). A formal basis for heuristic determination of minimum cost paths, *IEEE Trans. on SSC, 4, no. 2*: 100.

Hilditch, C. L. (1969). Linear skeletons from square cupboards, *Machine Intelligence* (Meltzer & Michie, ed.), Edinburgh University Press, Edinburgh.

Hofer-Alfeis, J. (1986). "Automated Conversion of Existing Mechanical-Engineering Drawings to CAD Data Structure: State of the Art," Proceedings of IFIP TC5 2nd International Conference on Computer Applications in Production and Engineering (CAPE '86), Copenhagen.

Kakumoto, S., Fujimoto, Y., and Kawasaki, J. (1978). "Logic Diagram Recognition by Divide and Synthesize Method," Proceedings of IFIP Working Conference on Computer Aided Design, Grenoble, France.

Kakumoto, S., Miyatake, T., Shimada, S., and Ejiri, M. (1983). "Development of Auto-Digitizer Using Multi-Color Drawing Recognition Method," Proceedings of CVPR '83, Washington, D.C.

Kakumoto, S., Miyatake, T., Shimada, S., and Ejiri, M. (1985). An automatic drawing recognition method using a real time color coding technique (in Japanese), *Trans. IECE of Japan, J68-D*: 4.

Kikkawa, W., Kitayama, M., Miyazaki, K., Asai, H., and Arato, S., (1984). "Automatic Digitizing System for PWB Drawings," Proceedings of 7th International Conference on Pattern Recognition, Montreal, p. 1308.

Masui, T., Shimizu, S., Yoshida, M., and Abe, T. (1980). "Recognition System for Design Charts Drawn on Section Paper," Proceedings of 5th International Conference on Pattern Recognition, Miami Beach, p. 127.

Miyatake, T., Matsushima, H., and Ejiri, M. (1985). Extraction of roads from topographical maps using a parallel line extraction algorithm (in Japanese), *Trans. IECE of Japan, J68-D*: 2.

Pavlidis, T. (1986). A vectorizer and feature extractor for document recognition, *Computer Vision, Graphics, and Image Processing, 35*: 111.

Sakurai, H. and Gossard, D. C. (1985). Solid model input through orthographic views, *Computer Graphics, 17*: 3.

Shimada, S., Kakumoto, S., and Ejiri, M. (1985). Representation and interpretation of overlapped line segments for automatic recognition of LSI cell drawings (in Japanese), *Trans. IECE, J68-D*: 8.

Shimada, S., Kakumoto, S., and Ejiri, M. (1986). A recognition algorithm of dashed and chained lines for automatic inputting of drawings (in Japanese), *Trans. IECE of Japan, J69-D*: 5.

Shimada, S. and Ejiri, M. (1986). GENTLE: Multi-media topographical expert system with natural language interface (in Japanese), *Trans. Information Processing Society of Japan, 27*: 12.

Takagi, M., Konishi, T., and Yamada, M. (1982). "Automatic Digitizing and Processing Method for Printed Circuit Pattern Drawings," Proceedings of 6th International Conference on Pattern Recognition, Munich, p. 713.

Tsutsui, K., Hayashi, H., Takagi, T., Matsushima, H., and Kakumoto, S. (1986). Computerized flow network mapping system for service maintenance in water and wastewater systems (in Japanese), *The Hitachi Hyoron, 68*: 9.

Yamamoto, K., Mori, A., Mori, S., and Shimizu, S. (1976). Machine recognition of handprinted Japanese characters KATAKANA and numerals—topological line segment method and automatic formation of masks" (in Japanese), *Trans. IECE of Japan, J59-D*: 6.

4

Image-Analysis Techniques for Geographic Information Systems

Rangachar Kasturi

Pennsylvania State University
University Park, Pennsylvania

4.1 INTRODUCTION

Information representation and management in heterogeneous databases containing text, one-dimensional signals, graphics, and images requires intelligent techniques to manipulate such data. Efficient query processors, preferably with natural-language interfaces, are also essential to retrieve and manipulate information contained in such databases. In most of the applications, it is necessary to access data based on their information content. Thus, algorithms and techniques to understand pictorial data and facilitate query processing and information management are very useful. The design of a general-purpose image-interpretation system, however, is a very difficult and challenging task. Thus, to design practical systems, it is essential to restrict the system's domain to a particular class of image data. In this chapter, we restrict our attention to the problem of information extraction from images of maps. However, most of the image-processing techniques described here are also applicable for analysis of graphics such as engineering drawings, flowcharts, circuit diagrams, block diagrams, etc.

Maps are excellent examples of the abstract symbolic representation of spatial data. Humanity's use of maps has a long and remarkable history. From the first known map engraved on a clay tablet over 4,000 years ago by the Mesopotamians (Throwers/1972) to today's comprehensive atlases, humanity has shown an increasing reliance on the information contained in maps. Usually, to update a map or to show the results of changes made to the data, a new map must be drawn, making the whole operation time-consuming and very laborious. Often, it is necessary to combine the information contained in several maps to generate a new map. During the past 20 years, computers have been used extensively to automate some of these processes. Computers have been used in automated cartography (Starr/1984), aerial and satellite image analysis (McKeown/1983), and digital database generation (Mahoney/1981). While automating the creation of maps and digital databases is important, it is even more challenging and useful to automate the process of query processing and data management in geographic information systems (GIS). In Sec. 4.2, we present an overview of the techniques used in commercially available geographic information systems. In Sec. 4.3, we discuss recent research trends that are directed toward designing systems that operate on geographic data presented in a raster-image format. In Sec. 4.4, we describe in detail our efforts in the design of a system for query processing and information extraction from images of paper-based maps.

4.2   OVERVIEW OF GEOGRAPHIC INFORMATION SYSTEMS

Geographic information systems are designed to allow users to collect, manage, and analyze large volumes of spatially referenced data (Guptill and Nystrom/1986). The emphasis here is on the word *analysis*, which distinguishes GIS from other forms of spatial data processing such as automated cartography. As pointed out by Goodchild, "The ability to manipulate spatial data into different forms and to extract additional meaning from them is at the root of current GIS technology" (Goodchild/1987). GIS technology integrates concepts of database management, computer graphics, and spatial modeling into software for managing geographic features (Robinson/1989). The characteristic feature of a GIS is that the data are spatially indexed, allowing the information from maps, graphs, and tabular data to be combined and displayed in a variety of formats and media.

The first step in the building of a GIS is the collection of data. The data may be from map overlays, satellite/aerial images, hand-drawn sketches, population census, survey data, etc. An overlay is a special form of a geographical map, with each layer having only one attribute. Figure 4.1 shows the map overlays of the Elizabeth River Geographic Information System Project, which is a joint effort of the Environmental Protection Agency and the United States Geological Survey to clean up the Chesapeake Bay in Virginia.

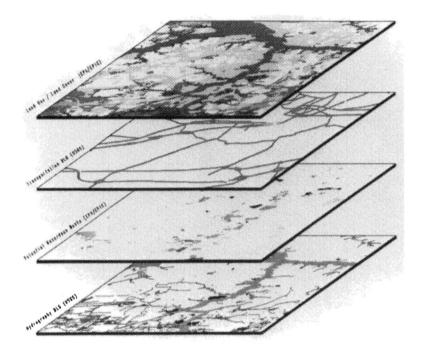

Figure 4.1   Geographic data in the form of digitized layers. Source: (United States Geological Survey/1987).

The four layers of the information in Fig. 4.1 are transportation, hydrography, potential hazardous-waste data, and photointerpreted land-use and land-cover data. After all the information has been entered into the system, each attribute can be viewed independently or can be combined together to create images such as the one shown in Fig. 4.2.

Computer software can analyze the information in various overlays and in other data files to assist the user in decision-making tasks. For example, information such as the size, ownership, pollutants, and vegetation of a potential hazardous site can be easily obtained by identifying the site on the screen (USGS/1987).

Typical GIS software, such as ARC/INFO of the Environmental Research Institute of Redlands, California, contains components to input, code, and display cartographic data and a relational database to analyze and describe attributes of various geographical features. Each feature may contain as many attributes as necessary. For example, a polygon represent-

Figure 4.2 A composite map obtained by combining individual layers. Source: (United States Geological Survey/1987).

ing a lake may have attributes like depth, the type of bed (sandy, rocky, etc.), fish population, mineral content, and the level of contaminants in the water.

Geographic information systems are finding many applications in every walk of life. For example, ARC/INFO has been used to increase the efficiency of collection of delinquent taxes in Kansas City, Kansas; to predict and manage the growth of communities around the country; and to monitor rising sea levels, decimation of African elephants by ivory hunters, and other global projects for the United Nations (Robinson/1989). MAPINFO software marketed by MapInfo Corporation of Troy, New York, has been used to assist the police department of Syracuse, New York, in tracking individuals targeted as prime suspects; by the Youth Environment Study of San Francisco, California, to help stem the spread of AIDS; and by the U.S. Army Recruiting Command to track the success of some 6,000 army recruiters. Other well-known systems include GeoVision of Ottawa, Canada; Intergraph of Huntsville, Alabama; and ERDAS of Atlanta, Georgia.

Although there are many applications that have already benefited from GIS technology, Goodchild observes that "because of the huge investments involved, the development of GIS tends to have been driven by application rather than by more abstract principles.... To an outsider GIS research appears as a mass of relatively uncoordinated material with no core of theory or organizing principles" (Goodchild/1987). He points out that no current commercial product comes close to matching the spatial analyst's view of the ideal GIS.

GIS traditionally has been organized as a four-part system: input, output, storage, and analysis. Input functions remain a major impediment to the adoption of GIS technology; manual digitizing is error-prone and extremely slow, whereas ambiguities of input documents plague the effectiveness and efficiency of automatic digitizing methods. Traditional maps are designed to match the abilities of the human map reader and are not ideally suited for automatic scanners. Data, once digitized, must be represented in a form suitable for computer processing. This is an area of great concern and debate in the design of geographic information systems.

Two of the basic spatial-data models used are vector (polygonal) and raster (grid) representation. In vector representation, geographic objects are represented by points, lines, and regions (polygons). Usually, a vector system is chosen when the data is interpreted in the form of features, so that the components of the feature can be linked together for easy access. In raster representation, the x-y location of the array corresponds to the spatial location of the feature point, and the array element contains the attribute feature for that point. Vector representation organizes the data by objects, whereas raster representation organizes it according to spatial address (Goodchild/1987). It is clear that both raster and vector representations are useful in a GIS, depending on the particular analysis task. Attempts have also been made to use a representation that places neighboring pixels close together, such as the Morton order and Peano scan order (Peuquet/1984). In addition to creating a suitable representation for the primitive objects (points, lines, and regions), it is essential to provide mechanisms to allow a full range of relationships between pairs of objects, whether they belong to the same class or are from different classes. Often, it is necessary to create a new class of data that is a mixture of point, line, and polygonal data. For example, a road that is represented by lines may also form the boundary between two geographic divisions that are represented by polygons. An excellent review of various data-representation techniques, including the so-called spaghetti model, the topological model, the POLYVRT (polygon converter) model, and various tessellation models, and their implication in spatial-data management, is given by (Peuquet/1984).

Although vector-manipulation algorithms for performing cartographic manipulations are more advanced than are raster algorithms, and vector representation is in general more compact than is raster representation, raster representation has many unique advantages. In particular, raster data matches the array structure of data representation in programming languages: two-dimensional images created from raster data are excellent abstractions of geographical data, and efficient manipulation of region-type features are possible using data in raster format. Often, many applications such as remote sensing, autonomous navigation, etc., require interaction between aerial/satellite images presented in raster form and digitized maps. In such cases, it is very convenient if both images and maps use the same representation. Although processing raster-type data in a sequential-processing computer is extremely slow, with the advent of parallel and pipeline processing technology it is becoming realistic to design systems that operate efficiently on data represented in raster form. Thus, recently, researchers have concentrated their efforts on developing algorithms for manipulating geographical data presented as a raster image. In the next section, we present recent research work in the area of map processing.

4.3 RECENT RESEARCH TRENDS IN MAP-DATA PROCESSING

Several works in the field of map-data processing have been reported in the literature. A collection of papers on map-data processing has been edited by (Freeman and Pieroni/1979). Shapiro and Haralick/1978 proposed an intelligent map-data processing system using a standard commercial database system and a query system to directly translate the query into a set of processing operations. Vaidya et al./1982 described a spatial information system with results on watershed data for a portion of the state of Virginia. Siekierska/1983 has described an experimental graphic workstation with different cartographic functions that enable the users to manipulate existing maps and to create new ones. A three-schema architecture and a query-by-example language are described (Barrera and Buchmann/ 1981) for a geographic database system intended for planners. A cartographic pattern-recognition system using homogeneous parallel algorithms and application of digital raster methods, and a road-river network generalized by automatic simplification, widening, thinning, emphasizing, and selecting using vector methods have been described in (Starr/1984). A database-management system and a query language have been developed by (Lien and Harris/1980). Harris et al./1982 have described a system for interpreting binary images of specially prepared polygon maps. An automatic recognition method to extract map components from large-scale maps has been described by (Suzuki et al./1987). Olsson/1984 has described a method for identifying different road types from map overlays. A geographical information overlay method that is a combination of the overlay of geographic maps and the distribution, integration, and representation of socioeconomic information is suggested as an alternative to conventional generation and calculation of data in different collection units (Tsurutani et al./1986). An automatic digitizer and a topological land-use expert system have been described in detail by Ejiri et al. in Chapter 3 of this book.

In this section, we present a summary of several papers describing research projects in which image processing, pattern recognition, and artificial-intelligence techniques have been applied. In particular, we describe the AUTONAP expert system for name placement in maps (Freeman and Ahn/1987), a system for creation of lower resolution maps from higher resolution data (Nickerson and Freeman/1986), and a map-oriented information system for urban planning (Jungert/1984).

4.3.1 Automatic Placement of Names in Maps

AUTONAP is an expert system designed by Freeman and Ahn/1987 that contains a set of rules that are applied by the cartographers when they manually perform the task of name placement. The system places names such that there is an unambiguous association between each name and the feature to which it refers, and there is no overlap among names and features. The placement of names for features of one category is dependent on the placement of names of features of other categories. In addition, names cannot be placed before the boundaries have been selected. Point-feature names are placed so that the distance between the name and the point feature is small. For line features, names are placed alongside the features such that their curving is minimized and the names are oriented similar to other line-feature names in the vicinity. Size, orientation, and location of names for area features are chosen such that they reflect the shape and extent of the area they describe. Names for area features are placed first, then point-feature names, and, finally, line-feature names. Area-feature names may sometimes have to be repositioned to permit a point-feature name to be placed. A map with names automatically placed by AUTONAP is shown in Fig. 4.3.

Figure 4.3 Example of automatic name placement using the AUTONAP System.
Source: (Freeman and Ahn/1987).

4.3.2   Map Generalization

One of the most challenging problems in computerized cartography is that of map generalization—the problem of generating a map of a reduced scale from a given map. The rule-based system described by Nickerson and Freeman/1986 is intended to generate maps at 1:250,000 scale from an original map of 1:24,000 by deleting features considered insignificant at the reduced scale, simplifying features, combining features, and converting some features from one type to another (e.g., area feature to point features). Features are deleted so that the average feature density remains the same in both source and target maps. Feature are then simplified by reducing the number of points used to describe them. Some features are combined to form new features, and some features are converted from one form to another. Features and symbols are also altered such that they are of appropriate size and detail in the new map. Feature names and other annotations are selected and placed to complete the map-generalization task.

4.3.3   A Map-Oriented System for Urban Planning

A rule-based map information system for urban planning is presented by (Jungert/1984). The information system is made up of three parts. The main part, called MASK, supervises all the activities in the information system, including the interaction with the user. It contains a map supervisor and a knowledge-based system, both written in LISP. By means of the knowledge given as user-defined rules, information is inferred from an internal database. The second part of the system includes the database handler and the geographic database. The database is divided into two parts: the first contains application-dependent information for urban planning; the second part contains information about the extensions of all available geographical objects; e.g., lakes, roads, and cities. The third part of the system is the workstation, where the maps are displayed. A knowledge-based method for finding the shortest path in a digitized map using a tile-based data structure is described by (Holmes and Jungert/1988).

In the following section, we describe our work toward obtaining a system for information extraction from images of paper-based maps.

4.4   INFORMATION EXTRACTION FROM IMAGES OF ·PAPER-BASED MAPS

Paper-based maps are designed to exploit the heuristic reasoning and world knowledge of the human user. Automated query processing to extract information from paper-based maps is a very challenging task. Often, a query contains several subqueries. Clearly, the number of possible queries, all of which may be answered using the information contained in a map, are enormous. An intelligent map-understanding system should be capable of analyzing and interpreting any query, extract relevant information from the map, and present the answer to the user in an appropriate format. The system should infer certain attributes and parameters using standard conditions and pattern behavior. It must also segment the map into meaningful entities, while maintaining their spatial and measurable attributes. All these processes in map analysis and understanding should be performed on data relevant to the query being processed, rather than performed on all the data contained in the map. This selective processing approach helps to minimize the size of the database and is essential to obtain a computationally efficient system. These issues are addressed, at least in part, in this research.

Our approach to information extraction from maps is based on the same logical operations that a human would do to extract information from maps. An individual, if asked to find the shortest path between two cities, for example, would proceed in the following logical manner: read the city names and locate their corresponding symbols, define a search space and find the network of roads within it, note the distance of each segment, and compute the shortest path. All this requires knowledge about symbols and lines and their association with physical entities, distance metric, etc. To automate such a process, an intelligent system would need to perform several tasks. For example, to answer the same query, the system should perform the following operations:

> Recognize characters to locate names
> Define small windows centered around the desired city names to locate the city symbols
> Search for symbols that represent the cities within these windows
> Define a larger window that encloses both city symbols
> Discriminate lines that represent roads from other line features
> Estimate distances between each node
> Determine the shortest path

Each of these operations may be decomposed into several primitive image-processing operations. For example, to determine the road network, several operations such as line thinning, line following, line-thickness computation, line-length determination, etc., must be performed.

It is clear from the above example that to answer a query, the system should be knowledgeable about the data representation in the particular map, should be capable of logically dividing the query into basic processing operations, and should be able to execute these operations to generate the desired response. The principal components of our system, which has been designed to answer queries similar to the one discussed above, are a *preprocessor*, a *query processor*, a *natural-language interface*, and an *image processor*. A dynamic database is used to store intermediate results and for communication among various processors. These components are described in detail in the following sections.

4.4.1 The Preprocessor

A paper-based map is converted into a digital image using a scanner. Several preprocessing operations are then performed to separate the text strings and orient the map image. These preprocessing operations are described in this section.

A typical map consists of a large number of entities, such as cities, rivers, railroads, time zones, country borders, state borders, etc. Usually, these entities are represented by various symbols and line structures with specific features and attributes. The symbols and lines in a map are intermixed with textual characters. Thus, an algorithm to segment a mixed text/line-structure image into text and line structures is necessary. Such an algorithm has been developed and is described in (Fletcher and Kasturi/1988). This algorithm is capable of separating text strings from an image containing graphical patterns and text strings, irrespective of their size, font style, and orientation. The algorithm outputs two images, one containing the character strings in the map and the second containing the line structure and graphical symbols. The image containing the text strings is intended to be processed by a character-recognition system to generate a table containing the names and their coordinates. In our work, a simplified map of the western U.S. was digitized at a resolution of 120 pixels/cm using a MicroTek MS300A image scanner connected to an IBM-PC/AT to obtain a 2,048 × 2,048 pixel image. The image in compressed format was sent to a VAX 11/785 for further processing. The graphics image obtained after text separation is shown in Fig. 4.4.

Note that this image contains important cities, state and national borders, and interstate and U.S highways. Since an OCR system is not readily available in our lab, a table listing the coordinates of the location of character strings is manually created.

The next step is to orient the map to be processed with respect to the system coordinates (Alemany/1986). Orientation is an operation that establishes the relationship between the longitude-latitude coordinates and the image coordinates by performing a coordinate transformation. As can

Figure 4.4   A simplified map of the western United States.

be seen in Fig. 4.4, the map contains crosshairs that are the points of
reference of the map. These cross hairs or other known points on the map
are used for orientation. The system prompts the user for the longitude and
latitude of the reference points. The following equations are then solved to
find the coordinate-transformation parameters:

$$
\begin{pmatrix}
1 & U_1 & V_1 & U_1V_1 \\
1 & U_2 & V_2 & U_2V_2 \\
1 & U_3 & V_3 & U_3V_3 \\
1 & U_4 & V_4 & U_4V_4
\end{pmatrix}
\begin{pmatrix}
A_1 \\
A_2 \\
A_3 \\
A_4
\end{pmatrix}
=
\begin{pmatrix}
X_1 \\
X_2 \\
X_3 \\
X_4
\end{pmatrix}
\tag{4.1}
$$

$$\begin{pmatrix} 1 & U_1 & V_1 & U_1V_1 \\ 1 & U_2 & V_2 & U_2V_2 \\ 1 & U_3 & V_3 & U_3V_3 \\ 1 & U_4 & V_4 & U_4V_4 \end{pmatrix} \begin{pmatrix} B_1 \\ B_2 \\ B_3 \\ B_4 \end{pmatrix} = \begin{pmatrix} Y_1 \\ Y_2 \\ Y_3 \\ Y_4 \end{pmatrix} \tag{4.2}$$

Here, U_is are the longitude coordinates and V_is are the latitude coordinates of the reference points. X_is and Y_is are the corresponding X and Y coordinates. The parameters A_i and B_i are the unknown constants of the coordinate transformation.

After obtaining the values of the unknown constants, any point (U_i, V_i) in the longitude and latitude format will correspond to the point (X_i, Y_i) in the image coordinate cartesian system, per the following equations:

$$X_i = A_1 + A_2U_i + A_3V_i + A_4U_iV_i \tag{4.3}$$
$$Y_i = B_1 + B_2U_i + B_3V_i + B_4U_iV_i \tag{4.4}$$

Note that these equations represent a nonlinear coordinate transformation.

If the coordinates of the crosshairs are not known, the preprocessor also allows the user to interactively specify the longitude-latitude coordinates of several reference points, which are then used to orient the map.

4.4.2   The Query Processor

The query processor (Amlani/1989) analyzes the query presented in a predefined syntax, translates the query into a sequence of operations for execution by the image processor, computes and passes the necessary parameters to the image processor, and generates the required response to the user. The query processor is written in Common LISP and executes on a VAX 11/785 computer.

(A)   Query Syntax

The general syntax of the query is shown below:

[Entity]{Constraint(&Constraint&...)}

where each constraint is of the form

Qualifier(⟨Query(&Query....)⟩)

The following discussion clarifies this query syntax:

All quantities enclosed by ( ) are optional.

Any phrase enclosed within [and } constitutes a valid query. The words *Show* or *Display* may be prefixed at the beginning of the query for clarity.

The *Entity* is any one of the displayable objects that are present in a map. Examples of *Entity* are road, city, state, etc.

The *Qualifier* specifies an attribute of the *Entity*, thereby providing additional information about it. Some of the qualifiers may require a value, as in the example Name = *New York*. Often, a *Qualifier* specifies the relational description among entities. Examples of *Qualifier* are largest, shortest, west of, within, containing, etc.

The delimiters in the query, [ ] { } ⟨ ⟩ and &, are helpful in the classification and translation of a query. They enable the query processor to identify an *Entity, Qualifier, Constraint, Sub-Query,* etc. All the delimiters, except &, are used in pairs.

(B) Types of Query

The interpretation and execution of a complex query is simplified by classifying the queries into different types. This classification is described below with the help of examples.

Basic query Basic query has one *Entity* and only one *Constraint*. This is the simplest form of a query. Examples of basic query are:

Show [state] {largest}
Show [state] {name = "Washington"}

Query with nested basic queries This type of query has a nested structure. A basic query forms the innermost part of the nested structure. A larger query is formed around the basic query by prefixing a relational descriptor and an *Entity*. Each subquery may have only one constraint. In such a query, the execution of the inner query generates the constraint and/or parameter for the outer query. Hence, the translation of such a query begins at the innermost basic query and proceeds outward. Examples of nested basic queries are:

Show [State] {Containing ⟨[City] {Name = "Seattle"}⟩}
Show [capital] {of ⟨[state] {containing ⟨[city] {name = "Seattle"}⟩}⟩}.

Query with multiple constraints Queries with multiple constraints are classified into a separate class, since each constraint may require the execution of one or more (possibly nested) subqueries. At each level of nesting, all the subqueries must be executed to obtain the necessary parameters before the execution of the next level of query. The following examples illustrate queries with multiple constraints:

Show [road] {type = "all" & within ⟨[state] {name = "Washington"}⟩}
Show [cities] {on ⟨[road] {type = "inter_state" & west_of
  ⟨[city] {name = "Harrisburg"}⟩} & within ⟨[state] {name =
  "Pennsylvania"}⟩}⟩}.

In the first example, it is evident that *road*, which is an *Entity*, has two constraints: (type = "all") and (within ⟨[state] {name = "Washington"}⟩). These constraints are separated by "&". The second example contains three constraints, two of which require the evaluation of subqueries. This concept of nesting and multiple constraints is similar to the evaluation of algebraic expressions with rank-ordered operators.

Query with special relational descriptor   The delimiter "&" in the above example separates the constraints and is used to denote the logical AND operation. The same delimiter is also used to convey a different meaning in queries that contain special relational descriptors. For example, in the query

Show [road] {type = "Interstate" & between ⟨[city] {name =

"Los Angeles"} & [capital] {of ⟨[state] {name = "Arizona"}⟩}⟩}},

the symbol "&" on the second line represents the separation of two parameters for the relational operator *Between*. Note that the two parameters for the relational descriptor *Between* are enclosed in ⟨ ⟩.

(C)   Query Translation

The translation of the query into executable calls to the image processor is preceded by the classification of the query into one of the query types discussed above. This classification is done on the basis of the presence or absence of particular delimiters and special relational descriptors. The order of query classification and translation is: basic query, query with nested basic queries, query with special relational descriptors, and query with multiple constraints.

After identifying the type of the query, the translation procedure relevant to that query type is followed. The translation procedure begins by dividing a complex query into subqueries until no further breakup is possible. Each of the subqueries is then translated into appropriate image-processing call routines by using a table-lookup procedure. Necessary parameters to these routines are also provided. This process is illustrated with the help of the following examples:

Example 1 (Basic Query)   Show [city] {name = "Seattle"}:

This query contains one entity and one qualifier. Each combination of an entity and a qualifier defines a unique set of operations for information extraction. These operations are performed by invoking a particular sequence of image-processor routines. For the above example, the query processor generates the following sequence of calls: Read_Image, Approximate_Coordinates, Define_Search_Window, Detect_Symbols, Nearest_City_Symbol, and Display. The approximate coordinates of the city "Seattle" is obtained by estimation of the position of the corresponding label in the map. A search window centered around this approximate point is defined and all the city symbols within the search window are found. The city symbol that is nearest to the approximate location is then displayed. During the course of this execution, the image processor returns certain flags and values that enable the query processor to continue (for example, to enlarge the search space, if no symbols are found within the window) or to stop the search for the city symbol.

Example 2 (Query with Nested Basic Queries)   Consider the query:

Show [Road] {within ⟨[state] {containing ⟨[city] {name
  = "Bismarck"}⟩}⟩}.

The innermost subquery in this query is the same as in Example 1, except for the city name. This subquery is translated into image-processing operations as described above. The execution of this subquery generates a parameter—the coordinates of the city symbol—for the next level of query. At the next level, we have an entity (state) and a relational descriptor (containing) pair. There can be various such combinations; viz., capital-of-state, road-within-state, etc. Corresponding to each of these routines, the query translator invokes a particular sequence of image-processor routines. In this example, the coordinates of Bismarck are passed as parameters to the following routines, which identify the state boundary that encloses the city of Bismarck: Read Image, Define Window, Skeletonize, Find Line Thickness, Track Dashed Lines, and Find Closed Contour. Note that some of these routines are the same as the ones called by the lower level subquery. This is because the query translator considers each subquery as a separate process and, hence, starts all over again, while retaining only the necessary parameters and results. Further, some of the routines behave differently depending on the parameters and flags passed on to them. For example, the Define Window routine defines search windows of different sizes, depending on whether a city or a state is being searched. The final step in answering this query involves the call to routines that display the lines representing the roads within the enclosed region.

Queries that contain the special relational descriptors (operations requiring more than one entity; e.g., *Between*) and queries with multiple constraints are handled in a similar manner. The query processor is thus responsible for the overall execution of the query through its interaction with the image processor.

4.4.3 A Natural-Language Interface to the Query Processor

It is clear from the discussions in the previous section that the queries presented to the query processor must confirm to a rigid syntax. To make the system more user friendly, a natural-language front end has been designed (Feng and Kasturi/1988) that converts the natural-language input to the rigid syntactic form required by the query processor. This interface is briefly described in this section.

(A)   Selection of a Natural-Language Processing Approach

Natural-language processors are typically classified according to the method that is adopted for representing the knowledge (Barr and Feigenbaum/ 1981). Based on these criteria, natural-language processors can be classified into four basic types: (1) keyword-scanning method, (2) text-based method, (3) limited logic method, and (4) knowledge-based method.

Keyword-scanning method   An example of the keyword-scanning approach is Green's BASEBALL (Tennant/1981). Its parsing system consists of 14 classifications of parts of speech. It adopts right-to-left scanning to break the input into functional phrases. This is combined with the keywords that are identified to convert the input phrase into a specification list. Although this method is very simple to implement, it is not powerful enough to process the complex queries that are expected in a map-based information system.

Text-based approach   Natural-language processors using the text-based approach essentially store a representation of the text itself in their databases. An example of the text-based approach is PROTOSYNTHEX by Simmons, Burger, and Long, in which an index of the text is constructed after deleting the prepositions, conjunctions, article adjectives, and pronouns. Each indexed word points to a list of all the memory locations in the text where the word occurs. Any time an indexed word is used, PROTOSYNTHEX searches all the memory locations associated with the indexed word for a match. This system, however, lacks any deductive abilities. While this method is suitable for general processing of natural-language text, it is not very attractive for use in a natural-language interface to the map query processor.

Limited-logic approach   Limited-logic systems, such as Raphel's semantic information retrieval (SIR) system (Raphel/1976), are able to answer questions that are not explicitly stored in the database. SIR contains 24 templates of the form "How many * does * have?" where * represents a noun, possibly modified by a qualifier such as *a, the, every, each*, or a number. Such a system, although possessing deductive inferring abilities, is rather limited in scope, since it is governed by the template-matching technique and thus lacks flexibility. In particular, different templates would be needed for representing the same query that is presented in slightly different manners.

Knowledge-based approach   Knowledge-based approaches are, of course, the most powerful of the natural-language query-processing systems. Such systems are built on a set of rules that make up the grammar of the language.

Some examples of these rewrite rules for English are:

$$\langle S \rangle \longrightarrow \langle NP \rangle \langle VP \rangle$$
$$\langle NP \rangle \longrightarrow \langle ART \rangle \langle N \rangle$$
$$\langle VP \rangle \longrightarrow \langle V \rangle \langle NP \rangle$$
$$\langle ART \rangle \longrightarrow a \mid an \mid the$$
$$\langle N \rangle \longrightarrow any\ noun$$
$$\langle V \rangle \longrightarrow any\ verb,$$

where S, NP, and VP refer to sentence, noun phrase, and verb phrase, respectively. Knowledge-based systems are very powerful and are suitable for use in general-purpose natural-language interpretation systems. The number of rewrite rules and the vocabulary in such a general-purpose system could be quite large. For interpretation of queries in a limited domain, such as map-data processing, such a general-purpose system is not appropriate. Thus, a hybrid of the limited-logic approach and the knowledge-based approach has been adopted for our implementation. This system, which we call the limited-knowledge (i.e., semantic-based) approach is described in the next section.

Limited-knowledge approach   The rewrite rules used by the limited-knowledge-based system are capable of interpreting queries that are more complex than those that can be handled by the templates of the limited-logic approach. However, it uses a semantic grammar rather than the syntactic grammar used by the knowledge-based systems. Like a syntactic grammar, a semantic grammar consists of a lexicon and a set of rewrite rules; but instead of using the general word classes like ⟨VERB_PHRASE⟩ and ⟨NOUN_PHRASE⟩, it uses the general semantic classes such as ⟨ACTION⟩, ⟨DISPLAY_OBJECT⟩, and ⟨QUALIFIERS⟩. The set of rewrite rules used by our system is shown in Table 4.1. A top-down, depth-first parsing technique has been used to match the input query with the rewrite rules. The general features of the natural-language interface are described in the following section.

(B)   General Features of the Natural-Language Interface

The natural-language interface contains a preprocessor and a top-down parser. The preprocessor converts all the letters to upper case and also verifies that each word in the query is a part of its lexicon. It interacts with the user to correct any spelling errors and to identify action phrases that are not a part of its vocabulary. The top-down parser attempts to match its semantic grammar to the input query. It begins with the top-most rewrite rule (see Table 4.1) and then recursively moves toward lower, more specific rewrite rules using a semantic-based transition network (Obermeier/1987).

The queries are parsed from left to right. Based on the rewrite rules in Table 4.1, the ⟨ACTION⟩ and ⟨DISPLAY_OBJECT⟩ nonterminals are extracted from the input to form a template. This query represents the main query of the natural-language input. If there are no ⟨QUALIFIERS⟩, the parsing is complete and the corresponding output is generated. Otherwise, additional parsing is necessary and an asterisk is placed at the end of the current output to denote this situation. The examples shown in Fig. 4.5 illustrate these concepts.

The remaining part of the input query is then parsed recursively by the processor. After these recursive processes, the final output, which corresponds to the queries 2 to 4 in Fig. 4.5, is as follows:

(show [state] {containing ⟨[city] {name = "Bismarck"}⟩})

(show [road] {type = "all" & within ⟨[state] {containing ⟨
  [city] {name = "Fresno"}⟩}⟩})

(show [road] {type = "interstate" & between ⟨ [city]
  {name = "Dallas"} & [city] {name = "Reno"}⟩}).

The output of the natural-language interface is input to the query processor, which then generates the appropriate calls to the image processor to generate the final response to the user.

4.4.4   The Image Processor

The image-processing system (Alemany/1986, Kasturi and Alemany/1988) consists of pattern-recognition and image-processing routines that are called by the query processor. These routines are designed to process map images to extract spatial and structural information. The image-processing routines include operations such as skeletonization, line tracking, closed-contour detection, symbol identification, etc. Each of these routines requires additional parameters and constraints to specify the processing area. These parameters are supplied by the query processor. To decrease the image-processing time, only relevant portions of the image—relevant to the query being processed—are processed. These routines are written in Pascal and FORTRAN. The image processor is described in detail in this section.

To illustrate the operation of the image processor, the following two queries will be used as examples:

1. Show [state] {Containing ⟨[city] {name = "Seattle"}⟩}
2. Show [road] {Within ⟨[state] {Containing ⟨[city] {name = "Seattle"}⟩}⟩}

Table 4.1   Rewrite Rules of the Natural-Language Interface

The following conventions are used:
 (1)   Nonterminal symbols are enclosed in ⟨ ⟩.
 (2)   Semantic classes are in upper case.
 (3)   Literals (i.e., terminal symbols) are in lower case.
 (4)   The logical operator "or" is represented as |.
 (5)   Any entity enclosed in ( ) is optional.

⟨SENTENCE⟩	→	⟨VP⟩ | ⟨VP⟩ ⟨QUALIFIERS⟩
⟨VP⟩	→	⟨ACTION⟩ ⟨DISPLAY-OBJECT⟩
⟨QUALIFIERS⟩	→	⟨SPECIFIERS⟩ ⟨DISPLAY-OBJECT⟩ (⟨CONJUNCT⟩) | ⟨SPECIFIERS⟩ ⟨DISPLAY-OBJECT⟩ (⟨CONJUNCT⟩) ⟨QUALIFIERS⟩
⟨ACTION⟩	→	show | display | give | present | exhibit | find | print | list | ⟨ACTION-QUESTION⟩
⟨ACTION-QUESTION⟩	→	what is/are | where is/are
⟨DISPLAY-OBJECT⟩	→	⟨STATE-PHRASE⟩ | ⟨CITY-PHRASE⟩ | ⟨CAPITAL-PHRASE⟩ | ⟨ROAD-PHRASE⟩
⟨STATE-PHRASE⟩	→	⟨STATE-NAME⟩ (state) | the state (of) ⟨STATE-NAME⟩
⟨CITY-PHRASE⟩	→	⟨CITY-NAME⟩ (city) | the city (of) ⟨CITY-NAME⟩
⟨CAPITAL-PHRASE⟩	→	the capital (city) (of) ⟨CAPITAL-NAME⟩ | the capital of ⟨STATE-PHRASE⟩
⟨ROAD-PHRASE⟩	→	(the) (state) ⟨ROAD⟩ | (the) (state) ⟨INTERSTATE⟩ (⟨ROAD⟩)
⟨STATE-NAME⟩	→	any state name in the state database
⟨CITY-NAME⟩	→	any city name in the city database
⟨CAPITAL-NAME⟩	→	any capital name in the capital database
⟨INTERSTATE⟩	→	interstate(s) | inter-state(s) | expressway(s) | freeway(s) | highway(s) | throughway(s) | thruway(s)
⟨ROAD⟩	→	road(s) | roadway(s) | route(s) | path(s)
⟨SPECIFIERS⟩	→	(⟨SUBORDINATE⟩) ⟨RELATION-VERB⟩ (⟨POSSESS⟩) | (⟨SUBORDINATE⟩) (⟨EXTRA⟩) ⟨RELATION-PREP⟩ | (⟨EXTRA-PROGRESSIVE⟩) ⟨RELATION-PREP⟩ | (⟨SUBORDINATE⟩) (⟨EXTRA⟩) ⟨RELATION∗⟩ ⟨DISPLAY-OBJECT⟩ ⟨CONJUNCT⟩ ⟨DISPLAY-OBJECT⟩ | (⟨EXTRA-PROGRESSIVE⟩) ⟨RELATION∗⟩ ⟨DISPLAY-OBJECT⟩ ⟨CONJUNCT⟩ ⟨DISPLAY-OBJECT⟩ | ⟨RELATION-PRGRSS⟩
⟨SUBORDINATE⟩	→	which | that

Table 4.1   (*Continued*)

⟨EXTRA⟩	→	is | are | go(es) | run(s) | pass(es) | lie(s) | lay(s) | lead(s)
⟨EXTRA-PROGRESSIVE⟩	→	being | going | running | passing | lying | laying | leading
⟨RELATION-PRGRSS⟩	→	containing | consisting | having | holding | including leaving | exiting | arriving | entering | whose
⟨RELATION-VERB⟩	→	contain(s) | consists(s) | has | have | hold(s) | include(s) | leave(s) | exit(s) | arrive(s) | enter(s)
⟨RELATION-PREP⟩	→	within | near | on | through | from | to | with | beside | close | closest | nearby | nearest | next | north | south | east | west | in | inside
⟨RELATION*⟩	→	between
⟨POSSESS⟩	→	of | to
⟨CONJUNCT⟩	→	and | but

To answer the first query, the query processor segments the query into two parts: first, the subquery "Find [city] {name = "Seattle"}"; second, the query "Find [state] {Containing ⟨Response from first part⟩}." To answer the first part, the system should locate the city symbol representing Seattle. To identify the state limit, the system should recognize the pattern of lines that identifies state borders and also the closed contour formed by this line pattern. The call routines generated by the query processor to answer the first subquery are the following:

1. Call City Index: This routine locates the city name "Seattle" from the City Index table. Note that this table was manually created after separating text strings from the map image. The table gives the approximate coordinates of the text string in the map. The actual coordinates of the symbol representing the city of Seattle on the map, however, will be in the vicinity of the point corresponding to these coordinates. Thus, a search is necessary to locate the city symbol corresponding to Seattle.

2. Call Window: This routine defines the spatial working area where the search for the city symbol is performed. The nearest city symbol to the transformed coordinates is identified as the symbol for the city of Seattle. The following three routines perform the symbol-identification operation.

3. Call Connected Component: This routine identifies potential city symbols within the windowed area of the map.

TEMPLATE 1:   < ACTION > < STATE-PHRASE >   (no qualifiers present)

TEMPLATE 2:   < ACTION > < STATE-PHRASE > *   (qualifiers present)

TEMPLATE 3:   < ACTION > < ALL > < ROAD-PHRASE > *   (qualifiers present)

TEMPLATE 4:   < ACTION > < INTERSTATE > < ROAD-PHRASE > *   (qualifiers present)

Natural Language Queries

(1)   *Show the state of Washington*   ⇒   MATCHES TEMPLATE 1
Associated, format-dependent query (complete):
          (show [general-entity] { name = place-entity })
Output
          (show [state] { name = "Washington" })

(2)   *Show the state containing Bismarck*   ⇒   MATCHES TEMPLATE 2
Associated, format-dependent query (partial):
          (show [general-entity]
Query construction after the main query is processed:
          (show [state]

(3)   *Display all the roads in the state which contains Fresno*
              ==> MATCHES TEMPLATE 3
Associated, format-dependent query (partial):
          (show [general-entity] { type = descript-entity &
Query construction after the main query is processed:
          (show [road] { type = "all" &

(4)   *Display the interstate highways from Dallas to Reno*
              ==> MATCHES TEMPLATE 3
Associated, format-dependent query (partial):
          (show [general-entity] { type = descript-entity &
Query construction after the main clause is processed:
          (show [road] { type = "interstate" &

Figure 4.5   Examples of query parsing.

4. Call Symbol: This routine examines the potential symbols and recognizes all the symbols contained within the window.
5. Call Short/Far: This routine identifies the city symbol nearest to the transformed coordinate as the symbol for the city of Seattle.

Once the city symbol has been identified, the system displays the symbol on the screen and then proceeds to execute the second part: finding the state containing this symbol. The call routines generated to answer this part are as follows:

6. Call Window: This routine redefines the spatial working area in which the search for a closed contour (representing the state) surrounding the city of Seattle is performed. This spatial working area is increased, if necessary, until the closed contour of the state is found. The following three routines perform the boundary-detection operation.
7. Call Skeleton: This routine generates a skeleton of the lines within the window.
8. Call Dashed Line: This routine identifies different line patterns contained in the window and tracks dashed lines.
9. Call Closed Contour: This routine finds a closed contour formed by the dashed lines identified in the previous step.

The second query, which requests the display of all the lines representing roads within the state in which the city of Seattle is located, is answered by executing the same set of call routines as above until the state of Washington is found. Finally, to display the roads, the following call is performed:

10. Call Trace: This routine searches the dynamic database for lines representing roads within the state boundary and displays them on the monitor. Note that the road information is available in the dynamic database at the end of the skeleton operation.

To execute all these routines, several parameters are necessary. For example, the *dashed-line* routine requires the line-segment length and the distance between the segments, the *skeleton* routine requires the location and extent of the spatial window, etc. These parameters are supplied by the query processor and by the output of other routines. In the following paragraphs, a brief discussion of some of these routines is presented.

(A) City Symbol Detection
The image-processing system has been designed to detect several symbols such as circle, square, and triangle. It has been assumed that the symbols are hollow and enclose a region of black pixels (note that in the image, the lines are white on black background). The shape of this black pixel region

is used to identify the symbol. These black pixel islands (enclosed by white pixel lines) are located using a connected component-generation algorithm.

The *connected-component* algorithm groups together black pixels that are four-connected to each other. The algorithm encloses each four connected regions by a circumscribing rectangle; i.e., each rectangle identifies a single connected component. A separate entry of each black pixel island and the size of its enclosing rectangle is maintained in a table. Thus, although these rectangles are overlapping, it is possible to process each island individually using the information contained in the table. For example, the circumscribing rectangles (overlapping) generated by applying the connected-component algorithm to the image shown in Fig. 4.6 (a window of the map image shown in Fig. 4.4) is shown in Fig. 4.7.

Each rectangle generated by the connected-component routine contains a potential symbol, since it encloses black pixel islands. However, since all symbols are limited in size by a predetermined range of values, it is possible

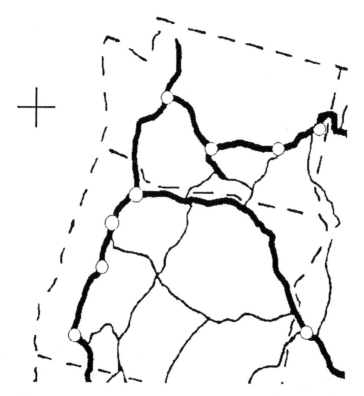

Figure 4.6   A portion of the map shown in Fig. 4.4 after enlargement.

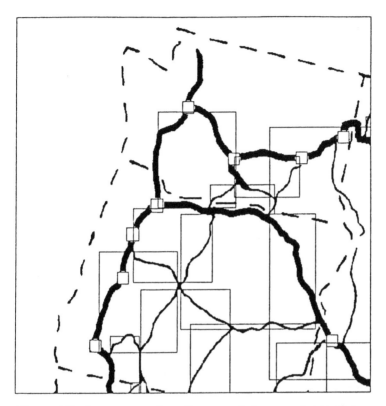

Figure 4.7   Image output from the connected-component routine.

to exclude the larger rectangles from further consideration as candidates for potential symbols. The rectangles enclosing potential symbols are analyzed to determine the area and perimeter of the black pixels. The ratio between the area and the perimeter of the closed contour is calculated. Ratio values approximately equal to 0.07 are identified as circles. The squares and triangles are identified in a similar manner. The symbol-detection routine is being enhanced to detect other symbols, such as highway symbols. Once all the city symbols are detected, the *short-far* routine, which identifies nearest and farthest symbols from a specified point or line, is applied to locate the symbol nearest to the approximate coordinates of the city Seattle.

(B)   Skeletonization
To decrease the volume of data in the digitized image, while maintaining the spatial relationship among various symbols, a skeletonization operation is

performed. The skeletonization algorithm described by (Harris et al./1982) has been adapted for our application. The algorithm operates on binary images in which the pixel values are either 3 (line pixels) or 0 (background). The algorithm uses a series of 3 × 3 operators to process the image. A buffer is created to process the image from top to bottom. The operation takes place by inserting two unprocessed lines at the top and obtaining the thinned lines at the bottom. The resulting image of this algorithm is a marked skeleton. In the skeletonized image, the pixels are coded as follows:

Intensity	Description
0	Background
1	Terminal pixel
2	Intermediate pixel
3	Junction pixel

This algorithm has been implemented in such a way that any rectangular window of the map (as opposed to the entire map) can be specified and skeletonized, depending on the descriptors and qualifiers specified in the query. Skeletonization is a primitive image-processing operation that is an essential component of many other operations. For example, the skeletonization algorithm, when applied to the image shown in Fig. 4.6, generates the skeleton image shown in Fig. 4.8. This routine creates several files that contain the data needed for further processing of the map.

(C)   Thickness Routine

After the skeleton routine is executed, the algorithm to detect the average thickness of line segments is executed. This algorithm works in the spatial work area defined by the *window* routine and uses the files generated by the skeleton routine. The *thickness* routine calculates the thickness at every point in the core line in four directions—vertical, horizontal, and the two diagonal directions. The minimum value is retained as the thickness of the line at that particular point. After this procedure is performed, the average thickness of the segment is calculated. At the end of the algorithm, the files generated by the skeleton routine are updated to include the thickness information. The thickness information is used to distinguish among various types of lines.

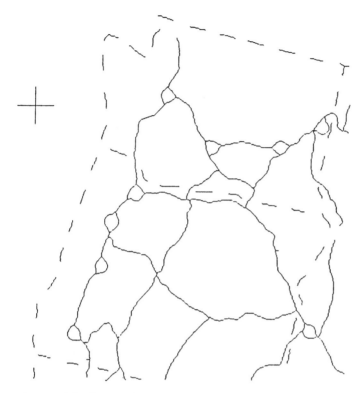

Figure 4.8   Skeletonized map.

(D)   Dashed-Line-Detection Algorithm

The *dashed-line-detection* algorithm, described in detail in Chapter 2, is applied to locate segments that belong to the state boundaries.   The algorithm consists of the following steps:

1.   Segment and gap-length computation
2.   Neighboring-segment detection and sorting
3.   Segment linking and dashed-line detection
4.   Dashed-line-segment interconnection

The dashed lines detected by applying the dashed-line-detection algorithm to the image shown in Fig. 4.6 are shown in Fig. 4.9.

(E)   Contour Following

For road-map analysis, the identification of a closed contour is important; the identification of the state containing Seattle is a good example of the use

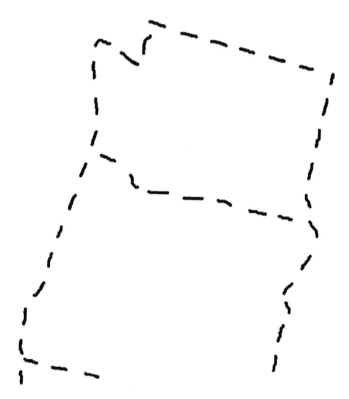

Figure 4.9 Dashed-line segments detected by the dashed-line-detection algorithm.

of the *contour-following* algorithm. This algorithm is capable of detecting the internal contour that surrounds a specific point—in this case Seattle—without following any external contour. This routine is similar to the algorithm described by Jimenez and Navalon/1982.

The contour-following algorithm operates within the spatial work area defined by the window, which may be as large as 2,048 × 2,048 pixels. A search is performed in a horizontal direction, starting from an internal point, by incrementing the x coordinate by 1 until a point $P_1(x_1,y_1)$ on the contour is found. At each pixel on the contour, a rotational order search (described below) is performed.

Consider the image shown in Fig. 4.10(a). The eight directions around $P_1(x_1,y_1)$ are labeled as shown in Fig. 4.10(b). The pixel $P_1(x_1,y_1)$ has been

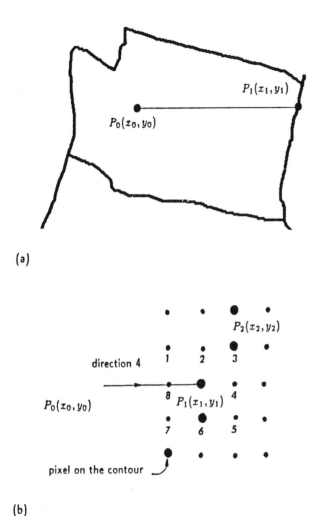

(a)

(b)

Figure 4.10   Contour-following technique: (a) locating a point, $P_1(x_1,y_1)$, on the contour; (b) enlarged view around pixel $P_1(x_1,y_1)$.

reached by moving from left to right; i.e., along direction 4 starting from an internal point $P_0(x_0,y_0)$. The next pixel along the contour is searched using the following rule: Add a value of 4 to the previous direction (in this case 4) using module 8 arithmetic (resulting in 0), and 1 to the result. The next pixel on the contour is searched starting from the direction specified by this result. In Fig. 4.10(b), pixel 1 is examined, followed by pixel 2, etc., until the next pixel $P_2(x_2, y_2)$ is found. This defines a new direction to the contour as 3. The next pixel is then searched starting from the pixel in direction 8. This method ensures that the correct internal contour is found.

The pixels or points detected along the search path each will be marked with a different value, always leaving the opportunity for the algorithm to use the same point several times, in case the search has to pass two or more times through the same point. This algorithm ends when the first two points of the contour are found again or if the window does not contain a closed contour.

The resulting image after the closed contour is detected is shown in Fig. 4.11(a). The system's response when "Seattle" is replaced by "Bismarck" in the query is shown in Fig. 4.11(b). If a closed contour is not found, the window routine is called again to increase the spatial work area.

(F)   Area Computation

After the contour-following algorithm is executed, two lists containing the x and y coordinates of the closed contour are transferred to the *area* routine. A common problem in image analysis is finding the area of a region enclosed by a contour. Pavlidis/1982 has described two basic approaches to solving the problem of contour filling, which can be modified easily to calculate the area enclosed by a contour. The first approach implements a *parity-check* algorithm and the second applies a *connectivity-criterion* algorithm. The parity-check algorithm uses the fact that a straight line intersects any closed curve an even number of times. The connectivity-criterion algorithm assumes that an interior point is provided, in addition to the contour, in which case a search is performed to find all pixels that can be reached from the given point without crossing the contour. The algorithm used in our system is the one based on the parity-check criterion. The area computation is performed on a row-by-row basis using this algorithm. The algorithm also sets the seventh bit of all the pixels enclosed by the closed contour. This bit is used by the trace routine to limit the display to the enclosed region.

(G)   Trace Routine

To answer the second query, presented earlier in this chapter, it is necessary to find all the roads that are within the state boundary. This algorithm reads the coordinates of all the pixels representing the roads present in the spatial work area and checks the seventh bit of these pixels. If the seventh bit is set, the system displays the pixel on the monitor. The same concept is used to show the symbols detected by the symbol routine. The image generated by this routine in response to this query is shown in Fig. 4.12(a). This figure contains the cities and roads within the state of Washington.

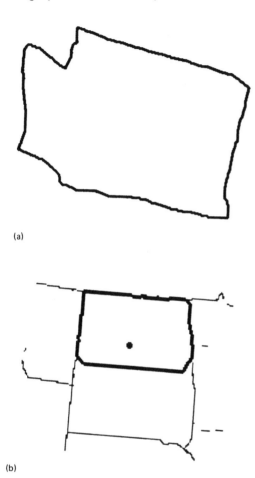

(a)

(b)

Figure 4.11   Image displayed by the system in response to the first query: (a) state containing "Seattle"; (b) state containing "Bismarck".

Table 4.2 gives a breakdown of the processing times needed to execute the image-processing routines to answer the second query. The second query, with a different city name (i.e., "Lubbock") generated the response shown in Fig. 4.12(b).

(H)   Shortest-Path Routine

As a final example, consider the query, "Show the shortest path between Seattle and Phoenix." The natural-language interface translates this query

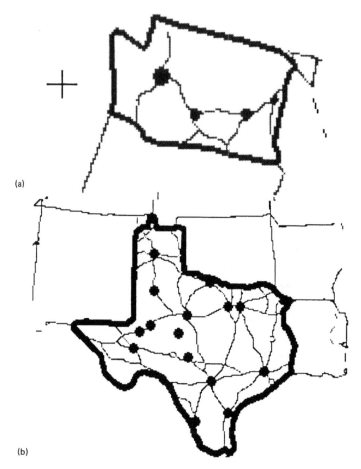

(a)

(b)

Figure 4.12 Image displayed by the system in response to the second query: (a) all roads within the state containing the city "Seattle"; (b) all roads within the state containing the city "Lubbock". (Note: These images were printed at a lower resolution than those in Fig. 4.11)

to the form:

> Show [spath] { type = "all" & between ⟨ [city] { name = "Seattle"} & [city] {name = "Phoenix"})}.

To answer this query, the system defines an elliptical search region whose foci are Seattle and Phoenix. The major diameter of this ellipse is arbitrarily set at 1.2 times the euclidean distance between the cities, as shown in

Table 4.2   Breakdown of CPU Times for
Processing the Second Query

Processing Step	CPU Time (sec)
1. Read 2K image	13.1
2. Skeletonization	37.21
3. Thickness	98.64
4. Dashed-line following	101.59
5. Closed contour	12.13
6. Trace	6.2
Total	268.87

Fig. 4.13. The image within the ellipse is thinned, and the shortest path between the cities using the road network within the ellipse is found using the bidirectional search algorithm (Barr and Feigenbaum/1981). The length of the shortest path is estimated. If this length is greater than the major diameter of the ellipse, the search space is increased to an ellipse with the current length of the shortest path as the major diameter and a shorter path, if it exists, is found. This procedure will ensure that the path found is the shortest, since the length of any path connecting the foci of an ellipse that includes a point outside the ellipse is always greater than the major diameter. The path found for our query is shown in Fig. 4.13 in bold lines.

4.4.5   Summary of Map-Based Information System

A system for extracting information from images of paper-based maps was described in this section. A natural-language interface converts the input queries to the rigid syntax that is expected by the query processor. The query processor classifies the query into one of the four types depending upon its complexity, specifies the operations to be performed on the image data by the image processor, and controls the information-extraction process. The operation of the natural-language processor and the query processor and their interaction with the image processor are illustrated with the help of several examples.

   In our present implementation, all the logical functions described above are present. However, the query processor has a very limited vocabulary of terms for spatial-data analysis. The image processor also has very limited symbol-recognition abilities. Additional enhancements are being added to enhance the capabilities of the system. The goal is to include all possible features and spatial relationships that are meaningful in answering queries

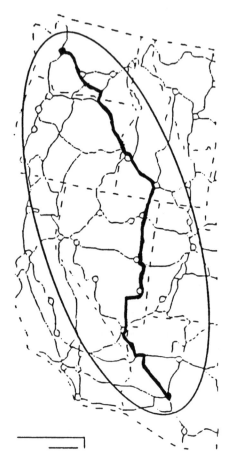

Figure 4.13   Finding the shortest path between Seattle and Phoenix.

in a geographical information system. These enhancements are being implemented using the same basic shell that we have already developed.

4.5   CONCLUSIONS

Geographic information systems present an excellent environment for the integration of many symbolic and image database-manipulation techniques. Many commercial systems are built using vector representation as a primary means of data abstraction. However, image-analysis techniques are well suited for processing queries on spatial data. Thus, recent research work has focused on designing systems that apply image-processing, pattern-recognition, and artificial-intelligence techniques for management of geographical information. In this chapter, we presented a brief description of the current state in GIS technology, described recent research trends in this area, and discussed our own efforts toward the design of a paper-map-based query-processing system. Powerful image-interpretation algorithms, especially if implemented on special-purpose computers, would have major impact on the growth of geographical information systems of the future.

ACKNOWLEDGMENTS

I would like to thank many individuals for their help and encouragement in writing this chapter. In particular, I would like to thank Mr. Kit Robinson of ComputerLand Corporation for permitting us to use much of the material on commercial geographical information systems that he had collected for use in his article in *ComputerLand Magazine* and Dr. Ejiri, Dr. Jungert, and Professor Freeman for their help in providing several of the references and figures used in this chapter. I would also like to acknowledge the dedication and hard work of several of my students including Juan Alemany, Mukesh Amlani, Hwa-tse Liang, Rodney Fernandez, and Wu-chun Feng in implementing and testing many of the algorithms used in our system. The work described in Sec. 4.4 was supported in part through a grant from the National Science Foundation (ECS 8307445).

REFERENCES

Alemany, J. (1986). *An Image Processing System for Interpretation of Paper-Based Maps*, M.S. Thesis, Dept. of Electrical Engineering, Pennsylvania State University, University Park, Pennsylvania.

Amlani, M. L. (1989). *A Query Processor for Information Extraction from Paper-Based Maps*, M.S. Thesis, Dept. of Electrical Engineering, Pennsylvania State University, University Park, Pennsylvania.

Barr, A. and Feigenbaum, E. A. (1981). *The Handbook of Artificial Intelligence*, Vol. 1, William Kauffman, Los Altos, California.

Barrera, R. and Buchmann, A. (1981). "Schema Definition and Query Language for a Geographical Database System," Proceedings of IEEE 1981 Conference on Computer Architecture for Pattern Analysis and Image Database Management, Hot Springs, Virginia, pp. 250–256.

Feng, W. C. and Kasturi, R. (1988). *ENLIM: A Natural Language Interface for Paper-Based Map Information System*, Computer Engineering Technical Report, Pennsylvania State University, University Park, Pennsylvania.

Fletcher, L. A. and Kasturi, R. (1988). A robust algorithm for text string separation from mixed text/graphics images, *IEEE Trans. on Pattern Analysis and Machine Intelligence*, *10*, no. *6*: 910–918.

Freeman, H. and Pieroni, G. (1979). "Map data processing", Proceedings of a NATO Advanced Study Institute on Map Data Processing, Academic Press, New York.

Freeman, H. and Ahn, J. (1987). On the problem of placing names in a geographic map, *International Journal of Pattern Recognition and Artificial Intelligence*, *1*, no. *1*: 121–140.

Goodchild, M. F. (1987). A spatial analytical perspective on geographical information systems, *International J. of Geographical Information Systems*, *1*, no. *4*: 327–334.

Guptill, S. C. and Nystrom, D. A. (1986). Geographic Information Systems: An Important Tool for Spatial Analysis, in *USGS Yearbook (1985)*, pp. 9–16.

Harris, J. F., Kittler, J., Llewellyn, B., and Preston, G. (1982). A modular system for interpreting binary pixel representation of line-structured data, *Pattern Recognition: Theory and Applications*, D. Reidel, Boston, pp. 311–351.

Holmes, P. and Jungert, E. (1988). "Shortest Path in a Digitized Map Using a Tile-Based Data Structure," Proceedings of 3rd International Conference on Engineering Graphics and Descriptive Geometry, Vienna, pp. 238–245.

Jimenez, J. and Navalon, J. L. (1982). Some experiments in image vectorization, *IBM J. of Research and Development*, *26*, no. *6*: 724–734.

Jungert, E. (1984). *A Map Oriented Information System for Urban Planning*, National Defence Research Institute (FOA), Linkoping, Sweden.

Kasturi, R. and Alemany, J. (1988). Information extraction from images of paper-based maps, *IEEE Trans. Software Engineering (Image Databases)*, *14*: 671–675.

Lien, W. Y. and Harris, S. K. (1980). Structured implementation of an image query language, *Pictorial Information Systems*, pp. 416–430.

Mahoney, W. C. (1981). Defence Mapping Agency (DMA) overview of mapping, charting, and geodesy (MCIG) applications of digital image pattern recognition, *Techniques and Applications of Image Understanding*, Washington, D.C. *Proc. SPIE*, *281*: 11–23.

McKeown, D. M. Jr. (1983). *Maps*, Report CMU-CS-83-117, Dept. of Computer Science, Carnegie-Mellon University, Pittsburgh, Pennsylvania.

Nickerson, B. and Freeman, H. (1986). "Development of a Rule-Based System for Automatic Map Generalization," Proceedings of 2nd International Symposium on Spatial Data Handling, Seattle, Washington.

Obermeier, K. K. (1987). Natural language processing, *Byte 12*: 225–232.

Olsson, L. (1984). *Development of a Road Database from Map Overlays*, National Defence Research Institute (FOA), Linkoping, Sweden.

Pavlidis, T. (1982). *Algorithms for Graphics and Image Processing*, Computer Science Press, Rockville, Maryland, pp. 167–193, 324–335.

Peuquet, D. J. (1984). A conceptual framework and comparison of spatial data models, *Cartographica*, *21*, no. *4*: 66–113.

Raphel, B. (1976). *The Thinking Computer: Mind Inside Matter*, W. H. Freeman and Company, New York.

Robinson, K. (1989). Computers make smarter maps, *ComputerLand Magazine, 4, no. 2*: 12–19.

Shapiro, L. G. and Haralick, R. M. (1978). "A General Spatial Data Structure," Proceedings of IEEE Computer Society Conference on Pattern Recognition and Image Processing, Long Beach, California, pp. 238–249.

Siekierska, E. (1983). "Toward an Electronic Atlas," Proceedings of 6th International Symposium on Automated Cartography, Ottawa, Canada, pp. 464–474.

Starr, L. E., ed. (1984). *Computer-Assisted Cartography Research and Development Report*, International Cartographic Association.

Suzuki, S., Kosugi, M., and Hoshino, T. (1987). Automatic line drawing recognition of large-scale maps, *Optical Engineering, 26, no. 7*: 642–649.

Tennant, H. (1981). *Natural Language Processing: An Introduction to an Emerging Technology*, Petrocelli Books, New York.

Throwers, N. J. W. (1972). *Maps and Man: An Examination of Cartography in Relation to Culture and Civilization*, Prentice-Hall, Englewood Cliffs, New Jersey, pp. 10–15.

Tsurutani, T., Yutaka, K., and Miyashita, T. (1986). A geographic information overlay method for regional analysis, *Systems and Computers in Japan, 17, no. 8*: 41–48.

USGS (1987). *United States Geological Survey Yearbook (1986)*, United States Government Printing Office.

Vaidya, P. D., Shapiro, L. G., Haralick, R. M., and Minden, G. J. (1982). Design and architectural implications of a spatial information system, *IEEE Trans. on Computers, C-31*: 1025–1031.

5

Digital Image Processing and Three-Dimensional Reconstruction in the Basic Neurosciences

Lyndon S. Hibbard*

The Milton S. Hershey Medical Center
Pennsylvania State University College of Medicine
Hershey, Pennsylvania

5.1 INTRODUCTION: BRAIN RESEARCH THEMES AND THE ROLE OF DIGITAL IMAGE PROCESSING

In one way or another, all neuroscience researchers seek answers to the question: How does the brain work? The most complex object in nature, a human brain, defines humans individually and as a species. In addition to its role in mentation and sensory processing, the brain plays important roles in regulating the health of the whole body. One example is its function as the topmost level of regulatory control in the endocrine system—the brain itself produces hormones that regulate the production of hormones generated by other tissues. The brain is capable of autoregulation as well. Since the brain requires a continuous and large (relative to other tissues) supply of oxygen and glucose for fuel, and since it can store no reserves of those substances, it adjusts the rate of transport of glucose across the blood-brain barrier and the rate of flow of the blood bringing the glucose and oxygen it needs. Blood flow and transport can be regionally up- or down-regulated, depending on the metabolic needs of the local tissues. Thus, in addition to the well-known structural complexity of the brain presented by neuroanatomy, the

---

*Current affiliation: Washington University School of Medicine, St. Louis, Missouri

brain's behavior as a consumer of energy and a processor of neurochemical information is likewise complex.

The correlation of brain structure and function is a task in which images take a natural part. Many of the techniques used to measure metabolic rate, blood flow, metabolite transport, and the localization of neurotransmitter substances (which actually do the work of transmitting a signal from one neuron to another) rely on chemical assays in thin sections of brain tissue that are recorded as images containing both anatomic and chemical information. These data can be efficiently acquired, examined, and reconstructed using digital image processing. Indeed, without computer methods for the storage, display, and analysis of these data, fundamental progress in many areas would not be possible.

5.1.1   Structure and Function in the Brain

Modern neuroscience had its beginning in the late 19th century, with the discovery by Golgi of a silver stain that is selectively absorbed by a small number of neurons in thin sections of the brain (Fig. 5.1). Under light microscopy, the Golgi stain rendered visible single neurons throughout the full extent of their dendritic processes. Cajal, using Golgi's stain, showed that the brain consists of two main classes of cells (neurons and glial cells), that the neurons are discrete cells, and that they are the basic signaling units of the brain (Nauta and Feirtag/1986). Neurons communicate with other neurons by releasing chemicals (neurotransmitters) that travel across gaps (synapses) that physically separate neurons and bind to specialized protein complexes (receptors) on the destination (postsynaptic) neuron. Within a neuron, the transfer of a signal from place to place occurs by means of the electrochemical action potential. Each neuron *feels* the contacts of hundreds or thousands of other neurons and projects dozens or hundreds of dendrites onto other neurons. How a neuron, feeling the inputs from many other cells, decides to transmit or not transmit a signal is a question of urgent interest and the subject of intense study currently (Kandel and Schwartz/1985).

Another general aspect of brain function is the regional specialization of cognitive, sensory, motor, and regulatory functions. Learning, memory, sensory perception, and voluntary motor activity, for example, originate in the cerebral cortex, sometimes with the localization of a very particular functionality to a small, discrete area. This was first reported in 1836 by a French physician, Marc Dax, who observed that of a group of his patients who had lost the ability to speak, all had, at autopsy, lesions in the left hemisphere (Bloom et al./1985). Broca later made the association of the loss of speech with lesions in a particular area of the left frontal lobe, now

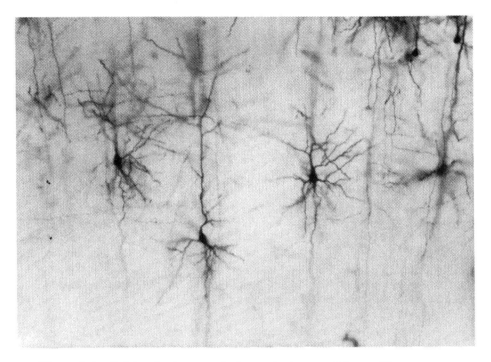

Figure 5.1 Golgi-stained neurons in the cerebral cortex of the rat. These neurons are pyramidal cells, named for the shape of the soma, or cell body. The somas appear here as the large dark spots, out of which the long thin dendrites and axons radiate to contact other neurons. Camillo Golgi (1844–1926) discovered the silver-containing stain that is absorbed by only a few percent of the neurons, permitting them to be observed in contrast to their surroundings. The reason for the selectivity of the stain for a subset of all neurons remains unknown today. Nonetheless, the capability of visualizing single neurons has made the Golgi stain an essential neuroscience tool for over 100 years. Magnification about 400 times. Photograph courtesy of Dr. Ian Zagon, Pennsylvania State University College of Medicine.

known as Broca's area. Since then, the regional functionality of the brain has been mapped in detail. The basal ganglia and the cerebellum control motor activity and the hypothalamus exerts control over the sympathetic nervous system and the endocrine system, affecting physiologic functions as diverse as blood-pressure regulation and sexual behavior. An excellent introductory review of the organization and function of the brain can be found in (Bloom et al./1985).

Thus, structure and function are linked in the brain, and the nature of these linkages defines how the brain works. To understand these connections, research is done in several broad areas: anatomy (structure of neurons, parts of neurons, and neuronal networks), chemistry (metabolism, chemical processing of information), and cell biology (transport of substances throughout, gene expression, and neuronal development). Each of these areas has benefited from the application of computer image analysis, and many of those applications will be described in this review.

5.1.2   Images Play a Natural Role in Neuroscience Research

Digital image analysis as a neuroscience research tool has become prominent in two particular applications: in the regional mapping of physiological and biochemical activity throughout the brain, and in the study of the 3-D morphology of neurons, assemblies of neurons, and other brain structures. In both cases, the brain must be cut into sections before it can be examined. The sampling of the brain in thin sections has a long history, and various conventions regarding the manner of sectioning have been elaborated in atlases, for example, of the rat brain (Pellegrino et al./1979, Paxinos and Watson/1986, Skinner/1971). The tissue may have been treated with indicator reagents before or after sectioning, or before or after the animal was sacrificed. In any event, the sections contain some kind of chemical signal that is recorded by autoradiography or photography, and the images are subsequently digitized. For 3-D morphological and ultrastructural studies, the images are obtained from light microscopy or transmission electron microscopy, followed by photography, film densitometry, or manual recording. This review will emphasize research done in animals rather than in humans for two reasons. First, most neuroscience research is done in animals (especially the rat) using techniques that require the sacrifice of the animal. The second is that the human central nervous system is studied by medical diagnostic imaging techniques that have been reviewed in many other places (Herman/1986, Mazziotta and Koslow/1987, Margulis et al./1987, Wong et al./1986, Wagner/1986).

5.2   IMAGING OF BRAIN METABOLISM AND FUNCTION

5.2.1   Autoradiographic Imaging of Brain Metabolism

Autoradiography is the generation of an image in a photographic or radiographic emulsion by the radiation emitted from a radioactive substance. In studies of metabolism, transport, and blood flow, the experiment involves the injection of a radiolabeled substance into the circulation, followed by sacrifice of the animal (most commonly the rat), sectioning of the brain, and incubation of the tissue sections with a radiographic film. After a period of time (a few days to a few months, depending on the isotope and the nature of the experiment), the film is developed, revealing life-size images of the tissue sections in which light and dark areas represent brain regions of low and high concentrations, respectively, of the tracer in the tissues. Depending on the nature of the molecule containing the radiolabel and on the design of the experiment, the observation of rates of glucose metabolism, blood flow, or transport across the blood-brain barrier may be quantitatively measured (Fig. 5.2). (The regional mapping of receptors will be described later.)

Autoradiographic imaging of brain function has its origins in the cerebral blood-flow measurements made by Kety and coworkers (Landau et al./ 1955, Freygang and Sokoloff/1958, Reivich et al./1969). These studies established that the variable rates of blood flow throughout different regions of the brain resulted in proportional deposition of radiolabeled substances, observable in the corresponding light and dark regions of the autoradiograph. Sokoloff et al./1977 presented a mathematical model of brain metabolism that permitted the association of autoradiograph gray-level with the rate of glucose utilization. Glucose is the primary energy source for the brain, and the rate of its use is (at least qualitatively) proportional to neuronal activity. In this method, the labeled compound is [2-^{14}C] deoxyglucose, a nonmetabolizable form of glucose that accumulates to varying extents throughout the brain, depending on local tissue metabolic rates. Like blood-flow autoradiography, high local glucose utilization will produce relatively large local tissue concentrations of the ^{14}C label and correspondingly high gray-levels in the autoradiograph. A second method, with certain operational advantages over the deoxyglucose method (Lu et al./1985), uses ^{14}C-labeled glucose, which is metabolized identically to ordinary glucose. In either case, the resulting autoradiograph can be digitized by a scanning microdensitometer (Sokoloff et al./1977, Goochee et al./1980) or solid-state video digitizer (Gallistel et al./1982, Alexander and Schwartzman/1984, Yonekura et al./1982).

5.2.2 Autoradiographic Imaging of Neurotransmitters and Receptors

Autoradiography, with its high sensitivity and spatial resolution, has become important for recording the localization and relative densities of neurotransmitters and their receptors. Receptors are transmembrane proteins found in pre- or post-synaptic membranes of neurons that specifically bind particular small molecules or peptides. The transmission of an impulse from one neuron to another occurs via the release of a quantity of neurotransmitter molecules from packets (vesicles) at the presynaptic membrane. The neurotransmitter molecules diffuse across the synapse and are very

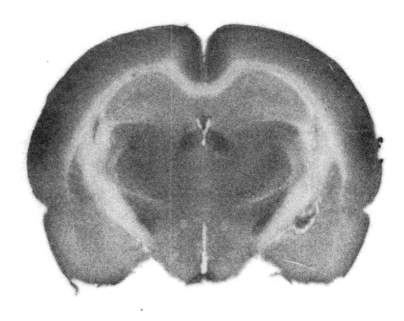

(a)

Figure 5.2 Autoradiography of rat brains reveals abundant structural and functional information. Shown here are ^{14}C-autoradiograms displaying five different chemical signals related to energy consumption, transport across the blood-brain barrier, and protein synthesis. (a) Glucose consumption to satisfy regional energy needs is pictured in a ^{14}C-glucose autoradiogram of a normal rat brain; (b) and (c) autoradiograms depicting the transport of the ^{14}C-labeled amino acids lysine and leucine, respectively, across the blood-brain barrier; (d) autoradiogram record of the transport of an alternative energy source for the brain, β-hydroxybutyrate (this substance is one of a group of substances—known as ketone bodies—produced by the metabolism of fatty acids when glucose is unavailable); (e) autoradiogram record of the incorporation of the labeled amino acid, ^{14}C-phenylalamine, into proteins synthesized by neurons and glial cells in the brain. The patterns in each autoradiogram are different from the others, demonstrating the individualized responses of the brain to each experiment. All sections are of the cerebral cortex at about the level of coronal section number 45 in the atlas of (Pellegrino et al./1979). Autoradiograms generously provided by Dr. Richard A. Hawkins, The Pennsylvania State University College of Medicine.

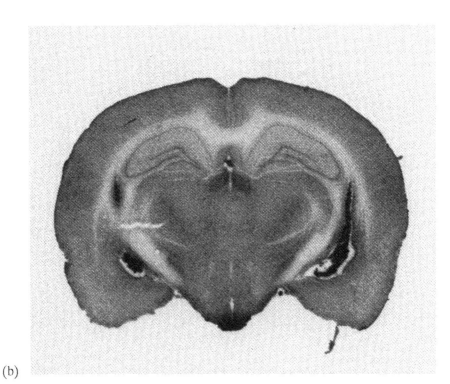

(b)

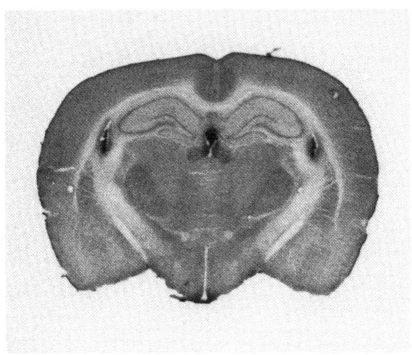

(c)

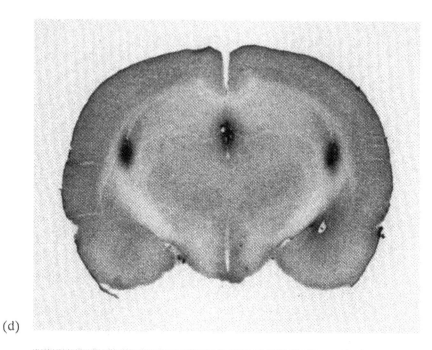

(d)

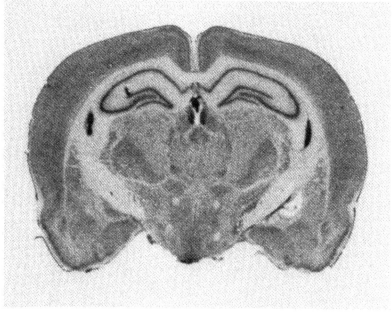

(e)

Figure 5.2 (Continued)

tightly bound by receptors in the postsynaptic membrane. Upon binding, the receptor initiates an intracellular signal that communicates to the rest of the cell the message that a signal has been received from a neighboring neuron. Interestingly, each neuron uses one kind (or, at most, a few kinds) of neurotransmitters, and it is generally accepted that only those neurons releasing a given neurotransmitter possess high-affinity receptors for the neurotransmitter (Bradford/1986). Thus, different regional neuronal networks use different neurotransmitters, presumably to convey information with more specificity and discrimination than is possible with a single neurotransmitter (Hokfelt et al./1984, Bloom/1984).

To answer questions about the organization of neurotransmitter networks, histochemical and immunochemical labeling of either the neurotransmitters or the enzymes that produce them in their tissue sections has been devised. Neurotransmitters are localized by the in situ formation of fluorescent derivatives of neurotransmitter molecules that are absorbed by groups of neurons. Other methods involve fluorescent or radiolabeled antibodies specific for the neurotransmitters, or the enzymes (either in the presynaptic or the postsynaptic neuron) that process the neurotransmitter molecules (Bradford/1986). Perhaps the most commonly used technique is the autoradiographic detection of receptors using ^{3}H-labeled neurotransmitter molecules (Young and Kuhar/1979, Unnerstall et al./1982, Herkenham and Pert/1982). The radiolabeled neurotransmitter molecules are tightly bound by the receptors and expose the emulsion to a greater or lesser extent, depending on the quantity of label present (Fig. 5.3). The resulting autoradiograph can be digitized in the same way as are the metabolic autoradiographs for quantitative comparisons with other binding experiments (Rainbow et al./1983, Altar et al./1984, Glaser et al./1985, Berck and Rainbow/1985).

5.2.3 Digital Autoradiography

One of the strengths of autoradiography as a method for recording brain function is at the same time an obstacle to its interpretation: the autoradiograph contains a vast amount of spatial and functional information. Autoradiographs offer a limiting spatial resolution of 25 to 100 microns, depending on the isotope used (Unnerstall et al./1982, Gallistel and Nichols/1983), and a gray-level resolution of 6 to 7 bits (Ramm and Kulick/1985). Rat brain coronal autoradiographs like those in Fig. 5.2, which are approximately 10×15 mm (at the tissue margins), digitized at 100-micron resolution, yield digitized images containing up to 15,000 pixels.

The conventional analysis of autoradiograph data depends on manual densitometry, or measuring the optical density of selected regions in the film

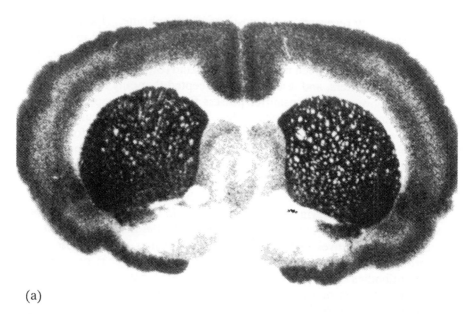

(a)

Figure 5.3 Autoradiography of brain neurotransmitter receptors. Neurotransmitters are small organic molecules that, by diffusing across the synaptic gaps between neurons, carry information from one neuron to another. Receptors are protein molecule complexes that specifically bind neurotransmitter molecules and communicate their arrival to the interior of the recipient neuron. The distribution of the receptors specific for various neurotransmitters is an important aspect of the organization of the brain. In addition, autoradiography can provide quantitative determinations of the binding affinity of receptors for their natural neurotransmitter

image by physically placing the film in the densitometer so that a reading can be made within a small circular spot (typically, less than 1 mm in diameter). There are many problems with this method. A single reading (or, at most, a few readings) is taken to sample the functional activity within a neural structure that may extend over a large volume relative to the densitometer spot diameter. The large numbers of readings pose a practical obstacle to the mapping of the functional heterogeneity of structures. Further, the variable contrast of autoradiographs may preclude the accurate sampling of small structures or structures whose boundaries are indistinct. Thus, manual densitometry is poorly suited to deal with the large quantities of data in autoradiographs.

Computer digitization of autoradiographs and their display as pseudo-color raster images on computer-graphics systems has become popular following the prototypical system described by (Goochee et al./1980). In

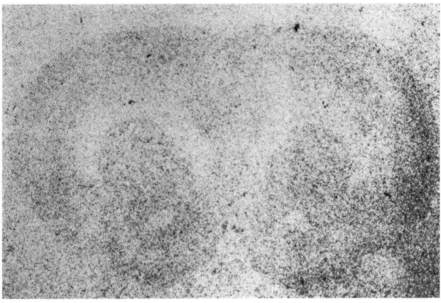

(b)

molecules, or neurotransmitter analogs. (a) Autoradiogram of a tritium-labeled neurotransmitter shows regional differences in the distribution of its receptor; (b) autoradiogram showing the nonspecific binding of the neurotransmitter used to produce (a). Here the labeled neurotransmitter was co-incubated with a large excess of an antagonist substance that competes with the neurotransmitter for the binding sites on the receptor. The nonspecific binding autoradiogram is a control for the autoradiogram in (a). These figures are from (Rainbow et al./1982) and are reproduced by permission.

this system, and in the dozens like it that have appeared in the neuroscience-methods literature since then, various display techniques are employed to enhance contrast or to selectively emphasize some aspect of the images. In addition, interactive sampling of neuroanatomic regions in the displayed image enables more accurate measurements of the physiological property of interest. The optimal performance requirements for computer-controlled densitometry systems have been reviewed in several places (Ramm and Kulick/1985, Alexander and Schwartzman/1984, Gallistel et al./1982, Goochee et al./1980).

An essential function that digital image processing can provide is the comparison of images from different experiments. Varied approaches to this task have been described, but almost all depend on some kind of interactive operator definition of regions of interest. Goochee et al./1980 used an on-screen cursor to define the outlines of a region whose mean glucose utilization rate was determined from the corresponding region of the digital image. As a kind of interactive densitometry, the size and shape of the region measured are unrestricted by pinhole size, as in manual densitometry. Comparison of experiments can be effected by measuring the same regions in equivalent serial sections from different experiments. The subjective element in the definition of regions remains, and the full sampling of structures extending over many serial sections is usually truncated to one or a few measurements in a few sections because of the labor involved.

An alternative approach is to combine images to produce difference images, where high-valued regions would display the larger differences between a pair of experiments. Before such a pointwise comparison can be carried out, the input images must be superimposed. Various approaches to image alignment have been put forward, and all but one depend on interactive input of the information by the experimenter. Gallistel et al./1982 associated autoradiograph regions with neuroanatomy using a split-screen video technique in which the digital autoradiograph was superimposed on a histochemically stained, adjacent-serial section. The histochemical section was manually repositioned so that its video image was aligned with that of the autoradiograph. The metabolic rate of a neural structure was obtained by interactively painting the histological region of interest, with a program retrieving the corresponding autoradiograph pixel values. A similar routine was followed by (Alexander and Schwartzman/ 1984), except that the experimenter interactively defined corresponding points in the histological image and the autoradiograph. The coordinates of these points in their respective images defined the linear transformation of one onto the other, but the images were not actually superimposed. Screen pixels within painted regions of the displayed histological image were carried over by the transformation into the corresponding pixels of the digital autoradiograph, and their gray-level values were summed. Isseroff and Lancet/1985 used essentially the (Gallistel et al./1982) method, except that they aligned successive serial autoradiographs by interactively moving one to match its outline with that of a fixed, aligned image.

A more elaborate, manual-alignment method was employed by Zilles, Glaser, and colleagues (Zilles et al./1986, Glaser et al./1985). To associate regions in digital autoradiographs with neural structures visualized in homologous Nissl-stained sections, the autoradiograph was output to a printer capable of representing 11 gray-levels. The image of the stained section was projected onto the autoradiograph print, and the neural regions of interest were traced by hand onto the print. The autoradiograph print, with neural structures outlined, was then mounted on a digitizing tablet and points were set to establish the transformation relating the autoradiograph print and the stored, digital autoradiograph. The neural structure contours were then traced with the digitizer tablet, and the values of the corresponding image pixels surrounded by these contours were summed.

The actual subtraction of one digital autoradiograph from another has been done by a small number of groups. Altar and coworkers (Altar et al./ 1985, Altar et al./1984) aligned images by an interactive video method like that of (Gallistel et al./1982), after conversion of pixel values from gray-level to units related to the actual number of moles of labeled neurotransmitter in the tissue (Fig. 5.4). Difference images and images of the percentage of specific binding map the differences throughout the tissue section in the highest detail possible. Isseroff and Lancet/1985 also reported using digital subtraction in the analyses of $[2\text{-}^{14}C]$ deoxyglucose autoradiograms.

One other approach to digital subtraction of autoradiographs has been worked out based on image-processing methods for serial image alignment, requiring no manual intervention (Hibbard and Hawkins/1984). Because rat brain coronal sections are roughly oval in shape, and because adjacent sections do not change very much in size and shape, the principal-axes transformation has been applied to calculate the combined rotation-translation transformation necessary to superimpose successive images' centroids and principal axes. From classical mechanics (Goldstein/1950), for any point in a rigid body, a set of orthogonal axes known as the principal axes can be calculated. Treating the digital autoradiograph as a rigid body, the principal axes are calculated for the image center of mass (centroid). The principal axis through the coronal section in its longer dimension corresponds to a best-fit line through the pixel data (Rosenfeld and Kak/1982). To superimpose serial digital autoradiographs, the images are separately translated so that the images' centroids are placed at a constant point in the image plane, and then the images are rotated about their centroids by an angle that makes the longer principal axis parallel with the horizontal (row) axis of the image coordinate system. The principal-axes transformation works well for rat brain coronal autoradiographs, which are decidedly longer in the medial-lateral direction than in the dorsal-ventral direction, and which do not change size and shape abruptly from section to section. The principal-axes method works less well when those conditions are not met, as in the case of damaged tissue sections or for well-cut sections whose principal moments have about the same value. In these instances, cross-correlation methods have been implemented to provide alignment (Hibbard et al./1987).

Here, the digital autoradiographs, aligned approximately by the principal-axes method, are smoothed by local averaging and passed through an

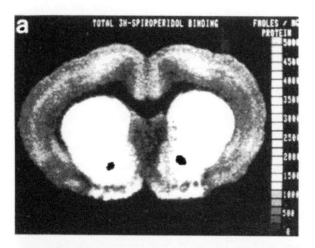

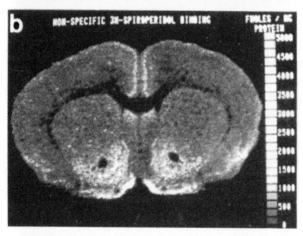

Figure 5.4 Direct comparisons of digital autoradiograms can be used to quantitate the binding of neurotransmitters and their analogs to receptors. Shown here is a study by (Altar et al./1984) of the binding of a tritium-labeled dopamine antagonist, ^{3}H-spiroperidol, to dopamine receptors in the cerebral cortex of the rat. (Antagonists block the biological effects accompanying neurotransmitter-receptor binding.) (a) The digitized autoradiogram of total (receptor-specific plus nonspecific) antagonist binding; (b) a digitized autoradiogram of the ^{3}H-spiroperidol incubated with a 700-fold molar excess of another substance, (+)-butaclamol, which also binds

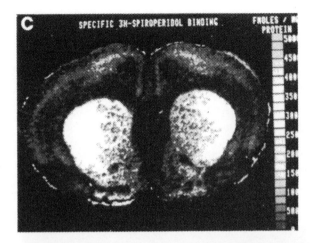

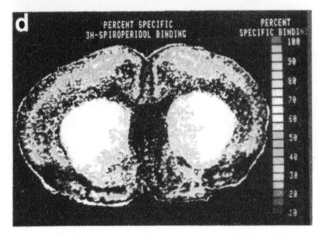

tightly to dopamine receptors. The resulting gray-level density is due to ^{3}H-spiroperidol binding to substances in the tissue not displaced by the (+)-butaclamol, and hence, not dopamine receptors. This autoradiogram was prepared using the section adjacent to that in (a). (c) Image resulting from the pixel-by-pixel subtraction of (b) from (a), representing the specific, spiroperidol-bound, dopamine receptors; (d) an image of the percent specific binding obtained by dividing the pixel values of (c) (specific binding) by those of (a) (total binding) and multiplying the quotients by 100. These figures are from (Altar et al./1984) and are used here by permission.

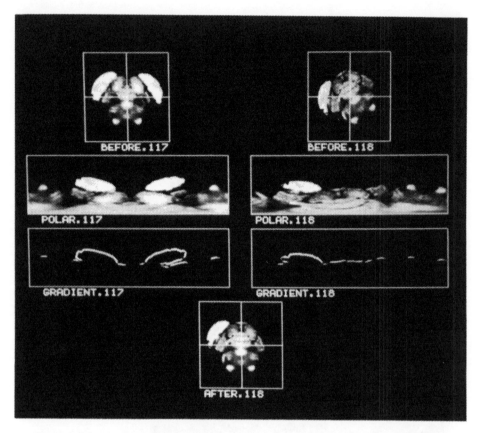

Figure 5.5 Application of polar-resampling and cross-correlation to align irregular-section autoradiograms. The top row shows two serial-adjacent autoradiograms in which BEFORE.118 is missing the left occipital lobe of the cerebral cortex present in 117. The second-row images are the polar-resampled versions of the top-row images. They were constructed by interpolating BEFORE-image pixel values along rays traced out from the BEFORE images' centers at equal angle intervals. These POLAR images were sampled around 360° by 2° increments. The edges of the highest contrast features were extracted by smoothing and by the use of Roberts gradient (Roberts/1965) to yield the GRADIENT images (third row). The GRADIENT images were actually input to the cross-correlation program, which iteratively calculated the translation/rotation transformation resulting in the aligned image AFTER.118. Figure taken from (Hibbard et al./1987), by permission.

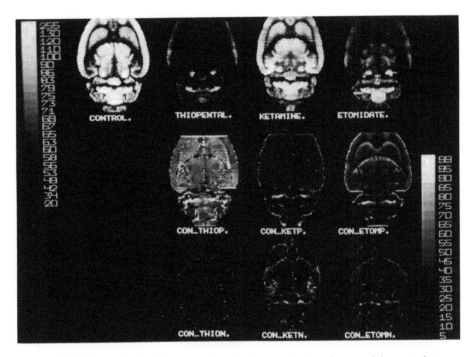

Figure 5.6 Horizontal images of rat-brain metabolism from serial coronal autoradiograms. Images of glucose utilization rates were prepared (as described in the text) for a normal rat (CONTROL in the top row) and for rats given anesthetic doses of three injectable anesthetics: thiopental, ketamine, and etomidate. The rest of the images are pixel-by-pixel differences, CONTROL minus anesthetic, in which the positive-difference images (second row) are displayed separately from the negative differences (third row). Interestingly, though all the anesthetized rats displayed the same exterior symptoms, their brain metabolic patterns varied widely. Figure adapted from (Hibbard et al./1987) by permission.

edge-detection operator. The resulting image, emphasizing the edges of the higher contrast features, is used for the calculation of the cross-correlation function (Moik/1980, Anuta/1970). That is, the edge-enhanced version of the image to be aligned is combined in the cross-correlation with that of an adjacent, aligned image. The location of the maximum in the correlation is related to the relative translation needed to superimpose the unaligned image onto the aligned image.

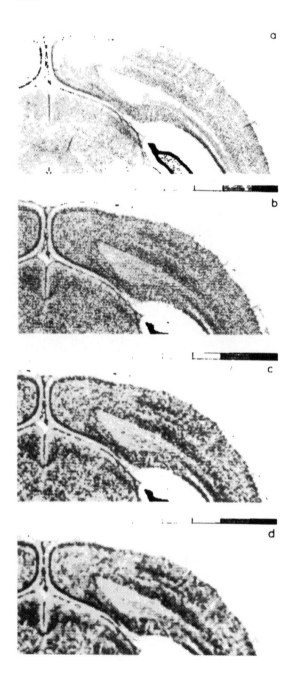

In general, the misalignment has both a translational and a rotational component, which are coupled such that the correction of one depends on the other. Therefore, the cross-correlation alignment of digital autoradiographs requires interactive, alternating refinements of the rotational and translational misalignments. To correct rotational misalignment, the images must be resampled in polar coordinates (Fig. 5.5) (Hibbard et al./1987, Pratt/1978). The location of the cross-correlation maximum in this case is related to the misalignment angle. In practice, the entire image does not need to be polar-resampled. Rather, a fan-shaped sector is chosen by the experimenter to include undamaged regions of both images with small, high-contrast features. This strategy is especially helpful in hindbrain coronal images, where the principal moments are about equal and where the various brain tissues occurring in that section may not be physically connected and, in some cases, may actually be missing from some sections. Cross-correlation of corresponding parts of images is sufficient to effect alignment.

Once aligned, the images in different serial sets can be compared by calculating point-by-point differences or ratios (Isseroff and Lancet/1985, Alter et al./1984, Hibbard and Hawkins/1984). Further, by combining similar data sets to form mean and standard-deviation images, the pixel values in the difference images can be screened according to their statistical significance (Hibbard and Hawkins/1984).

By treating the aligned digital autoradiographs as a three-dimensional reconstruction (discussed below), it is possible to refine the comparisons and extend them to images from planes other than the coronal section plane. Indeed, equivalent serial sections, in general, sample the brain in slightly different locations, producing an error in the correspondence of region edges. In the author's laboratory, three-dimensional reconstructions

---

Figure 5.7 Computer image analysis of the cytoarchitecture of the rat cerebral cortex. Schleicher et al./1986 used a digital imaging operator to enhance the regions of varying neuron density in Nissl-stained rat-brain sections. (a) Micrograph of a Nissl-stained section of the rat visual cortex. The gray density is proportional to the population density of neurons. (b) The same region digitized and thresholded into five gray-levels corresponding to five gray-level index (GLI) ranges. The GLI is simply the percent fraction of the sampling window area occluded by dark-stained neurons. (c) Data of (b) smoothed by one pass of a 3 × 3 median filter; (d) data of (b) smoothed by three passes of the same filter. The median filter has the effect of smoothing local variation in the Nissl density due to cell clusters or blood vessels without blurring edges between regions with different average gray-levels. Figure taken from (Schleicher et al./1986) with permission.

of rat-brain metabolic data from digitized autoradiographs are resampled in the horizontal plane (90° from the coronal plane) by locating the 3-D reconstruction within a defined coordinate system (Hibbard et al./1987). A constant row is taken from each coronal image, and that image of rows is scaled linearly about the brain coordinate-system origin to generate isotropically resampled, horizontal images like those of Fig. 5.6.

5.2.4   Digital Imaging in Cytoarchitectonics

Cytoarchitectonics is a branch of neuroanatomy concerned with the regional organization of the nervous system, based on local properties such as cell shape, size, and orientation. Region boundaries correspond to changes in these properties observed by histochemical staining of serial tissue sections. The regional organization of the brain thus observed can be related to local functional roles and to structural changes occurring during development and during the course of a disease. In the past, cytoarchitectonic studies were qualitative computations of the variation in staining throughout the brain. Recently, digital image processing methods have been employed to give these observations a quantitative basis.

The neuroanatomists Schleicher, Zilles, Wree, and colleagues in Cologne have used digital image-processing techniques (Schleicher et al./1984, Wree et al./1982) to study the volume densities of neuronal nuclei in Nissl-stained sections of the brains of mouse (Wree et al./1983), rat (Zilles et al./1980), and guinea pig (Wree et al./1981). Unlike the Golgi stain, which impregnates only a few percent of the neurons, the Nissl stain is absorbed by all neurons and none of the surrounding glial cells. The cytoarchitectonic analyses rest on the concept of a gray-level index (GLI), defined as the ratio of the area of the stained elements to the area of the entire field of measurement. The GLI is a measure of the density of stained cells within a rectangular solid whose square section is equal in area to the measurement field and whose length is equal to the section thickness. The stained cells' areas are determined by segmenting the cells from the rest of the image features and thresholding about a gray-level value determined from the gray-level distribution of the image. Nissl-stained sections of the cerebral cortex tend to have similar distributions over different areas in one section and from section to section. The stained cells, therefore, can be reliably isolated from the background. Maintaining constant section thickness and staining protocol, it is possible to compare cell density among several locations in the brain or among experiment and control brains. The result is a map of neuronal density over the entire brain in which the boundaries between anatomical regions are established by quantitative statistical criteria (Fig. 5.7).

5.3   IMAGING OF NEURONAL MORPHOLOGY AND BRAIN ULTRASTRUCTURE

From its beginning, neuroanatomy has proceeded from the implicit assumption that the shapes and dendritic branching of neurons determine their function. With modern techniques in staining, sectioning, and microscopy, new information has gradually turned that assumption into fact. Indeed, the nervous system consists of more different kinds of cells than does any other tissue, with each cell displaying regional specialization of subcellular parts. Differentiation of neurons into the various classes results from the expression of sets of genes characteristic of each cell type and from environmental influences. Neuronal specialization may take the form of specific neurotransmitters (described above) or a particular size and shape of cell body and dendritic tree. However expressed, neuronal form and function are correlated (Schwartz/1985).

Because of their link to function, neuronal structures have been the object of study by neurobiologists since the 19th century. Tissue-staining techniques permit the viewing of single neurons in thin sections of normally transparent brain tissue, and the information in single sections is sufficient to answer some functional questions. However, because neurons, or parts of neurons, may extend through many sections, a recreation of the original structure is produced by combining information from many planes into one 3-D reconstruction. In the sections that follow, neuronal shape and signaling properties will be reviewed briefly, followed by an example of the application of digital image processing applied to the assessment of nerve viability. Then, three-dimensional reconstruction will be discussed with respect to problems in its application and strategies for overcoming them.

5.3.1   Correlation of Neuron Structure and Function

The relation between form and function can be made evident in an examination of the electrical properties of the neuron. Information transfer occurs within a neuron by the motion of the action potential along the membrane of the cell. In its resting (nonsignaling) state, the neuronal membrane is the site of a charge separation equivalent to a potential at the inner membrane surface equal to -65 mV relative to the outside surface. The charge separation is created and maintained against passive dissipative losses by concentration gradients of Na^+ and K^+ ions across the membrane produced by the action of a protein molecular device within the membrane known as the Na-K pump. The lipid membrane does not permit the passage of charged, water-soluble ions, so their transmembrane transport can occur only through pores or channels that permit the passage of specific ions or molecules. Ionic gradients can be established by the cell using ATP (adenosine triphosphate) to provide the energy to pump ions across the membrane. The Na-K pump removes three Na^+ ions from the neuron for every two K^+ ions it brings in, producing Na and K concentration gradients and a net negative change on the inside of the membrane.

The action potential transiently depolarizes the membrane, traveling as a wave along the membrane from one part of the neuron to another. The rate of propagation is inversely dependent on the product of the axial resistance and the capacitance per unit length of the axon, and directly proportional to the diameter of the axon. Myelination (wrapping of glial cell membranes around neuron axons) also increases propagation rates by effectively decreasing the capacitance of the membrane. Conduction in myelinated axons is much faster than in nonmyelinated axons of the same diameter. Therefore, the length and diameter of dendrites and axons directly affect the conduction of action potentials (Koester/1985).

5.3.2   Nerve Viability Assessed by Digital Image Processing

Ellis et al./1978 described a program for the automatic measurement of axon diameter and myelin sheath thickness from images of peripheral nerve cross sections. Measurements of axonal diameter and the thickness of the myelin sheath encircling the axon were used to classify and gauge the condition of peripheral nerve fibers. This is useful to assess changes due to degenerative neuropathologies caused by drug studies or disease modeling in experimental animals (Ellis et al./1978 and references cited therein). The fiber cross sections were distinguished from the background by thresholding, using a gray-level distribution. The individual fibers were isolated in this way from the background, but further processing was required to separate fibers in contact with each other. By combining boundary-tracking and a measure of local curvature, their program found those points representing the interfiber contacts and located the full boundary of each individual fiber. Fibers passing the recognition test were measured to determine the cross-section areas of the axon itself and its surrounding myelin sheath, with the thickness of the sheath providing a measure of the viability of the nerves. Similar, though less sophisticated, approaches have been published by (O'Leary et al./1976 and Geckle and Jenkins/1983).

Gayer and Schwartz/1983 describe another interesting application of image processing to nerve viability. Neuritic outgrowth, or the sprouting of thin fibers by cells in culture, is a common bioassay in neurobiology. The greater the proliferation of neuritic outgrowth, the greater the potential for cell growth and regeneration. Conventional observations of outgrowth result in the subjective scoring of a culture on a scale of 0–4 based on the lengths and densities of the fibers. Gayer and Schwartz used digital image analysis (Fig. 5.8) to score the density of outgrowth more quantitatively, by imaging neurons grown in culture and counting the pixels belonging to outgrowth fibers (Fig. 5.8). Digital images of cells in culture were obtained with a video camera and a frame buffer. The fibers had higher gray-levels than did the background, but the background varied greatly, making direct thresholding impossible. An adaptive point operation was devised to suppress the background to an acceptable low and even gray-level range. From the value of each pixel, $N(i,j)$, the quantity $A(i,j) - KS(i,j)$ was subtracted, where $A(i,j)$ is the average gray-level in an 8 × 8 window about the point (i,j), $S(i,j)$ is the standard deviation over the same window, and K is a constant whose optimal value was determined to be 2. With the background flattened uniformly to near zero, the fiber features could be thresholded and counted as a percent area of the image. Used this way, computer imaging could efficiently screen neuritic outgrowths for studies of the effects of drugs on neuronal growth and regeneration.

5.3.3   Three-Dimensional Reconstruction—Problems and Strategies

Computer-aided 3-D reconstruction provides a mechanism for examining a complex object that is otherwise viewable only in sections. In neurobiology, 3-D reconstruction studies have examined single neurons in section, assemblies of neurons (nuclei), the regional organization of neuronal groups, and even the entire brain itself (e.g., the rat-brain metabolic reconstructions described above).

All reconstructions face the same problems, regardless of the subject under study: (1) how to extract that subset of all the features in the serial section images representing the structure of interest, (2) how to align the images to recreate the structure in three dimensions, and (3) how best to display the resulting reconstruction. Computer segmentation of images is hampered by the lack of reliable models for the features of interest in complex images. So far, models for complex images have been too difficult to design or too computationally intensive, or both. For example, transmission electron microscopy of almost any biological ultrastructure produces images of enormous complexity. Simply sorting the edges of the features of interest from those of all the other features present can tax current scene-analysis methodologies to the limit. By comparison, the problem of serial image registration has been solved, both by manual and programmed means, in a number of instances and is (usually) not dependent on prior segmentation of the features of interest. In some instances for electron micrographs, however, no single rotation/translation transformation produces registration everywhere in the field of an image-

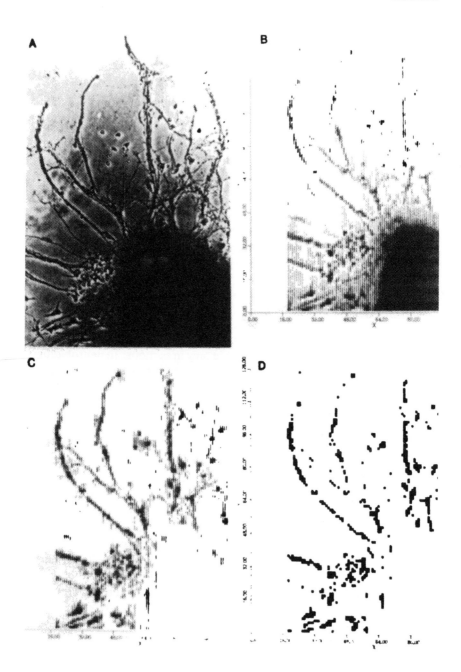

pair, due to stretching and warping of the tissue during fixation and mounting.

Because of these difficulties, most neurobiological reconstructions have used manual, interactive methods for data entry, segmentation, and alignment. The result has been proliferation of published computer 3-D reconstruction programs (Huijsmans et al./1986) based mostly on the interactive graphics capabilities of various microcomputers.

Because of the requirement for intelligible visual representations of 3-D reconstructions, the display problem has received more attention than have programmed solutions to image segmentation and alignment. Drawing on conventional computer-graphics technology (Foley and Van Dam/1982, Newman and Sproull/1979) for hidden-line and surface removal and lighting models, a number of laboratories have presented data as stacked polygons and as shaded solids using raster graphics (Huijsmans et al./1986).

5.3.4   The Computer Microscope and Neuron Tracing in Three Dimensions

The mapping of the structure of stained neurons began in the 19th century using light microscopy and *camera lucida* drawings. The camera lucida, invented in 1807 (Glaser et al./1985), is an optical attachment to the microscope that, by means of a half-silvered mirror, combines the images of the prepared specimen and the hand drawing of that specimen. More recently, computer imaging systems have replaced the camera lucida.

Glaser and Van der Loos/1965 introduced computerized tracing of dendritic processes with their computer microscope: a microscope with

---

Figure 5.8   The proliferation of dendritic outgrowths from neurons in cell culture is a common assay for the effects of drugs or disease states on neuronal viability. The more numerous and lengthy the outgrowths (A), the greater the potential for neuronal growth. Gayer and Schwartz/1983 used digital imaging methods to score the outgrowth density more quantitatively than conventional subjective scoring by visual inspection. (A) Micrograph of dendritic processes radiating from a neuronal soma in cell culture, (B) the digitized version of (A). Before the fibrous features could be identified and counted, the widely varying background had to be suppressed. (C) Background flattened by subtracting the standard deviation from the pixel average over a moving window; (D) thresholding removes the remaining nondendrite density from the operated image of (C). By counting the remaining pixels, the extent of outgrowth could be measured. Figure from (Gayer and Schwartz/1983) by permission.

a mechanical x-y stage in which the specimen being viewed is moved so that successive points on a dendrite are placed under a cross hair and the stage position is recorded (Fig. 5.9). Vertical motion of the stage to follow the vertical component of the displacement of the dendrite was also recorded. Recent descriptions of this system (Glaser et al./1983, Glaser and McMullen/1986), now called the *imaging combining computer microscope*, incorporate many technical advances in the interactive user interface, storage, and graphical outputs. Similar neural tracing systems have been devised by (Wann et al./1973, Hillman et al./1974, Llinas and Hillman/ 1975, Overdijk et al./1978, Paldino/1979, Capowski and Sedivec/1981, and

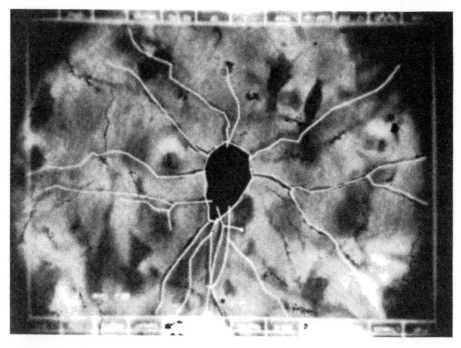

Figure 5.9 Golgi-impregnated neuron as viewed through the computer microscope of Glaser and McMullin/1986. The computer graphics display and menu is superimposed on the viewed image in the microscope using a miniature CRT in a camera lucida arrangement. The dendrites are traced by following them with a screen cursor and marking them at intervals. The (X, Y, Z) coordinates of each marked point are obtained from sensors attached to the microscope stage and focus mechanism. Once recorded, the dendrites can be displayed with a 3-D graphical representation, and morphology and connectivities can be assessed quantitatively. Figure from (Glaser and McMullin/1986) by permission.

Capowski/1983). All these authors describe their systems as semiautomatic, meaning that a human operator must guide the computer microscope along a dendrite, but the sampled cursor coordinates are recorded by the computer. Another common feature of these systems is that they sample a continuous neural structure in thick sections by moving the plane of focus. This method has been termed *optical sectioning* to distinguish it from the imaging of structure in discrete thin sections (described below). Several reviews of neural tracing and reconstruction systems have appeared recently (Huijsmans et al./1986, Capowski/1985).

The Golgi-stained neuron is a 3-D object, often compared to a tree, with its dendrites corresponding to a tree's branches. The body, or soma, of the neuron is the tree's trunk centered within the dendritic network. The dendrites describe complex paths through space, with branching, dendritic spines, fiber swellings, and end points to be recorded and distinguished. The semiautomatic computer-microscope provides the means to record this data, guided by a human operator. The tracking, however, is a laborious task, and programmed tracking would reduce the labor, enabling more extensive studies.

Successful tracking requires the implementation of edge detection, edge following, and branch-point decision making, and the search for algorithms appropriate for neural tracking has been under way since the early 1970s. Garvey et al./1972 reported a television-microscope scanner that tracked dendrites by searching for maximum darkness within an arc swept out in front of the cursor in the direction of its travel. The program also adjusted the microscope focus to maximize the dendrite darkness, providing at the same time the coordinate in the third dimension. Capowski/1983 described in detail the design and operation of an automatic dendrite-tracking system using that basic algorithm.

Two recent reports of neuron-tracking programs both involve aspects of Marr's theory of visual processing (Marr/1982). Reuman and Capowski/ 1984 use the Marr–Hildreth filter ($\nabla^2 g$) (Marr and Hildreth/1980) to detect dendrite edges as zero crossings in filtered images of Golgi–stained neurons. Several methods for following the dendrites are described and compared, including a variant of the semicircular-arc method of (Garvey et al./1972). Finally, decision-making is discussed, with the problem being how to tell if the current point on the dendrite-path belongs to one or another of several possible classes, including dendrite end points and several kinds of branch points.

The second recent study (Selfridge/1986) uses Marr's theory of vision, specifically the decomposition of a scene into *primal-sketch* primitives, or simple geometric figures easily encoded in spatial-domain operators. One example is a short line segment in different angular orientations that maximally map onto edges of neuronal features. The local line features' location and orientation data form a graph. The edge feature is a path that is grown within the graph by connecting local line elements according to simple empirical rules that reproduce the cell edges in test images.

Unfortunately, programs to track neurons as reliably as do human observers do not yet exist, for several reasons. The first is the complexity of the decision-making required to follow the long, thin dendritic processes of neurons when they appear to contact and cross each other, or when interrupted by an apparent gap due to incomplete staining. Distinguishing neural from nonneural features (glial cell processes, blood vessels, staining and sectioning artifacts) and tracking dendrites parallel to the optical axis of the microscope pose additional difficulties (Wann et al./1973). Finally, the use of global-scene information to provide more reliable dendrite tracking is too computationally demanding to be practical currently (Reuman and Capowski/1984).

One more application of the optical-section approach to reconstruction should be mentioned. Tieman et al./1986 presented a technique for producing stereo-pair images of Golgi-stained neurons in thick sections with the aid of digital image processing. Here, successive planes of a thick section were brought into focus with a microscope and digitized with a video camera and a frame buffer. Each digitized optical section was convolved with a simple 3×3 operator to increase the contrast of the dark, dot-like Golgi stain features with the surrounding light background. The result was an image in which only those features in sharpest focus were preserved. After interactive editing to remove any extraneous features not part of the neurons being imaged, the optical sections were recombined to form the final image. The final-image pixel values were simply the maximum of the values of the corresponding pixels in all the optical sections. Stereo-pairs of images can be prepared by recombining the digital optical images with appropriate offsets (see Tieman et al./1986 for details). Both the single and stereo-pair presentations of Golgi-impregnated neurons contain high detail with essentially infinite depth of field (Fig. 5.10).

5.3.5   Three-Dimensional Reconstructions from Serial Sections

The alternative to optical sectioning of thick sections for 3-D reconstructions is to image thin serial sections. This has been the most frequently used approach in mapping neural structures. The variety of applications has been wide: development of the visual systems in simple organisms (Levinthal et al./1976, Lopresti et al./1973, Macagno et al./1973, Rakic et al./1974), the spatial distributions of neurons containing particular neuro-

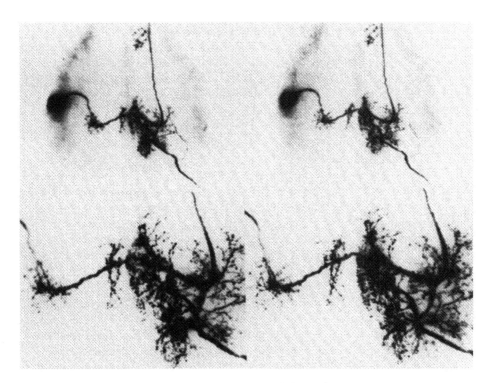

Figure 5.10   High-resolution stereograms of Golgi-stained neurons in thick sections prepared using digital image processing.   Because the depth of field of the microscope is much thinner than the full thickness of the dendritic "tree" of a neuron, one sees only a thin slice of the neuron in focus at any one time. By stepping through a stained neuron and digitizing the plane in focus, (Tieman et al./1986) built up a stack of optical sections. Each section was convolved with a simple 3 × 3 operator to increase the contrast of the dark, dot-like Golgi features with respect to the background. The result was that only those features in sharpest focus are preserved. The final stereo-pairs were constructed by combining the maximum-density features of all the optical sections, with appropriate offsets to produce the stereo effect. Figure from (Tieman et al./1986) by permission.

transmitters (Waterhouse et al./1986), reconstruction of the structures of and intercellular connections among retinal neurons (McGuire et al./1986, McGuire et al./1984, Stevens et al./1980), quantitative analyses of Golgi-stained cells (Yelnik et al./1981, Hillman et al./1974, Llinas and Hillman/1975), the morphology of neuronal groups (Augustine et al./1985), and the structural details of blood vessels serving the brain (Hibbard et al./1986).

The general method was defined in the early 1970s by Levinthal and coworkers (Levinthal and Ware/1972, Levinthal et al./1974) with the development of the well-known CARTOS system (computer-aided reconstruction by tracing of sections) (Macagno et al./1979, Sobel et al./1980, Kropf et al./1985). In studies of the development of the nervous systems of simple organisms, electron micrographs of many thin serial tissue sections were obtained. Aligned copies of the electron micrographs were produced using a video device that displays one aligned image and a second, serial-adjacent image to be aligned with it in rapid alternate views on a video display. By manually rotating and translating the second image, alignment with the first was achieved when the difference-flicker between the images was minimized. After alignment, each micrograph was rephotographed on successive frames of a 35-mm film strip. The outlines of the features of interest were then manually digitized from the aligned images, and the contours could be viewed in 3-D using a graphics display. Levinthal has called this system a 3-D notebook because it allows the investigator to record the contours and other morphologic details and then interact with the 3-D display of the reconstructed object. Recent reconstructions using CARTOS are shown in Fig. 5.11.

More recently, an automatic contour extraction (ACE) facility was added to CARTOS (Kropf et al./1985). Isodensity contours are examined in image regions where they are expected to occur based on extrapolated contour data in the preceding sections. An acceptable boundary contour is taken as one that is a non-self-intersecting closed curve and whose quantifiable shape features (area, perimeter, coordinate extrema, and others) are reasonably close in value to those of the accepted contours in the previous sections. The contours are represented as ordered sets of pixels lying on, or adjacent to, the visually correct boundary. ACE is capable of some heuristic decision-making regarding the acceptance/rejection of boundary pixels in contours degraded by gaps or noise.

Many other 3-D reconstruction systems have been reported that follow the basic plan of CARTOS, with differences dictated by the local needs of the scientific project to which the reconstructions were applied. Stevens, Sterling, and coworkers (Stevens and Trogadis/1984, Stevens et al./1980) have conducted extensive studies of the structures of single neurons in the visual system of the cat (McGuire et al./1980, McGuire et al./1984) with a system that relies on manual image alignment and contour tracing (Fig. 5.12). Woodward, Schlusselberg, Smith, and colleagues have developed the CARP (computer-aided reconstruction package) system (Woodward et al./1986, Smith et al./1983, Smith et al./1981), which has been applied to a very large number of scientific problems inside and outside neurobiology. This system incorporates a computer-microscope,

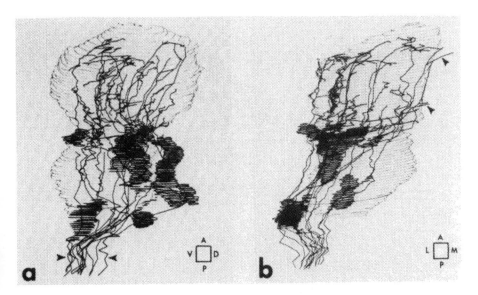

Figure 5.11   Computer three-dimensional reconstruction of neurons in the visual system of a small crustacean, *Daphnia magna*. Shown here are lateral and dorsal views, respectively, of one morphological class of neurons in the optic ganglion. Cell bodies are represented by the clusters of dark contours, and their dendritic processes are evident as single lines running along the A-P (anterior-posterior) direction. E. R. Macagno and colleagues have studied many aspects of this animal's visual system for a number of years (Macagno et al./1973, Sims and Macagno/1985), and reconstructions of the neurons involved in visual processing will be necessary to explain the animal's photostimulatory responses. Shown in this reconstruction are parts of the 17 neurons of the LS group, part of the approximately 200 neurons residing in this portion of the optic ganglion. Figure from (Sims and Macagno/1985), by permission.

manual tracing of serial section images, and processing of autoradiographs and radiological images. While it is a mostly interactive system, some programmed image segmentation is provided based on gray-level thresholding and boundary following (Woodward et al./1986). The properties of a large number of 3-D reconstruction systems have been reviewed by Huijsmans et al./1986).

Two reconstruction systems recently reported make use of digital image-processing methods for segmentation and/or image alignment to a greater extent than do other conventional systems. Augustine et al./1985 reported the reconstruction of the abducens nucleus in the baboon brain stem. This nucleus is a nerve bundle responsible (along with other nerve groups)

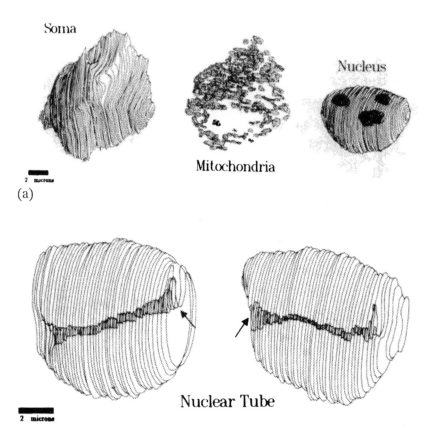

Figure 5.12   Three-dimensional reconstructions of single neurons. (a) Reconstructions of the soma and some interior parts (the mitochondria and the nucleus) of cells of the PC12 tumor cell line. These are cells that can be grown in culture and will undergo differentiation to become neurons with the addition of nerve growth factor to the culture medium. (b) Nuclear tubes in reconstructed PC12 nuclei. The purpose of these tubes or channels, which are invaginations running entirely through the nuclei, is unknown. The presence of these tubes would have gone unnoticed without 3-D reconstruction. Figures are taken from (Stevens and Trogadis/1986) by permission.

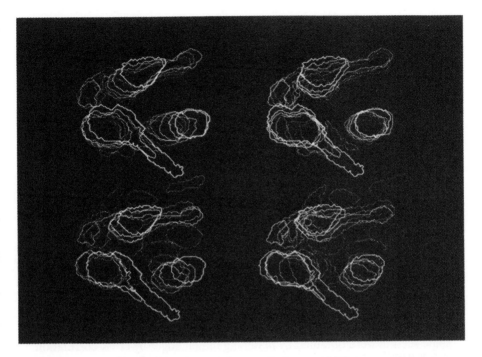

Figure 5.13 Digital image-processing methods can aid in generating three-dimensional reconstructions. Shown here are stereo views of a brain capillary module before (bottom) and after (top) the application of Fourier cross-correlation to match capillary edges in serial electron nuclei graphs. The serial images in the *before* reconstruction were themselves aligned by the principal-axes transformation from an even more disordered state. This is an example (described in Hibbard et al./1986) of a reconstruction carried out entirely by programs and demonstrates the potential for digital image-processing methods to contribute a reduction of labor and increased objectivity to three-dimensional reconstruction.

for the control of eye movements. Neurons in the nucleus were labeled with a histochemical marker, and the brain stem was cut into 40-micron sections. Viewed by light microscopy, the labeled neurons exhibited high contrast relative to the surrounding tissue, and digital images of the sections were obtained. The images were aligned manually before digitization by superposing anatomic landmarks outside the nucleus. Contrast in the digital image was further enhanced by spatial convolution with high-pass filters (not described), and the filtered images were thresholded to create a binary image. At this stage, the neuron features in the binary image had to be merged to represent the area of the entire nucleus. A region-growing algorithm fused the neuron features to fill in the nuclear area, and the outline of the nucleus was contoured. The contours of all the slices were combined to generate 3-D wire-frame and shaded-solid views of the reconstructed nucleus.

The second system is by (Hibbard et al./1986), who reported 3-D reconstructions of microcapillary modules in the median eminence of the hypothalamus of the rabbit. This portion of the hypothalamus physically connects the brain and the pituitary, and the capillary network within the median eminence is believed to play a role in controlling the two-way hormonal communication between the brain and the pituitary. To establish whether local control of blood flow within the median eminence is possible, a series of reconstructions was undertaken to study the placement of smooth-muscle sphincters within microcapillary loops and arbors. Hibbard et al./1986 report two reconstructions of small capillary bundles carried out entirely by programmed image segmentation, mosaicking, and two stages (coarse and fine) of alignment (Fig. 5.13). Thin tissue sections were imaged by transmission electron microscopy and digitized. The capillary interiors (lumens) were identified using an operator designed specifically to extract the lumens with their high gray-levels, low local variances, and long-range contiguity. Mosaics of multiple overlapping micrographs of the same section were obtained by matching lumen edges by cross-correlation (Moik/1980). The mosaic images were then aligned, first by superposing centroids and principal axes, and then by alternate cycles of translational and rotational cross-correlation, as described above for the metabolic 3-D reconstruction (Hibbard et al./1987).

5.4 FUTURE DIRECTIONS FOR COMPUTER IMAGING AND 3-D RECONSTRUCTION IN NEUROBIOLOGY

Since the brain and its parts are 3-D objects whose structure and interconnections determine how the brain functions, the use of computer imaging and 3-D reconstruction will become even more important in the future. In the past, most digital imaging applications have been largely qualitative comparisons of experiments and controls to support the assertion that the experiments produced an effect. With the development of metabolic models used to calculate a unit of function from an image gray-level, the images became quantitative tools, mapping function onto anatomy. Several laboratories then showed how it is possible to compare experiment and control, pixel by pixel, over the whole brain, enabling comparisons that are both global and highly detailed.

Imaging and reconstruction of structure are following that progress in autoradiography analysis and reconstruction. Until recently, virtually all reconstructions were hand-made, derived from manual alignment and feature tracing. The generation of efficient display algorithms has been useful, but the interactive encoding and manipulation of data impose a limit on the size and number of studies that can be practically undertaken (Macagno et al./1984). Indeed, digital image processing offers the only practical means both for generating large-scale reconstructions and for analyzing their properties.

ACKNOWLEDGMENTS

The author wishes to thank Judy Perry for assistance in preparing the manuscript and Susan Brecht Hibbard for editorial assistance. The author would also like to thank Dr. Richard A. Hawkins of the Pennsylvania State University College of Medicine at Hershey for introducing me to computer imaging and neurobiology and for providing essential support and direction during the early phases of development of our imaging program system. Support for the author's work described here has been provided by the National Science Foundation (BNS-8506479) and by The Whitaker Foundation.

REFERENCES

Alexander, G. M. and Schwartzman, R. J. (1984). Quantitative computer analysis of autoradiographs utilizing a charge-coupled device solid state camera, *J. Neurosci. Meth.*, *12*: 29.

Altar, C. A., Walter, R. J., Jr., Neve, K. A., and Marshall, J. F. (1984). Computer-assisted video analysis of [^{3}H] spiroperidol binding autoradiographs, *J. Neurosci. Meth.*, *10*: 173.

Altar, C. A., O'Neil, S., Walter, R. J., Jr., and Marshall, J. F. (1985). Brain dopamine and serotonin receptor sites revealed by digital subtraction autoradiography, *Science*, *228*: 597.

Anuta, P. E. (1970). Spatial registration of multispectral and multitemporal digital imagery using fast Fourier transform techniques, *IEEE Trans. Aerosp. Electron. Syst.*, *8*: 353.

Augustine, J. R., Huntsberger, T., and Moore, M. (1985). Computer-aided reconstructive morphology of the baboon abducens nucleus, *Anat. Rec.*, *212*: 210.

Berck, D. J. and Rainbow, T. C. (1985). Microcomputer-assisted densitometer for quantitative receptor autoradiography, *N. Neurosci. Meth.*, *13*: 171.

Bloom, F. E. (1984). The functional significance of neurotransmitter diversity, *Am. J. Physiol.*, *246*: C184.

Bloom, F. E., Lazerson, A., and Hofstadter, L. (1985). *Brain, Mind, and Behavior*. Freeman, New York.

Capowski, J. J. and Sedivec, M. J. (1981). Accurate computer reconstruction and graphics display of complex neurons utilizing state-of-the-art interactive techniques, *Comput. Biol. Res.*, *14*: 518.

Capowski, J. J. (1983). An automatic neuron reconstruction system, *J. Neurosci. Meth.*, *8*: 353.

Capowski, J. J. (1985). The reconstruction, display, and analysis of neuronal structure using a computer, *The Microcomputer in Cell and Neurobiology Research*, (R. R. Mize, ed.), Elsevier, New York, p. 85.

Ellis, T. J., Rosen, D., and Cavanaugh, J. B. (1978). Automated measurement of peripheral nerve fibers in transverse section, *J. Biomed. Eng.*, *2*: 272.

Foley, J. D. and Van Dam, A. (1982). *Fundamentals of Interactive Computer Graphics*, Addison-Wesley, Reading, Massachusetts.

Freygang, W. H., Jr. and Sokoloff, L. (1958). Quantitative measurement of regional circulation in the central nervous system by the use of radioactive inert gas, *Adv. Biol. Med. Physics*, *6*: 263.

Gallistel, C. R., Piner, C. T., Allen, T. O., Adler, N. T., Yadin, E., and Negin, M. (1982). Computer assisted analysis of 2-DG autoradiographs, *Neurosci. Biobehav. Rev.*, *6*: 409.

Gallistel, C. R. and Nichols, S. (1983). Resolution-limiting factors in 2-deoxyglucose autoradiography (I. Factors other than diffusion), *Brain Res.*, *267*: 323.

Garvey, C., Young, J., Simon, W., and Coleman, P. (1972). Semiautomatic dendrite tracking and focusing by computer, *Anat. Record*, *172*: 314.

Gayer, A. and Schwartz, M. (1983). The use of image analysis for measuring neuritic outgrowth from goldfish regenerating retina explants. *J. Neurosci. Meth.*, *7*: 275.

Geckle, W. J. and Jenkins, R. E. (1983). Computer-aided measurement of transverse axon sections for morphology studies, *Comput. Biomed. Res.*, *16*: 287.

Glaser, E. M. and Van der Loos, H. (1965). A semi-automatic computer-microscope for the analysis of neuronal morphology, *IEEE Trans. Biomed. Eng.*, *2*: 22.

Glaser, E. M., Tagamets, M., McMullen, N. T., and Van der Loos, H. (1983). The image-combining computer microscope—an interactive instrument for morphometry of the nervous system, *J. Neurosci. Meth.*, *8*: 17.

Glaser, E. M. and McMullen, N. T. (1986). Semiautomatic microscopy as a tool for cytologists, *Anal. Quant. Cytol. Histol.*, *8*: 116.

Glaser, T., Rath, M., Traber, J., Zilles, K., and Schleicher, A (1985). Autoradiographic identification and topological analyses of high affinity serotonin receptor subtypes as a target for the novel putative anxiolytic TVX Q 7821, *Brain Research*, *358*: 129.

Goldstein, H. (1950). *Classical Mechanics*. Addison–Wesley, Reading, Massachusetts.

Goochee, C., Rasband, W., and Sokoloff, L. (1980). Computerized densitometry and color coding of [^{14}C] deoxyglucose autoradiographs, *Ann. Neurol.*, *7*: 359.

Herkenham, M. and Pert, C. B. (1982). Light microscopic localization of brain opiate receptors: A general autoradiographic method which preserves tissue quality, *J. Neurosci.*, *2*: 1129.

Herman, G. T. (1986). Computerized reconstruction and 3-D imaging in medicine, *Am. Rev. Comput. Sci.*, *1*: 153.

Hibbard, L. S. and Hawkins, R. A. (1984). Three-dimensional reconstruction of metabolic data from quantitative autoradiography of the rat brain, *Am. J. Physiol.*, *247*: E412.

Hibbard, L. S., McGlone, J. S., Davis, D. W., and Hawkins, R. A. (1987). Three-dimensional representation and analysis of brain energy metabolism, *Science*, *236*: 1641.

Hillman, D. E., Chujo, M., and Llinas, R. (1974). Quantitative computer analysis of the morphology of cerebellar neurons (I. Granule cells), *Anat. Rec.*, *178*: 375.

Hokfelt, T., Johansson, O., and Goldstein, M. (1984). Chemical anatomy of the brain, *Science*, *225*: 1326.

Huijsmans, D. P., Lamers, W. H., Los, J. A., and Strackee, J. (1986). Toward computerized morphometric facilities: A review of 58 software packages for computer-aided three-dimensional reconstruction, quantification, and picture generation from parallel serial sections. *Anat. Rec.*, *216*: 449.

Isseroff, A. and Lancet, D. (1985). An inexpensive microcomputer based image-analysis system: novel applications to quantitative autoradiography, *J. Neurosci. Meth.*, *12*: 265.

Kandel, E. R. and Schwartz, J. H. (1985). *Principles of Neural Science*, 2d ed., Elsevier, New York.

Koester, J. (1985). Functional consequences of passive membrane properties of the neuron, *Principles of Neural Science*, 2d ed. (E. R. Kandel and J. H. Schwartz, eds.), Elsevier, New York, p. 66.

Kropf, N., Sobel, I., and Levinthal, C. (1985). Serial section reconstruction using CARTOS, *The Microcomputer in Cell and Neurobiology Research*, (R. R. Mize, ed.), Elsevier, New York, p. 265.

Landau, W. M., Freygang, W. H., Jr., Roland, L. P., Sokoloff, L., and Kety, S. S. (1955). The local circulation of the living brain; values in the unanesthetized and anesthetized cat, *Trans. Am. Neurol. Assn.* 80: 125.

Levinthal, C. and Ware, R. (1972). Three dimensional reconstruction from serial sections, *Nature* (London), *236*: 207.

Levinthal, C., Macagno, E., and Tountas, C. (1974). Computer-aided reconstruction from serial sections, *Fed. Proc.*, *33*: 2336.

Levinthal, F., Macagno, E., and Levinthal, C. (1976). Anatomy and development of identified cells in isogenic organisms, *Cold Spring Harbor Symp. Quant. Biol.*, *40*: 321.

Llinas, R. and Hillman, D. E. (1975). A multipurpose tridimensional reconstruction computer system for neuroanatomy, *Golgi Centennial Symposium Proceeding* (M. Santini, ed.), Raven Press, New York, p. 71.

Lopresti, V., Macagno, E. R., and Levinthal, C. (1973). Structure and development of neuronal connections in isogenic organisms: Cellular interactions in the development of the optic lamina of *Daphnia, Proc. Natl. Acad. Sci. USA, 70*: 433.

Lu, D. M., Davis, D. W., Mans, A. M., and Hawkins, R. A. (1983). Regional cerebral glucose utilization measured with [^{14}C] glucose in brief experiments. *Am. J. Physiol., 245*: C428.

Macagno, E. R., Lopresti, V., and Levinthal, C. (1973). Structure and development of neuronal connections in isogenic organisms: Variations and similarities in the optic system of *Daphnia magna, Proc. Natl. Acad. Sci. USA, 70*: 56.

Macagno, E. R., Levinthal, C., and Sobel, I. (1979). Three-dimensional computer reconstruction of neurons and neuronal assemblies, *Ann. Rev. Biophys. Bioeng., 8*: 323.

Margulis, A. R., Hricak, H., and Crooks, L. (1987). Medical applications of nuclear magnetic resonance imaging, *Quart. Rev. Biophys., 19*: 221.

Marr, D. and Hildreth, E. (1980). Theory of edge detection, *Proc. Roy. Soc. Lond., B207*: 187.

Marr, D. (1982). *Vision*. Freeman, San Francisco.

Mazziotta, J. C. and Koslow, S. H. (1987). Assessment of goals and obstacles in data acquisition and analysis from emission tomography: Report of a series of international workshops, *J. Cereb. Blood Flow Metab., 7*: 51–531.

McGuire, B. A., Stevens, J. K., and Sterling, P. (1984). Microcircuitry of bipolar cells in cat retina, *J. Neurosci., 4*: 2920.

McGuire, B. A., Stevens, J. K., and Sterling, P. (1986). Microcircuitry of beta ganglion cells in cat retina, *J. Neurosci., 6*: 907.

Moik, J. G. (1980). *Digital Processing of Remotely Sensed Images*, NASA SP-431, Washington, DC.

Nauta, W. J. H. and Feirtag, M. (1986). *Fundamental Neuroanatomy*, Freeman, New York.

Newman, W. M. and Sproull, R. F. (1979). *Principles of Interactive Computer Graphics*, McGraw-Hill, New York.

O'Leary, D. P., Dunn, R. F., and Kumly, W. E. (1976). On-line computerized entry and display of nerve fiber cross-sections using single or segmented histological records, *Comput. Biomed. Res., 9*: 229.

Overdijk, J., Uylings, H. B. M., Kuypers, K., and Kamstra, A. W. (1978). An economical, semi-automatic system for measuring cellular tree structures in three dimensions, with special emphasis on Golgi-impregnated neurons, *J. Microscopy, 114*: 271.

Paldino, A. M. (1979). A novel version of the computer microscope for the quantitative analysis of biological structures: Application to neuronal morphology, *Comput. Biomed. Res., 12*: 413.

Paxinos, G. and Watson, C. (1986). *The Rat Brain in Stereotaxic Coordinates*, 2d ed., Academic Press, New York.

Pellegrino, L. J., Pellegrino, A. S., and Cushman, A. J. (1979). *A Stereotaxic Atlas of the Rat Brain*, Plenum, New York.

Pratt, W. K. (1978). *Digital Image Processing*, Wiley-Interscience, New York.

Rainbow, T. C., Bleisch, W. V., Biegon, A., and McEwen, B. S. (1982). Quantitative densitometry of neurotransmitter receptors, *J. Neurosci. Meth.*, *5*: 127.

Ramm, P. and Kulick, J. H. (1985). Principles of computer-assisted imaging in autoradiographic densitometry, *The Microcomputer in Cell and Neurobiology*, (R. R. Mize, ed.), Elsevier, New York, p. 311.

Reivich, M., Jehle, J., Sokoloff, L., and Kety, S. S. (1969). Measurement of regional cerebral blood flow with antipyrine-^{14}C in awake cats, *J. Appl. Physiol.*, *27*: 296.

Reuman, S. R. and Capowski, J. J. (1984). Automated neuron tracing using the Marr–Hildreth zerocrossing technique, *Comput. Biomed. Res.*, *17*: 93.

Rosenfeld, A. and Kak, A. C. (1982). *Digital Picture Processing*, Academic Press, New York.

Selfridge, P. G. (1986). Locating neuron boundaries in electron micrograph images using "primal sketch" primitives, *Comput. Vision Graph. Image Process.*, *34*: 156.

Schleicher, A., Zilles, K., and Wree, A. (1984). A quantitative approach to cytoarchitectonics: Software and hardware aspects of a system for the evaluation and analysis of structural inhomogeneities in nervous tissue, *J. Neurosci. Meth.*, *18*: 221.

Skinner, J. E. (1971). *Neuroscience: A Laboratory Manual.* Saunders, Philadelphia.

Sims, S. J. and Macagno, E. R. (1985). Computer reconstruction of all the neurons in the optic ganglion of the *Daphnia magna*, *J. Comp. Neurol.*, *233*: 12.

Smith, W. K., Schlusselberg, D. S., and Woodward, D. J. (1981). A computer system for neuroanatomical data acquisition, analysis, and display, *Soc. Neurosci. Abstr.*, *7*: 418.

Smith, W. K., Schlusselberg, D. S., Culter, B. G., Woodward, D. J., and Lacey, E. R. (1983). Hierarchical database for biological modeling, *Proc. NCGA*, *4*: 106.

Sobel, I., Levinthal, C., and Macagno, E. R. (1980). Special techniques for the automatic computer reconstruction of neuronal structures, *Ann. Rev. Biophys. Bioeng.*, *9*: 347.

Sokoloff, L., Reivich, M., Kennedy, C., Des Rosiers, M. H., Patlak, C. S., Pettigrew, K. D., Sakurada, O., and Shinohara, M. (1977). The [^{14}C] deoxyglucose method for the measurement of local cerebral glucose utilization: theory, procedure, and normal values in the conscious and anesthetized albino rat, *J. Neurochem.*, *28*: 897.

Stevens, J. K., Davis T. L., Friedman, N., and Sterling, P. (1980). A systematic approach to reconstruction microcircuitry by electron microscopy of serial sections, *Brain Res. Rev.*, *2*: 265.

Stevens, J. K. and Trogadis, J. (1984). Reconstructive three-dimensional electron microscopy: A routine biological tool, *Anal. Quant. Cytol. Histol.*, *8*: 102.

Sundsten, J. W. and Prothero, J. W. (1983). Three-dimensional reconstruction from serial sections (II. A microcomputer-based facility for rapid data collection), *Anat. Rec.*, *207*: 665.

Tieman, D. G., Murphy, R. K., Schmidt, J. T., and Tieman, S. B. (1986). A computer-assisted video technique for preparing high resolution pictures and stereograms from thick sections, *J. Neurosci. Meth.*, *17*: 231.

Unnerstall, J. R., Niehoff, D. L., Kuhar, M. J., and Palacios, J. M. (1982). Quantitative receptor autoradiography using [^{3}H] Ultrofilm: Application to multiple benzodiazepine receptors, *J. Neurosci. Meth.*, *6*: 59.

Wagner, H. N. (1986). Clinical PET opens gates to *in vivo* biochemistry, *Diagn. Imag.*, June: 82.

Wann, D. F., Woolsey, T. A., Dierker, M. L., and Cowan, W. M. (1973). An on-line digital-computer system for the semiautomatic analysis of Golgi-impregnated neurons, *IEEE Trans. Biomed. Eng.*, *BME-20*: 233.

Waterhouse, B. D., Mihailoff, G. A., Baack, J. C., and Woodward, D. J. (1986). Topographical distribution of dorsal and medium raphe neurons projecting to motor, sensorimotor, and visual cortical areas in the rat, *J. Compar. Neurol.*, *249*: 460.

Wong, W. S., Tsuruda, J. S., Kortman, K. E., and Bradley, W. G. (1986). *Practical Magnetic Resonance Imaging: A Case Study Report*, Aspen Publishers, Rockville, Maryland.

Wree, A., Zilles, K., and Schleicher, A. (1981). A quantitative approach to cytoarchitectonics (V. The areal pattern of the cortex of the guinea pig), *Anat. Embryol.*, *162*: 81.

Wree, A., Zilles, K., and Schleicher, A. (1983). A quantitative approach to cytoarchitectonics (VIII. The areal pattern of the cortex of the albino mouse), *Anat. Embryol.*, *166*: 333.

Woodward, D. J., Smith, W. K., Schlusselberg, D. S., Chapin, J. K., Waterhouse, B. D., and German, D. (1986). Image analysis and computer graphics: Emerging applications in ethanol research, *Alcoholism: Clinical and Experimental Research*, *10*: 251.

Yelnik, J., Percheron, G., Perbos, J., and Francois, C. (1981). A computer-aided method for the quantitative analysis of dendritic arborizations reconstructed from serial sections, *J. Neurosci. Meth.*, *4*: 347.

Yonekura, Y., Brill, A. B., Som, P., Bennett, G. W., and Fand, I. (1982). Quantitative autoradiography with radiopharmaceuticals, Part 1: Digital film analysis system by videodensitometry: Concise communication, *J. Nucl. Med.*, *24*: 231.

Young, W. S., III and Kuhar, M. J. (1979). A new method for receptor autoradiography: [^{3}H] opioid receptors in rat brain, *Brain Res.*, *179*: 255.

Zilles, K., Zilles, B., and Schleicher, A. (1980). A quantitative approach to cytoarchitectonics (VI. The areal pattern of the cortex of the albino rat), *Anat. Embryol.*, *159*: 335.

Zilles, Schleicher, A., Rath, M., Glaser, T., and Traber, J. (1986). Quantitative autoradiography of transmitter binding sites with an image analyzer, *J. Neurosci. Meth.*, *18*: 207.

6

Applying Digital Processing Methods in the Analysis of Retinal Structure

Sunanda Mitra and Thomas F. Krile

Texas Tech University
Lubbock, Texas

6.1 INTRODUCTION

Fundus photographs are photographs of the retina taken through a dilated pupil of the examined eye. The retina of an eye is an essential part of the central visual pathways that enable us to visualize the real world (Davson/ 1976). The retina also provides us with a landmark that reflects many structural changes caused by retinal diseases that ultimately may lead to partial or complete loss of vision. Therefore, fundus photographs are used by every ophthalmologist to diagnose retinal diseases early and differentiate between them (Gass/1977). The quality of fundus photographs, however, is often poor due to various optical factors involved in such imaging procedures. To enhance and analyze fundus images captured either directly or by digitizing already available fundus images in the form of slides or negatives, conventional image-processing procedures have been employed (Peli et al./1986, Whiteside et al./1986, Mitra et al./1987, and Nagin and Schwartz/1980). Image-enhancement procedures involving spatial-domain operations, such as histogram equalization, contrast stretching, and gradient operations, and frequency-domain operations of high-pass homomorphic filtering are capable of improving the visibility of poor-quality images. However, detection of early changes in the retinal structure, such as the nerve-fiber layer, is extremely difficult due to the complex structure of the fundus and the high resolution required.

In this chapter, we state the complexity of fundus-image analysis, suggest several methods of analysis that might be useful in detecting such subtle changes, illustrate the feasibility of some of the procedures for detecting and differentiating specific retinal diseases, demonstrate the potential of a class of image-coding techniques for fast transmission of digital image data, and, finally, discuss the efficient execution of complex image-analysis procedures in PC-based systems.

6.2   PROBLEMS OF FUNDUS-IMAGE ANALYSIS

The appearance of the fundus viewed through a dilated pupil varies depending on the region of the retina viewed. Direct viewing of the fundus shows the macular area, which appears to have a simple view, yet in the inner layers, the macula has the most complex structure and is the most significant area for fine vision (Fig. 6.1). The posterior pole (20° to 45°) of the fundus reveals the complex blood-vessel pattern (Fig. 6.2) surrounding the optic disc (i.e., the surface of the optic nerve head at the retina). The optic nerve carries almost a million nerve fibers from the retina to higher visual pathways leading to the visual cortex. Included in this posterior area are a number of blood vessels crisscrossing the optic disc and branching into the retina, and whitish bands of nerve fibers forming thick layers in specific areas before converging into the optic disc. Since the optic disc is approximately 1.8 mm in diameter (Quigley/1983), into which a million nerve fibers converge in layers, the average size of each fiber is approximately a few microns.

Resolution of the fundus cameras used in most clinics is much worse than the above requirement. In addition to the resolution limitations, normal fundus cameras operate at a high light-intensity level required for proper illumination of the retina. Such a high intensity level not only is uncomfortable to the patient, but also may cause further damage to a diseased retina. Recently developed scanning-laser ophthalmoscopes (Bille et al./1984, Webb et al./1980) reduce the amount of light falling on the retina by allowing the laser beam to pass through a small portion of the central pupil and by collecting the reflected light beam through the rest of the pupil by required optical pathways. Further development, such as the confocal scanning-laser ophthalmoscope (Webb et al./1987), allows imaging of the retina without dilation of the pupil.

Another major problem of objective analysis of retinal images, irrespective of image quality and resolution, is the lack of registration and illumination normalization among sequential retinal images, even when

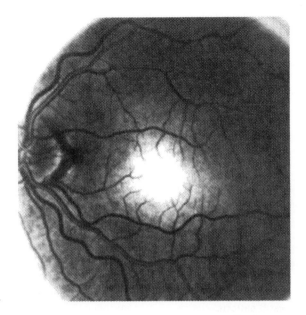

Figure 6.1 A typical fundus photograph. The central whitish section of the picture is the macular region, which is essential for fine vision. This area does not have any blood vessels.

these are taken on the same day. Most image-processing algorithms, including registration algorithms, are, in general, computationally intensive and were previously executable only by mainframe computers. However, recent progress in semiconductor manufacturing has led to quite inexpensive yet powerful microprocessors and other signal-processing chips, so that complex image-processing procedures can now be executed by PC-based systems for enhancing and extracting salient features from a poor-quality image. Even when better quality fundus images are available, image-analysis techniques can be employed for objective analysis of the salient fundus features and for obtaining 3-D information from stereo images.

Each retinal disease, however, is associated with typical structural changes in the retina, and specific image-processing and analysis techniques need to be developed for early detection of such retinal abnormalities. We have addressed two specific retinal diseases—glaucoma and one type of macular degeneration—both of which may result in at least partial loss of vision. Analysis of sequential fundus images of such diseased retinas enables one to detect early changes in the retinal structure, yielding a more accurate and objective evaluation of the stage of the disease. The direct benefit of

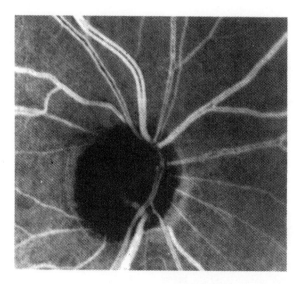

Figure 6.2 A magnified section of a fluorescein angiogram of a fundus near the optic disc showing the blood-vessel pattern.

such evaluation in the management and therapy of these diseases has yet to be determined through long-term studies (Quigley/1983, Gass/1977). However, the capability of precise analysis of fundus images will certainly make such evaluation an easier task for ophthalmologists.

We describe here our approach to analyzing sequential fundus images by proper registration and subsequent analysis of the subtracted image, which contains information regarding the change in fundus texture without being cluttered by other complex structures such as blood vessels. Since sequential fundus images of a patient are not always available and each fundus is unique in its blood-vessel structures, we are also developing techniques of segmentation and edge detection for analyzing single or stereo images. Lastly, attempts have been made to efficiently code physiological images for fast digital transmission of image data. All the above techniques have been implemented in a PC-based system and are routinely executed for fundus-image analysis and transmission.

6.3 METHODS OF FUNDUS-IMAGE ANALYSIS

6.3.1 Image Registration

Development of precise registration techniques is an essential requirement for many image-analysis applications. It is of particular importance in fundus-image analysis, since the fundus images obtained in a clinic even on the same day show marked rotational, translational, and scale differences. These differences must be corrected for quantitative evaluation of changes in two sequential fundus photographs taken over a period of time.

Currently used image-matching techniques range from classical correlation to comparison of symbolic representations of reference and test images (Ghaffary and Sawchuk/1983). The more useful methods of the classical technique for registering a pair of images are basic correlation measures, statistical correlation, and the sequential similarity detection algorithm (SSDA) (Pratt/1978, Barnea and Silverman/1972). The above methods are computationally intensive, and any reduction in redundancy may lead to nonconvergence. We have recently developed two different fast-registration techniques (Nutter et al./1987, Lee et al./1987). One is an automated and modified SSDA technique (Nutter et al./1987) that simultaneously corrects for translation, rotation, and scale errors. The other technique, which separately and more precisely determines the rotation and translation parameters, is based on manipulation of the cepstra of two images to be registered. The latter technique, however, does not address the scale-change problem (Lee et al./1987).

We will describe both registration techniques in some detail and with illustrative examples to demonstrate the versatility and accuracy of the algorithms employed.

Table 6.1 Parameters for Five-Stage Subtractive Correlation

Window	Solid	Frame	Frame	Composite	Composite
Correction	Translation	Rotation	Magnification	Coarse Fit	Fine Fit
X-Shift (Pixels)	±6	±1	±0	±2	±1
Y-Shift (Pixels)	±6	±1	±0	±2	±1
Rotation (Degrees)	±0	±6	±0	±1.5	±0.75
Magnification (%)	±0	±0	±10	±2	±1
Number of Operations	169	117	21	81	81

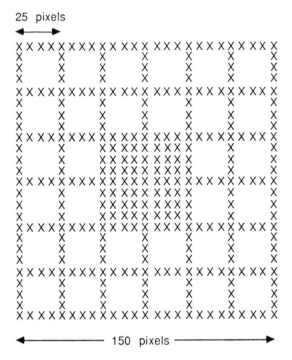

25 pixels

150 pixels

Figure 6.3 A composite mask combining a solid mask with a frame mask for better registration in the latter stages.

(A)   An Automated Multistage SSDA Algorithm

Accurate registration of the test image with the reference requires a multistage subtractive-correlation process for translation, rotation, and magnification, as exemplified by the typical five-stage procedure shown in Table 6.1. This five-stage registration process uses different masks for each process. A combination of solid and frame masks, as shown in Fig. 6.3, yields better registration in the later stages (Mitra et al./1988). An increase in the number of stages will result in greater accuracy and convergence at the cost of more computation time.

The subtractive-correlation error factor is determined from the following modified SSDA (Barnea and Silverman/1972):

$$E_{ij} = \sum_{l=1}^{M} \sum_{m=1}^{M} A(l,m)[S_{i,j}(l,m) - W(l,m)]^2, \tag{6.1}$$

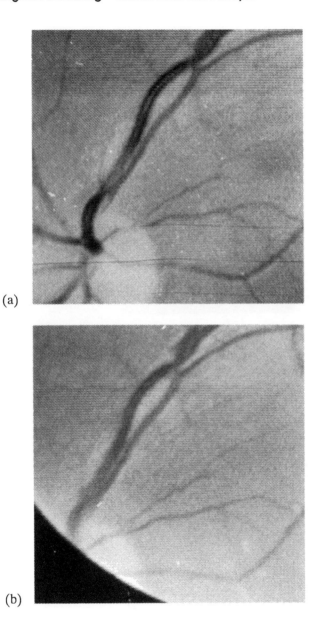

(a)

(b)

Figure 6.4 (a) and (b) are typical unregistered fundus images of the same person taken on the same day.

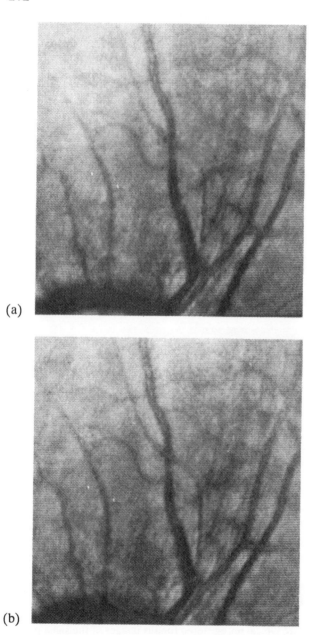

(a)

(b)

Figure 6.5 (a) and (b) are subsections of the fundus images of an eye with optic atrophy taken on the same day after the images have been illumination normalized and registered. (c) and (d) show similar fundus-image subsections of the same eye taken two weeks apart.

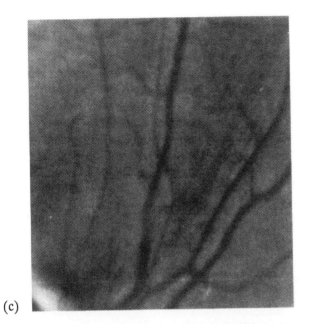

(c)

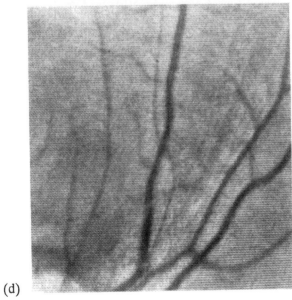

(d)

where $A(l, m)$ denotes the mask, while S and W refer to the search area and the window to be matched, respectively.

Before using an automatic five-stage registration procedure, each image is put through several preprocessing operations: (1) gross geometric matching of the test and reference images, (2) normalization of illumination (see Sec. 6.3.2), and (3) contrast stretching of the test and reference images. After satisfactory registration of a test image with the reference image has been completed, the images are subtracted. The difference image is then further processed by filtering and contrast stretching, resulting in an enhanced difference image.

Since the fundus image has many complex structures, to identify any textural change (particularly in the nerve fiber-bundle areas), spectral and gray-scale scans are performed on the difference image. Figure 6.4 shows typical unregistered fundus images of a normal subject's eye taken on the same day. Figures 6.5(a) and 6.5(b) show the subsections of two illumination-normalized and registered images of an optic-atrophy patient taken on the same day, when the difference image should serve as a norm with no change. Figures 6.5(c) and 6.5(d) show the subsections of the fundus images of the same eye taken two weeks apart. Figures 6.6(a) and 6.6(b) show the difference images of Figs. 6.5(a) and 6.5(b) and Figs. 6.5(c) and 6.5(d), respectively. The difference image of the two optic-atrophied fundus images taken two weeks apart demonstrates a changed texture as opposed to the normal difference image of Fig. 6.6(a), although visual inspection of the original optic-atrophied fundus images themselves does not reveal any changes.

To objectively identify the changes in the difference image, further spectral and gray-level analyses are performed, as explained in Sec. 6.3.3. The preprocessing and postprocessing procedures are explained in detail elsewhere (Nutter et al./1987).

(B) Image Registration Using Power Spectrum and Cepstrum Techniques

This method corrects rotational and translational shifts separately. Basic concepts involved with this method are (1) computation of the Fourier power spectra of the images to make the identifying features parallel to each other, thus converting the rotational shift into a translational shift; and (2) the power cepstra of the images are computed to correct for the translational shift.

The cepstra technique is usually used to detect and remove echoes from one-or two-dimensional signals (Bogert et al./1962, Dudgeon/1977). We have extended this concept to correct translational differences in sequential images (Lee et al./1987, 1988).

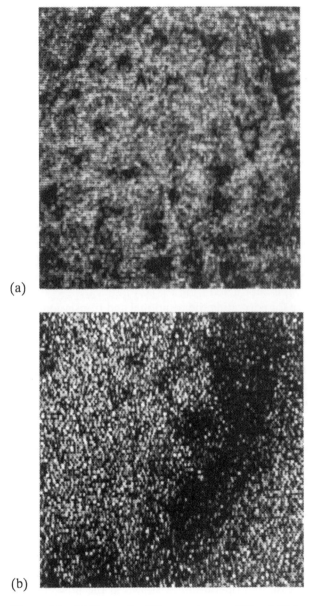

(a)

(b)

Figure 6.6   (a) Shows the difference image of Figs. 6.5(a) and 6.5(b); (b) shows the difference image of Figs. 6.5(c) and 6.5(d).

The power cepstrum of a one-dimensional function of time is defined to be the power spectrum of the logarithm of the power spectrum of that function (Kemerait and Childers/1972). When this definition is extended to an image-intensity function $I(x,y)$, the power cepstrum of $I(x,y)$ is expressed as $P[I(x,y)]$, where

$$P[I(x,y)] = F\left(\ln\left\{F[I(x,y)]^2\right\}\right)^2 \qquad (6.2)$$

where F stands for the Fourier transform of a function. If we let $I(x,y) = R(x,y) + S(x,y) = R(x,y) + A_oR(x-x_o, y-y_o)$, where $R(x,y)$ = a reference image and $S(x,y)$ = a shifted image, then

$$P[I(x,y)] = F[\ln|R(u,v)|^2]^2 + K_1\delta(x,y) + K_2\delta(x\pm x_o, y\pm y_o) + \cdots \qquad (6.3)$$

where the K_i's are constants.

The resulting cepstrum, therefore, will consist of the power cepstrum of the reference image plus a series of impulses occurring at integer multiples of the translational differences (x_o, y_o). Therefore, the translational peaks in the power cepstrum of the reference plus shifted images can be easily detected when $P[R(x,y)]$ is subtracted from $P[I(x,y)]$.

The actual procedure for correcting the rotational and translational differences in fundus images is somewhat more complex, since the fundus images are two-dimensional projections of curved retinal surfaces, and the dominant features of two sequential images cannot be made perfectly parallel. Therefore, the rotational correction cannot be completely accurate. The rotational and translational corrections for simple images such as two shifted T's are shown in Fig. 6.7. The similar procedure for fundus images is shown in Fig. 6.8. The details of these procedures have been described elsewhere (Lee et al./1987, 1988). This registration technique is capable of fast correction of rotational and translational shifts. For noisy signals, the cepstrum technique has been found to be a robust estimator of rotation/translation parameters (Lee et al./1987, 1988).

6.3.2 Illumination Normalization

In many cases, particularly in fundus images, variations in illumination are also a significant cause of artifacts in difference images. Therefore, in an automated registration method (as in Sec. 6.3.1), the normalization of illumination is also an essential operation, as shown in the flow chart of Fig. 6.9.

We assume a multiplicative model for the image, which is a two-dimensional intensity function $f(x,y)$ given by (Gonzalez and Wintz/1987):

$$f(x,y) = I(x,y)\cdot R(x,y) \qquad (6.4)$$

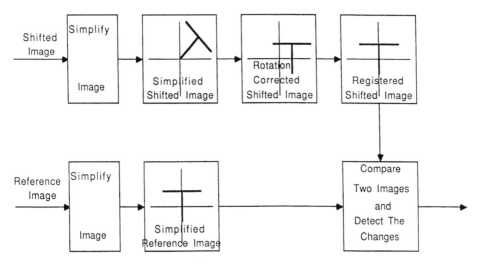

Figure 6.7 Block diagram of the registration algorithm procedure with two shifted T's as illustrative examples.

where $I(x,y)$ is the intensity function due to the source illumination and $R(x,y)$ is the intensity function due to the object reflectivity. Since we are interested in characterizing the reflectivity of the nerve fibers in a fundus image, the image intensity is separated into additive components of illumination and reflectance by taking the natural logarithm of the intensity function $f(x,y)$. The images acquired under different conditions of source illumination can then be normalized by partially eliminating the slowly varying $\ln[I(x,y)]$ component through use of the following relationship:

$$F_o(x,y) = F_I(x,y) - \alpha\mu_w + \alpha\mu_G \qquad (6.5)$$

where $F_o(x,y)$ is the transformed image as defined by Eq. 6.5, $F_I(x,y)$ is the image created by taking the natural logarithm of each of the pixels of the original image (i.e., $\ln[f(x,y)]$), μ_w is the image created by convolution of a specified (e.g., 48×48) two-dimensional, unweighted window $W(x,y)$ with $F_I(x,y)$, μ_G is a constant proportional to the average intensity of $\ln[f(x,y)]$, and α is a multiplicative constant to be determined for each input image.

Noting that all the operations in Eq. 6.5 occur with the natural logarithms of the images, Eq. 6.5 transforms the input image term $\ln[f(x,y)]$ into $F_o(x,y)$ by subtracting a fraction of the convolved input image, μ_w, which is essentially a low-pass-filtered version of the original input image. Since μ_w represents the slowly varying illumination component of the input image, the resulting image will contain only the more rapidly varying reflective

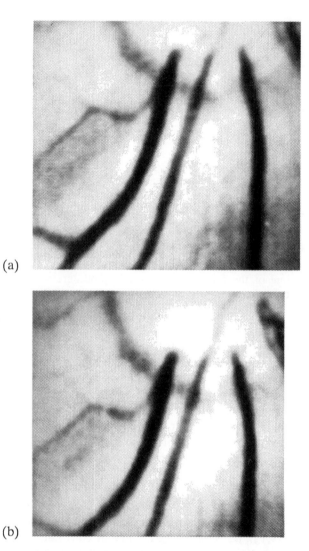

(a)

(b)

Figure 6.8 (a) Reference subimage, (b) shifted subimage, (c) binarized reference subimage, (d) binarized shifted subimage, (e) block diagram of rotation correction, (f) rotation-corrected edge of shifted subimage, (g) superimposed edge of (f) and edge of reference image, (h) block diagram of translation correction, (i) power cepstrum of (g), and (j) registered shifted subimage.

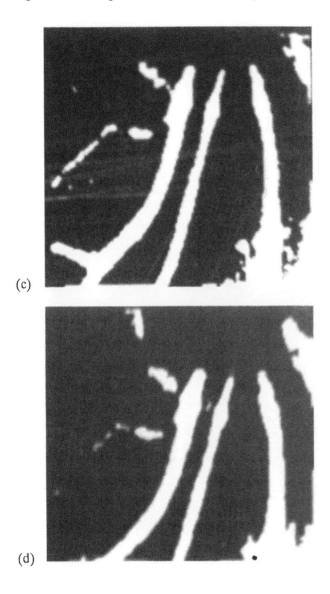

(c)

(d)

component of the input. The term μ_G, the global mean of the logarithm of the image, is added to preserve the DC bias in the output image. Since $\mu_w = W(x,y)^*[F_1(x,y)] = \ln(I) + \mu_G$ in reality, and $\ln[f(x,y)] = F_I(x,y) = \ln(I) + \ln(R)$, Eq. 6.4 can be rewritten as

$$F_o(x,y) = \ln(I) + \ln(R) - \alpha[\ln(I) + \mu_G] + \alpha\mu_G$$

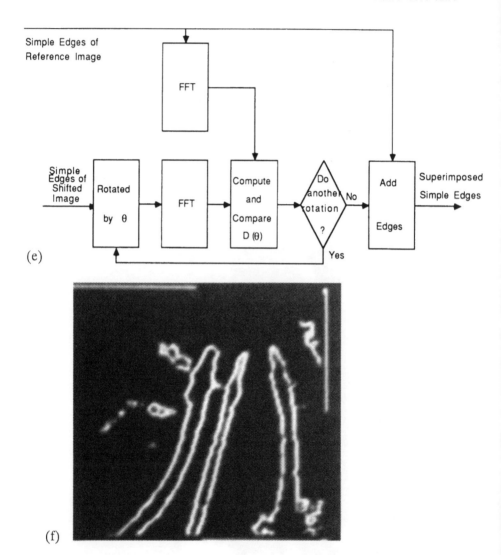

Figure 6.8 (Continued)

$$\text{or } F_o(x,y) = \ln(R) + (1-\alpha)\ln(I) \tag{6.6}$$

Writing $F_o(x,y)$ as $\ln[f_o(x,y)]$, Eq. 6.6 reduces to

$$f_o(x,y) = R(x,y) \cdot I(x,y)(1-\alpha) \tag{6.7}$$

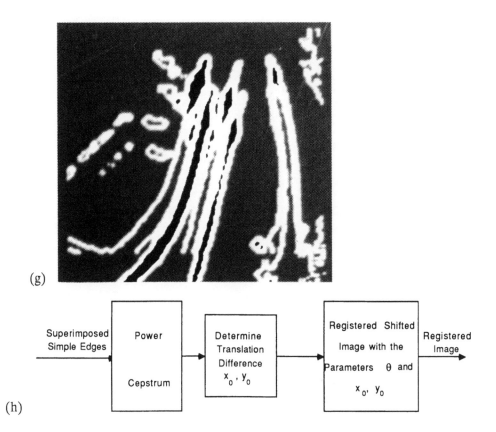

(g)

(h)

Therefore, the transformed image $f_o(x,y)$ has the essentially unaffected reflectance component of the original image and a reduced part of the illumination component, which keeps the variation due to source illumination to a minimum. α is calculated for each input image by searching the entire image for the largest degree of the illumination component that can be physically realized and removed by using a spatial convolution equivalent to low-pass filtering. The operations represented by Eq. 6.5 can be considered as an effective high-pass homomorphic filtering operation in the spatial domäin, since a portion of the low-pass image μ_w has been subtracted from the original image. Such a high-pass-filtered image should contain relatively more information regarding the reflectance components of the image (including the edges, discontinuities, textures, and other high-frequency terms), since the slowly varying low-frequency source illumination has been significantly removed. Total removal of the source illumination cannot be physically realized, thus yielding an exponent of $I(x,y)$ between zero and

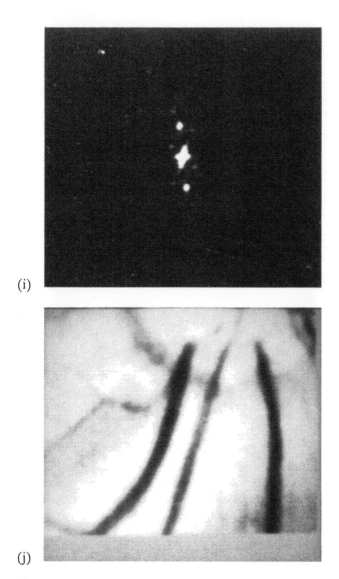

(i)

(j)

Figure 6.8   (Continued)

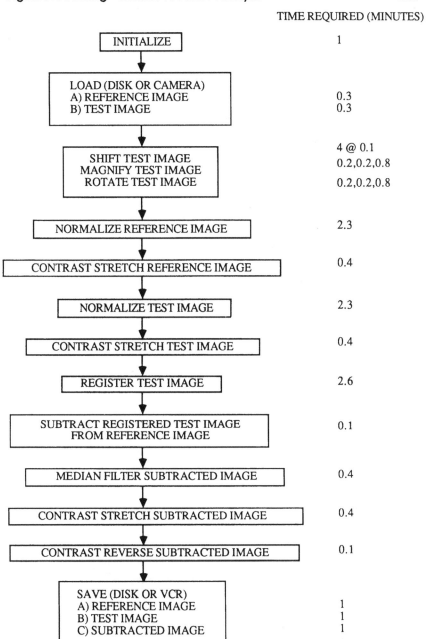

TIME REQUIRED (MINUTES)

INITIALIZE	1
LOAD (DISK OR CAMERA) A) REFERENCE IMAGE B) TEST IMAGE	0.3 0.3
SHIFT TEST IMAGE MAGNIFY TEST IMAGE ROTATE TEST IMAGE	4 @ 0.1 0.2,0.2,0.8 0.2,0.2,0.8
NORMALIZE REFERENCE IMAGE	2.3
CONTRAST STRETCH REFERENCE IMAGE	0.4
NORMALIZE TEST IMAGE	2.3
CONTRAST STRETCH TEST IMAGE	0.4
REGISTER TEST IMAGE	2.6
SUBTRACT REGISTERED TEST IMAGE FROM REFERENCE IMAGE	0.1
MEDIAN FILTER SUBTRACTED IMAGE	0.4
CONTRAST STRETCH SUBTRACTED IMAGE	0.4
CONTRAST REVERSE SUBTRACTED IMAGE	0.1
SAVE (DISK OR VCR) A) REFERENCE IMAGE B) TEST IMAGE C) SUBTRACTED IMAGE	1 1 1

16.4

Figure 6.9 Flow chart of a typical automated registration algorithm including illumination normalization.

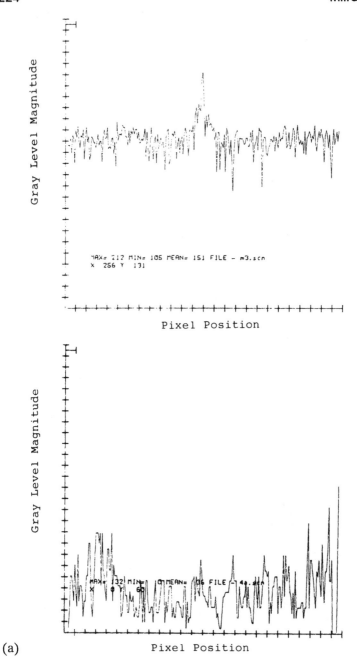

(a)

Figure 6.10 (a) Spectral scans of Figs. 6.6(a) and 6.6(b) along the directions shown; (b) spatial line scan of Figs. 6.6(a) and 6.6(b) along the directions shown.

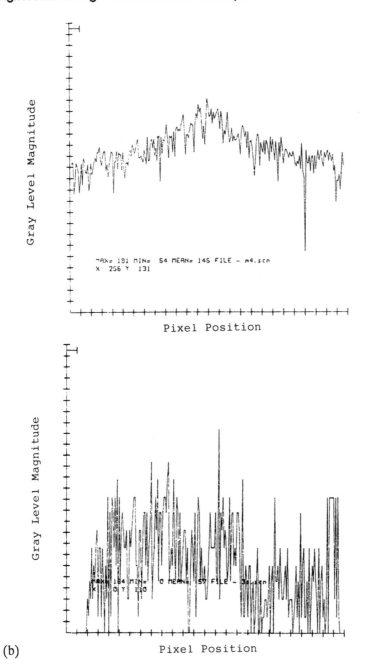

MAX= 191 MIN= 54 MEAN= 145 FILE - m4.scn
X 256 Y 131

Pixel Position

MAX= 184 MIN= 0 MEAN= 57 FILE - 3b.scn
X 0 Y 110

(b)   Pixel Position

one. The maximum value of α approaches 1, and a typical value is near 0.7 (Nutter et al./1987).

6.3.3 Spectral and Gray-Level Scans

In an attempt to evaluate the difference images such as shown in Figs. 6.6(a) and 6.6(b) quantitatively, spectral and gray-level scans can be taken along any horizontal or vertical lines. Figure 6.10 shows the spectral and gray-level scans of Figs. 6.6(a) and 6.6(b) along the lines indicated. In this particular case, spectral variations do not appear to be significant. The gray-level scans, however, show significant differences between the two subtracted images.

6.3.4 Segmentation and Edge Detection

A combination of various segmentation and edge-detection techniques may provide further descriptive information regarding the textural changes in different regions of the fundus image (Ballard and Brown/1982). The use of a simple edge-detection operator, such as the Sobel operator for obtaining simplified fundus images for registration, is illustrated in Figs. 6.8(f) and 6.8(g). Image segmentation by histogram thresholding is also used for better registration of fundus images (Nutter et al./1987).

For efficient pattern recognition, the human visual system appears to use a Laplacian of Gaussian (LOG) type of operator or multiscale zero-crossing operators (Marr/1982). LOG of the zero-crossing operators can be used for multiscale segmentation as well as for edge detection, as shown in Figs. 6.11 and 6.12. The algorithms for LOG operators are briefly discussed and referenced in Sec. 6.4.

Recent research efforts suggest the validity of zero-crossing descriptors as unique representatives of a wide class of images or signals in general (Logan/1977, Curtis et al./1987). We are presently investigating a modification of zero-crossing descriptors for 3-D image reconstruction from stereo images (Grimson/1981). Such a reconstruction scheme has a variety of applications, including satellite imaging and stereo fundus imaging. Stereo fundus photographs are routinely taken in ophthalmology clinics to evaluate the contour and topography of optic discs that have been damaged by glaucoma. At present, however, most clinical evaluation is done by the subjective perception of the change in the optic-disc depth by an experienced ophthalmologist. Our present research effort includes a better evaluation of zero-crossings by inspecting their spatial width.

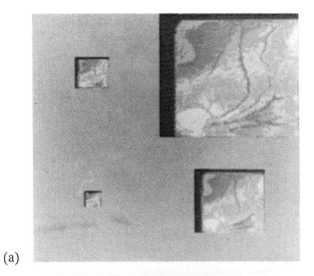

(a)

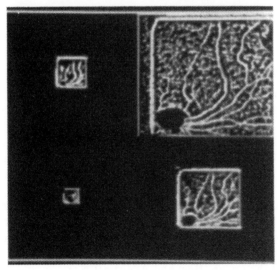

(b)

Figure 6.11   (a) Three levels of compression of a fundus image; (b) the error function of each level.

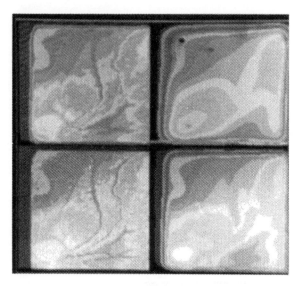

Figure 6.12   Reconstruction of a fundus image by progressive transmission.

6.4   IMAGE CODING AND DATA COMPRESSION

Physiological images, in general, contain an enormous amount of data
that require a very large number of bits for digital representation. Since
adjacent image data are highly correlated, the goal of image coding is to
reduce the number of representative bits from which the original image can
be reconstructed without noticeable degradation. The saturation limit of
compression ratio of 10:1 attainable by information-theory guided-coding
techniques (Netravali and Limb/1980) has been surpassed recently by
second-generation image-coding techniques based on human visual-system
models (Kunt et al./1985). We have successfully implemented in our PC-
based image-processing system a highly efficient and recursive method of
image compression based on a pyramid of reduced-bandwidth and low-
resolution versions of the image (Burt and Adelson/1983).

   The above-mentioned technique is considered a second-generation coding technique due to its LOG pyramid coding, which is assumed to be
also a coding mechanism in a human visual-processing model. The pyramid coding, however, can also be considered as a hybrid first-generation
coding combining predictive and transform coding techniques. The pyramid coding uses a hierarchical discrete correlation (HDC) method, where
the correlation of a function with certain large kernels is computed as a

weighted sum of correlations with smaller kernels. The kernels at each iteration of the HDC computation differ in size by a factor r.

Let $f(x,y)$ be an image function defined only at integer values of x and y, i.e., $f(m,n)$. Let $w(x,y)$ be a discrete weighting function defined at integers x and y and nonzero only for $-k/2 < x, y < k/2$. Then the HDC is defined as a set of correlation functions $g(x,y)$ that can be obtained from f and w as follows:

$$g(m,n,0) = f(m,n)$$
$$g(m,n,1) = g(rm,rn)$$
$$= \sum_{i=-k/2,\, j=-k/2}^{k/2} \sum^{k/2} w(i,j) \cdot g(m+ir, n+jr) \tag{6.8}$$

A Laplacian pyramid can be defined as a sequence of error images. Each error image is the difference between two levels of the Gaussian pyramid, whereas the Gaussian pyramids can be viewed as a set of low-pass-filtered copies of the original image, as illustrated in Figs. 6.11 and 6.12.

The detail of operations and the constraints of normalization, symmetry, unimodality, equal contribution, and separability are discussed elsewhere (Burt/1981, Wu/1987). Figure 6.11 shows typical pyramid coding for a fundus image.

Figure 6.12 shows the progressive image-transmission process, in which the images show reconstruction through three pyramid levels. Therefore, any image can be transmitted from one computer to another in an efficient manner by starting with a compressed image and using progressive transmission for reconstructing the image at the other end.

6.5 APPLICATIONS

6.5.1 Glaucoma

Although persistent high intraocular pressure (IOP) is considered an indicator of the onset of glaucoma, the actual diagnosis of glaucoma is based on the documentation of damage to the optic-nerve fibers and optic-nerve head (disc) and on the corresponding loss in vision usually detected by visual-field testing (Iwata/1983). Recent studies indicate that persistent abnormalities of nerve-fiber layers can be detected as early as five years before conventional vision tests can indicate a loss of vision (Sommer et al./ 1984). These results have been based on the evaluation of a large number of fundus photographs of normal and glaucomatous eyes. A major problem in glaucoma management and therapy is the lack of dependable and objective evaluation of optic-nerve damage (Quigley/1983).

The usefulness of the registration technique in analyzing sequential fundus images to detect textural changes in the nerve fiber-layer area is shown in a typical case of optic atrophy in Fig. 6.6. Since the nerve-fiber damage in glaucoma appears to precede a loss of vision by quite a few years, early detection of nerve-fiber loss may result in better control of further progression of the disease. However, sequential fundus images are not always available; therefore, other techniques such as segmentation and edge detection must be developed simultaneously to analyze single fundus images for early detection of glaucoma.

6.5.2 Macular Degeneration

Macular degeneration is another retinal disease (Gass/1977) that causes loss of fine vision, although peripheral vision remains intact in this case. Conventional image-enhancement techniques improve the quality of fundus images for better evaluation and classification of a particular group of macular diseases—namely, macular holes (Mitra et al./1986). The diagnosis of macular hole refers to either a full thickness or a partial thickness (lamellar) hole in the macular area. The most commonly occurring full-thickness macular hole is of unknown cause, although a number of factors, such as myopic degeneration, contraction of an epiretinal membrane, trauma, and solar retinopathy, have been implicated as etiologic factors in some cases (Gass/1977, Maumenee/1967). Differential classification of macular holes, macular cysts, and pseudomacular holes (hole-like appearance due to contraction in the epiretinal membrane) is not always readily available (Gass/1975). Fluorescein angiography may be of some help in differential diagnosis of macular holes from pseudomacular holes (Allen and Gass/ 1978).

Exact classification of the above macular problems is of importance because an eye with a macular hole (full thickness or partial) suffers a permanent loss of vision, whereas an eye with a pseudomacular hole may regain normal vision, if spontaneous peeling of the epiretinal membrane occurs. A modified visual-field testing technique, spatial-contrast-threshold perimetry, is found to be sensitive enough for some of the above classifications (Mitra/1985). Figure 6.13 shows how simple gray-level scans can differentiate the condition of the macular tissue by its reflectivity. The image-analysis techniques described here may provide better characterization and localization of the affected tissue areas, thus aiding in differential classification of macular degeneration.

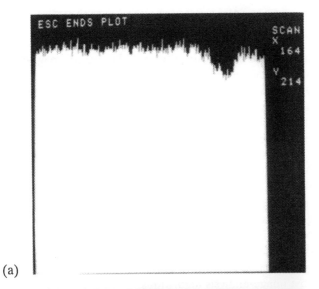

(a)

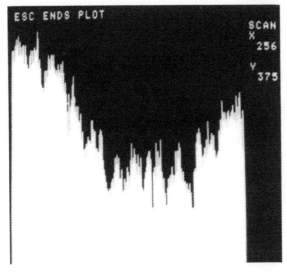

(b)

Figure 6.13 (a) Shows spatial scan along a normal macula showing uniform reflectivity; (b) spatial scan along a degenerated macula indicating major changes in reflectivity in the localized area that is suspected of degeneration.

6.6 IMPLEMENTATION AND EXECUTION OF IMAGE-ANALYSIS ALGORITHMS IN PC-BASED SYSTEMS

We have implemented all the above algorithms in PC-based image-processing systems. During the development stage, the main system used was a COMTAL VISION ONE/20 interfaced with a VAX 11/780 computer. However, all the algorithms except data-compression algorithms are now efficiently executed by a system organized around the Dimension 68000 microcomputer with custom-designed 8-bit digitizer card, a motherboard with 2 Mbytes of RAM (random access memory), a 20-Mbyte hard-disk drive, 368-Kbyte and 1.2-Mbyte floppy-disk drives, and an 8-MHz clock (Whiteside et al./1986, Nutter et al./1987). A Laplacian–Gaussian pyramid-coding algorithm for data compression and transmission has been implemented and executed using TI-PCs (Wu/1987). The software written to implement the above algorithms is in C language, with digitizer access and other subroutines written in Assembly. The software can be modified for an IBM-PC-compatible system. At present, faster microcomputers are available for easy execution of most image-analysis algorithms.

6.7 CONCLUSIONS

In 1986, when we started developing image-analysis algorithms with the aim of improving the status of evaluation of fundus images, powerful PC-based systems had to be custom-designed. At present, many powerful systems are available on the market for various applications. Considering the present problems of fundus photography, which is still considered an art developed by skilled personnel, a fundus image-analysis system appears to be equipment essential for any ophthalmology clinic. Therefore, cost-effective systems should be developed along with software for specific applications. This new area of application also creates exciting problems requiring further basic research in pattern recognition and 3-D image reconstruction.

ACKNOWLEDGMENTS

The authors would like to acknowledge funding from the Microcraft Co., Dallas, Texas, for carrying out some of the work reported here. One of the authors, S. Mitra, would also like to acknowledge support from Texas Instruments for making part of the research possible by the use of some equipment and partial financial support. The authors are extremely grateful to Drs. Reay Brown, Mary Lynch, and Zuhair Shihab for their encouragement in the fundus image-analysis research, their clinical input, and their generosity in loaning the fundus images. The authors are very appreciative of the contributions of graduate students Brian Nutter, Dah-Jye Lee, Steven Whiteside, and Jerry Wu. The dedication and hard work of Pansy Burtis in the preparation of this manuscript is also appreciated.

REFERENCES

Allen, T. M. and Gass, J. D. M. (1978). Contraction of a perifovial epiretinal membrane simulating a macular hole, *Amer. J. Ophthal.*, *69*: 555.

Ballard, D. H. and Brown, C. M. (1982). *Computer Vision*, Prentice-Hall, New Jersey.

Barnea, D. I. and Silverman, H. F. (1972). A class of algorithms for fast digital image registration, *IEEE Trans. Computers*, *C-21, no. 2*: 179.

Bille, J., Jahn, G., and Dreher, A. (1984). Active optical focussing of the laser scanner ophthalmoscope, *Proc. Soc. Photo-Opt. Instrum. Eng.*, *498*: 76.

Bogert, B. P., Healy, M. J. R., and Tukey, J. W., (1963). The frequency analysis of time series for echoes: cepstrum and saphe cracking, *Proceedings of the Symposium at Brown University, July 1962* (M. Rosenblatt, ed.), John Wiley and Sons, New York, 209.

Burt, P. J. and Adelson, E. H., (1983). The Laplacian pyramid as a compact image code, *IEEE Trans. Comm.*, *COM-31, no. 4*: 532.

Curtis, S. R., Shitz, S., and Oppenheim, A. V. (1987). Reconstruction of nonperiodic two-dimensional signals from zero crossings, *IEEE Trans. ASSP, ASSP-35, no. 6*: 890.

Davson, H. (1976). *The Eye*, Academic Press, New York.

Dudgeon, D. E. (1977). The computation of two-dimensional cepstra, *IEEE Trans. ASSP, ASSP-5, no. 6*: 148.

Gass, J. D. M. (1975). Lamellar macular hole, *Trans. Amer. Ophthal. Soc.*, *73*: 231.

Gass, J. D. M. (1977). *Stereoscopic Atlas of Macular Disease*, Mosby, St. Louis.

Ghaffary, B. K. and Sawchuk, A. A., (1983). A survey of new techniques for image registration and mapping, *Proc. SPIE 432*: 2.

Gonzalez, R. C. and Wintz, P. (1987). *Digital Image Processing*, Addison Wesley, Reading, Massachusetts.

Grimson, W. E. L. (1979). *From Images to Surfaces*. The MIT Press, Cambridge, Massachusetts.

Iwata, K. (1983). The earliest finding of primary open-angle glaucoma (POAG) and the mode of progression, *Glaucoma Update II*, G. K. Krieglstein, and W. Leydhecker, (eds.), Springer–Verlag, p. 133.

Kemerait, R. C. and Childers, D. G. (1972). Signal detection and extraction by cepstrum techniques, *IEEE Trans. on Inform. Theory, IT-18, no. 6*: 745.

Kunt, M., Ikonomopoulos, A., and Kocher, M. (1985). Second-generation image coding techniques, *Proc. IEEE, 73*: 549.

Lee, D. J., Krile, T. F., and Mitra, S. (1987). Digital registration techniques for sequential fundus images, *Proc. SPIE 89.*

Lee, D. J., Krile, T. F., and Mitra, S. (1988). Cepstrum and spectrum techniques applied to image registration, *Applied Optics*, 7: 1099.

Logan, B. F., Jr. (1977). Information in the zero crossings of bandpass signals, *Bell Syst. Tech. J.*, *56*: 487.

Marr, D. (1982). *Vision*. Freeman, San Francisco.

Maumenee, A. E. (1967). Further advances in the study of the macula, *Arch. Ophthal.* *78*: 151.

Mitra, S., Whiteside, S., and Krile, T. F. (1987). Differential retinal structural damage exhibited by image enhancement of fundus photographs, *Low Vision: Principles and Applications* (G. C. Woo, ed.), Springer–Verlag, New York, p. 125.

Mitra, S., Nutter, B. S., Krile, T. F., and Brown, R. H. (1988). Automated method for fundus image registration and analysis, *Appl. Optics*, 7: 1107.

Mitra, S. (1985). Spatial contrast sensitivity in macular disorder, *Doc. Ophthal. 59*: 247.

Nagin, P. and Schwartz, B. (1980). Approaches to image analysis of the optic disc, *IEEE Computer Soc., Fifth Int. Con. on Pattern Recognition*, 948.

Netravali, A. N. and Limb, J. O. (1980). Picture coding: a review, *Proc. IEEE*, *68*: 336.

Nutter, B. S., Mitra, S., and Krile, T. F. (1987). Image registration algorithm for a PC-based system, *Proc. SPIE*, 829.

Peli, E., Hedges, T. R., and Schwartz, B. (1986). Computerized enhancement of retinal nerve fiber layer, *Acta Ophthalmologica*, *64*: 113.

Pratt, W. (1978). *Digital Image Processing*, John Wiley & Sons, New York.

Quigley, H. A. (1983). The value of the retinal nerve fiber layer examination in the diagnosis of glaucoma, *Glaucoma Update II* G. K. Krieglstein and W. Leydhecker, eds.), p. 129.

Sommer, A., Quigley, H. A., Robin, A. L., Katz, N. R., and Arkell, S. (1984). Evaluation of nerve fiber layer assessment, *Arch. Ophthalmol.*, *102*: 1966.

Whiteside, S., Mitra, S., Krile, T. F., and Shihab, Z. (1986). Fundus image processing system for early detection of glaucoma, *Proc. SPIE*, *697*: 123.

Webb, R. H. and Hughes, G. W. (1981). Scanning laser ophthalmoscope, *IEEE Trans. Biomed. Eng.*, *BE-28*: 488.

Webb, R. H., Hughes, G. W., and Delori, F. C. (1987). Confocal scanning laser ophthalmoscope, *Appl. Optics.*, *201*, 26: 1492.

Wu, Z. L. (1987). Laplacian pyramid coding for video data transmission, M.S. Thesis, Dept. of Electrical Engineering, Texas Tech University.

7

Visual Perception Using a Blackboard Architecture

Terry E. Weymouth and Amir A. Amini

University of Michigan
Ann Arbor, Michigan

7.1 INTRODUCTION

A computer system for the automatic interpretation of a television camera signal relies on knowledge of the world. This knowledge can be either general or specific. The interpretation of groups of image features requires general knowledge, such as the facts about geometry that assist in the interpretation of line drawings as three-dimensional figures. Image understanding also relies on domain-specific knowledge, such as the facts that associate particular relations among features with an object. An interpretation system must have a means for effective and efficient access to a rich and varied knowledge resource. It must access information about objects: their interrelations, their function, and their visual characteristics.

The study of scene interpretation using intensity images is a field that has seen a great division of effort. On one hand, many researchers have

Supported in part by AFOSR grant F33615-85C5105, NASA-AMES grant NAG-2-350, and NSF grant IRI-8711957.
Note: Portions of this chapter were previously published in proceedings and Journal papers (Weymouth and Amini/1987, Amini and Weymouth/1987, Amini et al./1988, Amini et al./ 1989), and in a dissertation by Weymouth (Weymouth/1986). The figures have been taken from (Amini and Weymouth/1987, Weymouth/1987a, Weymouth/1986, Weymouth and Amini/ 1987, Amini et al./1989, Amini et al./1988). Permission to reproduce the relevant portions of text and figures has been obtained from the authors and publishers.

concentrated on developing sound theoretical underpinnings for methods of deriving information from image sequences (Brady/1982, Horn/1986). Their efforts have led to mathematical models that, although powerful as tools of explanation, are applicable only in restricted cases. This is because the models are usually not general enough to encompass many realistic situations. On the other hand, much research effort has gone into attempting automatic interpretation with the tools and techniques at hand. Such systems also are generally fine-tuned so that they are applicable to only a few images of scenes from restricted domains. The systems for the interpretation of aerial photographs by (Nagao and Matsuyama/ 1980, Glicksman/1982, Brooks/1981) were each designed to use general knowledge about image features and object characteristics, along with specific knowledge about the objects expected in the images, to guide interpretation and construct a description of the scene. The interpretation of natural outdoor scenes has also been an area that has yielded similar systems; for example, see (Ohta/1980, Weymouth/1986). In each of these studies, the argument has been made that the type of scene being interpreted and the system doing the interpretation should be as general as possible. However, the experiments and results are usually restricted to only a few examples of scenes from a very small class of scenes (Binford/ 1982).

The development of powerful mathematical models of the process of image formation and its relation to surface shape and other object characteristics is an important area for research in computer vision. Important engineering questions also must be answered, however. Using the best methods available, systems can be developed that interpret scenes in specialized settings. What is needed to better understand the engineering of scene interpretation is a series of detailed studies in restricted domains.

We propose the use of blackboard architecture (Nii/1986) in capturing both the general knowledge about the visual world and, at the same time, the high-level knowledge available for tackling the specialized domain at hand. The general knowledge of the visual environment suits the blackboard paradigm quite well, since the structure of the blackboard is hierarchical in nature and the visual environment is hierarchical as well (see, for example, Hanson and Riseman/1978). Furthermore, we claim that knowledge of any specialized visual domain can be incorporated in the blackboard hierarchy. This statement stems from the fact that any such knowledge necessarily has both a local and a global character. Expert knowledge having a local character may be programmed into processes at the lower levels of the blackboard hierarchy, whereas the more abstract and, thus, global knowledge may be made available to such a system at the higher levels of the blackboard hierarchy. In a later section (Sec. 7.4), we will discuss an example from a medical domain involving the signal-to-symbol transformation, where organization of the expert knowledge in a blackboard hierarchy is illustrated.

Blackboards have been used in a variety of problems involving the signal-to-symbol transformation (Erman et al./1980, Hanson and Riseman/1978, Nagao and Matsuyama/1980). The blackboard architecture is composed of three parts: a blackboard data structure, a set of knowledge sources, and a control mechanism. The blackboard is a hierarchical, global database, where knowledge sources (usually procedures) that capture the domain knowledge operate on a given level and produce hypotheses at the same or a different level of abstraction. Blackboards are desirable to use because:

> Knowledge sources can interact only through the blackboard, allowing for uniformity in system development. Knowledge sources cannot pass messages directly to one another. They take as input hypotheses present on the blackboard and write their outputs on the blackboard.

> Blackboards support top-down and bottom-up problem solving. Invoking knowledge sources that produce hypotheses on higher levels (lower levels) than they are located in amounts to bottom-up (top-down) problem solving. In the bottom-up approach to interpretation, based on image data, hypotheses are made about object locations. In the top-down approach, based on contextual information, locations of related objects are hypothesized. The control component opportunistically applies top-down or bottom-up problem solving.

> The modular structure of knowledge sources on blackboards facilitates parallel problem solving. One way to achieve modularity is to decompose the problem into loosely coupled subproblems. We will discuss such ideas in later sections.

> Finally, blackboards provide an excellent environment for system development. Changing knowledge sources, removing knowledge sources, and adding new knowledge sources can be done without any need to change the underlying architecture. For example, if a new knowledge source is developed that can abstract part of the data more effectively, it can replace the old knowledge source without affecting the underlying architecture of the system.

The next section of this chapter (Sec. 7.2) presents an examination of the characteristics that make a domain of study a reasonable one for research in scene interpretation. From this analysis, we construct a guide for selecting domains of study. Careful consideration of domains is required if we are to develop systems that will yield new and useful results in the engineering of vision systems.

Section 7.3 shows how different programs can interact to perform an entire interpretation. Because of separations evident in the data or the structure of knowledge, portions of the interpretation can be performed in parallel. In addition, a rough partial interpretation can provide contextual information that will limit the need for complex interpretation strategies. When such contexts are verified, specific interpretation strategies can be opportunistically applied. These techniques are illustrated using results from (Weymouth/1986) on the interpretation of natural outdoor scenes.

In Sec. 7.4, we will examine the utility and effectiveness of the blackboard architecture as a general framework for the representation of knowledge and for the control of interacting processes using that knowledge for scene understanding. The blackboard architecture provides both the flexibility for incremental design and the mechanisms for controlling the interactions among detailed procedures from which a description of a scene can evolve. We will see the merits of the blackboard framework through the discussion of an experimental domain. The domain is that of the automatic interpretation of a microscope slide containing preparations of the hair cells of the inner ear in the organ of Corti (Yost and Nielsen/ 1977). An understanding of the medical domain is not germane to the discussion in this section; however, certain characteristics of the process will be discussed in terms specific enough so that points can be illustrated. Human interpretation, as well as computer interpretation, will be used as a basis for the discussion. In addition, we will describe some extensions to the blackboard framework that are necessitated by the requirements for scene interpretation. These extensions permit the simultaneous activation of the same procedure on several hypotheses.

In Sec. 7.5, we have selected several knowledge sources for description. In developing these knowledge sources, we kept general settings in mind. A knowledge source is discussed that locates 2-D object positions using incomplete information about their boundaries. This knowledge source is focused on in a subsection. Also, a top-down algorithm for locating object boundaries in an image is summarized.

7.2   RESTRICTING DOMAIN KNOWLEDGE*

Many laboratory and industrial inspection tasks depend on a small amount of expert knowledge. Knowledge-based interpretation is especially useful

---

*This section first appeared as part of a paper in the proceedings for Vision87 (Weymouth and Amini/1987).

for those tasks that have eluded automation attempts based on image-processing and pattern-recognition methodologies, yet are routinely done by trained human technicians. In these tasks, where a small amount of well-structured knowledge seems to support the information used in controlling perception, research on the engineering of vision systems might well be able to make some advances. This also seems to be the case in many of the types of vision-oriented tasks for robots, where robots would be working in restricted domains with a few objects using well-defined procedures that, nevertheless, require some amount of perception and reasoning. Such situations seem tailor-made for research in computer vision.

Research in restricted domains, such as medical and industrial image interpretation, will lead to useful applications. More important, however, are the advances that can be made in our understanding of the types of knowledge necessary for scene interpretation and the advances that can be made in the area of vision algorithms by the use of real data.

What characterizes a good domain for study? We have identified a few characteristics that we list here; in the following paragraphs we will examine each of these listed items:

Real data are readily available.
There are known methods for getting features from the data.
The system will be part of the environment.
Human expert knowledge or protocol for interpretation is available.
There is a clearly identifiable structure to the interpretation process.
There are clear criteria for evaluation.

7.2.1   Real Data

It would seem obvious that we should use data from the environment that we seek to model. This is problematic in the construction of an interpretation system, for we wish to know what changes are due to choices in design and what changes are due to changes in the environment. So, a typical approach is to use either very select data, or to synthesize the data, so that the results of the changes to the processes being studied are clearly isolated. In addition, many of the problems of applying the results of past research arise from the seemingly reasonable presumption that an earlier stage in the process can yield a perfect description. A typical statement might be "Assume that we have perfect boundary information." Such assumptions remove many of the interesting issues; it is a very difficult task to design a process that produces a perfect description at *any level of abstraction*. However, pressure continues to be applied to reduce the problems of interpretation to problems of dealing with the data. Today the theme is echoed in decrying the lack of good segmentation routines, error-free edge detectors, and reliable surface-approximation routines. The reality of building machines for scene interpretation is that the original data and the results from all intermediate steps are errorful and uncertain. To meet the requirements of having enough control over the experimental setup, we select restricted domains. To keep ourselves honest about the nature of the problem, we work with the data and the faults of current feature-extraction technology as we find it.

7.2.2 Image Features

One serious problem in dealing with image data is sheer volume. In the design of any interpretation system, there are data-abstraction routines that extract features from the data. These processes are usually based on a mapping of the image to some feature space, a selection of elements in that feature space, and a labeling of those selected elements. Frequently, research is necessary to do this development. Thus, research on interpretation can be aided if there are already methods for extracting features that relate to the goals of the interpretation system. For example, if the interpretation of the scene relies on tracking important boundaries, it is helpful to have on hand edge-extraction routines that work in the setting being studied. Most often, however, one can realize the need for new algorithms as one sees fit. In fact, in the process of designing and building interpretation systems, researchers often develop new feature-extracting algorithms applicable in quite general settings.

There is another reason that the problem should have come under the scrutiny of research on feature-based processing. Many inspection problems in restricted domains can be handled by statistical methods. For example, see Besl et al./1985 for an inspection method of solder joints based on nearest neighbor classification. Such methods may render knowledge-based techniques unnecessary or, if the classification techniques solve only part of the problem, provide a stepping stone to get the interpretation system off the ground.

7.2.3 System in the Environment

An interesting aspect of past studies of interpretation is that most of them have been done off-line. That is, images are digitized as a separate process, and the analysis is performed on the image rather than on the scene. This is driven, in some part, by technology. Until recently, the cost of getting good-quality digitized images was prohibitive; also, the cost of storing large numbers of images was a problem. These problems are less of a concern now. Even when the system cannot be tested directly in the environment, it is often possible to simulate the interaction of the system

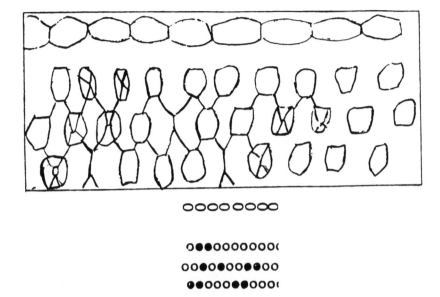

Figure 7.1   This photograph of the organ of Corti illustrates a possible configuration of scarred cells. These are recorded on the hair-cell sheet (shown below
photograph). This is an inspection task that requires expert knowledge.

with the environment by storing several images representing the scene from
several viewpoints and selecting the image that most nearly corresponds to
the desired viewpoint when the system requires new information.

The assumption that the interpretation system is part of the environment
has an important consequence, which relates directly to the idea of using
real data: because the environment can be resampled, many problems of
error recovery and dealing with uncertainty can be handled in a reasonable
way.

7.2.4   Human Expert Knowledge

An expert is a valuable source of information about the high-level knowledge that can be brought to bear on a problem. While it is obvious that
vision is not a problem that can be solved by a rule-based expert system, it
is also the case that expert knowledge, such as methods for training human
inspection technicians, can contribute to an understanding of the types and
structure of the knowledge necessary for an automatic interpretation system. However, even when expert knowledge is available, the duplication of
human performance is not necessarily an obtainable goal. Still, since one

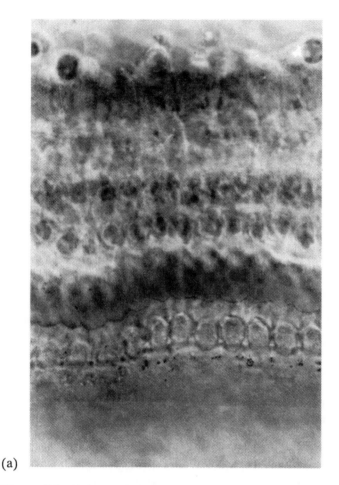

(a)

Figure 7.2 This set of pictures is taken from the same scene at various cross-sections. The pictures are ordered alphabetically. (a) corresponds to the lowest focal plane, and (d) corresponds to the highest. Interpretation of the defects in the cell sheet depends on an integration of information from each focal plane. This is a multiple-image inspection task.

of the problems facing the designer of a knowledge-based machine vision system is the cataloging and understanding of the types of knowledge that are needed for any particular task, when some of this knowledge is available in any form, such information provides a clear starting point.

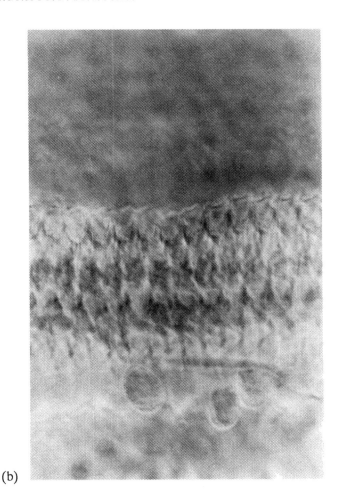

(b)

7.2.5 Knowledge Structure

If the interpretation process and the knowledge necessary for it have a clearly defined structure, this also can be used as a starting point for the internal structure of the interpretation system. In the architecture section that follows, we discuss the usefulness of being able to identify levels of abstraction and processes that transform hypotheses at one level into hypotheses at a higher level by grouping. The levels of abstraction are possible because of the inherent structure of the knowledge that can be brought to bear on the problem.

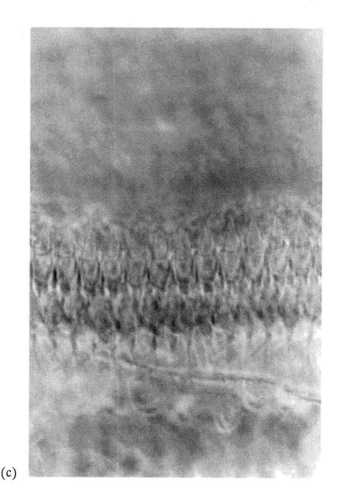

(c)

7.2.6 Evaluation

A clear means of evaluating the performance of the interpretation system
enables the qualification of the results of interpretation experiments. This is
an essential requirement for testing possible design changes for their effect
on interpretation. Unless we can say which interpretations are good and
which better, we cannot measure the results of our experiments.

We have selected two experimental domains that have the characteristics
described above. One is the inspection of inner-ear hair-cell microscope
slides. To interpret such slides, trained observers first diagram the observed
cell structure and classify the cells (see Fig. 7.1); they then count cells by
class and chart the density of cells over predefined areas of the sample

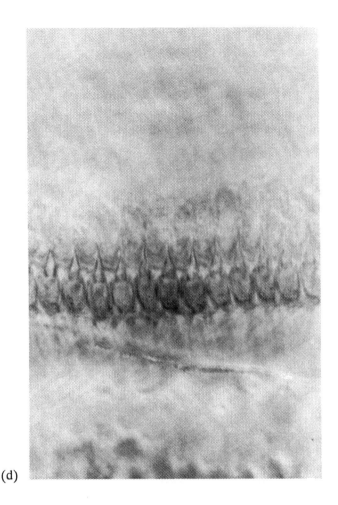

(d)

tissue. There are rules for filling in the structure and extending the count of cells into areas in which the cells cannot be observed clearly in the sample. During this inspection process, the observer has to adjust the focal plane of the microscope to see different features of the cells and cell structure. These variations are seen in Fig. 7.2. which shows images of slides taken at four settings of the focal plane. It is clear from these images that the individual cells are indistinct. Perception of the cells depends on the structure in the images and on the expert's knowledge. In addition to the structure inherent in the image, knowledge is available from descriptions of training procedures and methods used by the experts.

Hair-cell interpretation is an inspection task routinely done by trained experts and results in a measure of cell density. Thus, we have a clear measure of system performance. Given an expert's interpretation, we can measure system performance at two levels: the number of misclassified cells and the error in the overall measurement. We are not attempting to duplicate human performance; rather, our goal is to investigate issues in interpretation. Nevertheless, this scoring metric provides a way to select among alternative design choices and evaluate experimental runs.

The interpretation of simple geometric objects that move also provides a domain in which experiments can be performed. This domain is especially attractive because the position of the camera and of the objects in the scene can be precisely measured. As with the interpretation of hair-cell scenes, interpretation of these scenes involves interactions between partial interpretations from earlier views of the scene and the data. This environment is constantly changing. Thus, perception is not only the incremental construction of a description of the environment, but also the construction of a dynamic description. Consider the geometrical objects shown in Fig. 7.3. Interpretation of the objects as geometric forms depends not only on clustering moving lines, but also on constructing and matching *moving* models of the objects.

During the interpretation of dynamic scenes, elements at any level of abstraction can be selectively grouped to form elements at higher levels of abstraction. Geometric information guides this grouping process. The models of objects, the expected configurations of surfaces, and general knowledge about the types of surface and edge conjunctions all are used to make inferences among levels of abstraction.

Each primitive of inference is represented by a knowledge source. Knowledge sources produce updated hypotheses in response to changing situations. They are, in effect, procedures for knowledge-based mainte-nance of partial interpretations. At the lowest levels of abstraction (see Fig. 7.4), interpretation maintenance is the process of tracking image fea-tures over time.

While tracking can be initiated by changes in image intensity or similar image-level features, it can be continued only with knowledge from higher levels of abstraction. For the contribution from those higher levels of abstraction to be compatible, the knowledge sources at those levels must also be tracking the movement of entities there.

In overall character, an interpretation system for dynamic scenes is a multilevel tracking system. Knowledge sources at the lowest levels are simple procedures that maintain local hypothesis consistency by matching moving picture elements. This process is aided by projected model infor-mation from the higher levels, when it is available. In contrast, knowledge

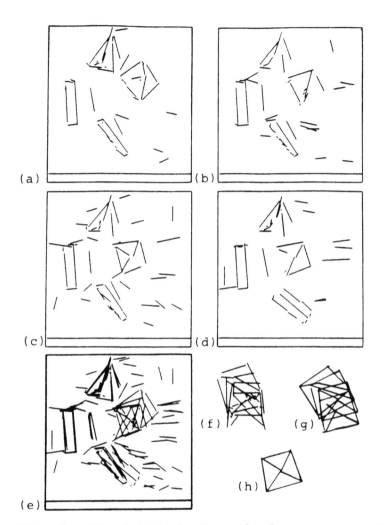

Figure 7.3   Shown in (a)–(d), four frames of motion-sequence processes to extract straight lines (Burns et al./1986). Shown in (e), a superimposing of those four frames to illustrate relative motion of linear features. Moving lines can be grouped (f), on common motion features and extended to form closed figures (g), shown here still superimposed. The hypothesis after four frames (h) shows the projection of a pyramidal solid that can be used for tracking. Matching in further frames will reveal three-dimensional characteristics.

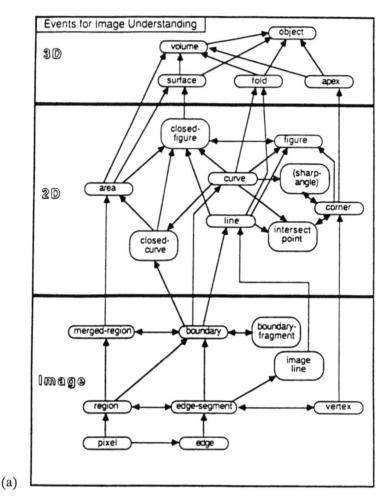

(a)

Figure 7.4 (a) The active elements for tracking visual events over time can
be distributed over several levels of abstraction. During processing, these levels
interact. (b) The information propagates from the lowest level of abstraction to
higher levels under the control of knowledge sources; verification of hypothesis,
however, causes interactions that are directed by higher levels of abstraction toward
lower levels. The frequency of updating necessary to maintain consistency with the
data decreases at higher levels of abstraction, where processing is more complex
and global.

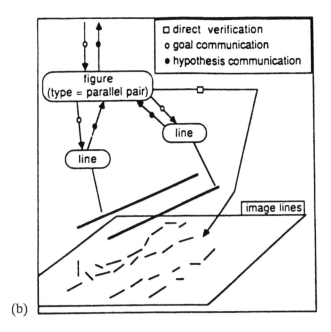

(b)

sources at the highest levels are qualitative, using a geometric reasoning process to maintain global information about object placement and motion. The propositions made by researchers in computer vision for using tokens, or groups of features, in motion correspondence (Ullman/1979, Barnard and Thompson/1980, Lowe and Binford/1985, Anandan/1984) may be thought of as a special case of the methodology described herein.

7.3   INTERPRETATION STRATEGIES

Before describing the specific experimental domains introduced in the first section, we digress to illustrate the effectiveness of using specific knowledge for interpretation. In this section, we describe some research done during 1984 and 1985 as part of the research of the VISIONS group at the University of Massachusetts (Hanson and Riseman/1975, Hanson and Riseman/1978, Hanson and Riseman/1983, Hanson and Riseman/ 1986). More extensive details are given in Weymouth/1986. The examples discussed in this section will illustrate the importance of specific knowledge in interpretation. Although most work on the application of knowledge to interpretation places an emphasis on the type of knowledge and the details of inference over that knowledge, we will use the results from Weymouth/ 1986 to emphasize the possibilities for parallelism in control.

There are several sources of potential parallelism in interpretation. They all stem from separations between the paths of inference and can be categorized by the type of separation:

Within the scene, events can be separated in space and time.

Within feature spaces, events can be separated by distance metrics between feature hypotheses or by selection according to predefined criteria.

Within solution space, events can be separated based on inference relations among the events and between partial solutions.

In this section, we will illustrate how knowledge sources can be active in parallel by taking advantage of these types of separation.

Computer interpretation of a single static image requires a large amount of detailed knowledge. The structure and form of that knowledge determine the effectiveness of the image interpretation. In the system described in this section, the primary representation of knowledge (within long-term memory) is via a set of object knowledge sources, called schemas, each of which contains two types of information: a declarative description of the object and pointers to interpretation strategies that encode control information.

The activation of any schema is treated as a separate invocation of its interpretation strategies. They each build a partial interpretation, perhaps combining previous partial interpretations that are joined into a network reflecting the current interpretation (stored in short-term memory). The resulting network associated with a scene schema instance interprets the scene.

Interpretation requires the opportunistic application of detailed knowledge, while the description of the objects in any particular domain necessitates the accumulation of large amounts of such information. Thus, it is necessary to organize that information so that context can play a role in its application.

For example, the system developed by (Nagao and Matsuyama/1980) contains object-recognition subsystems consisting of a set of specialized procedures that control interpretation of the image for the recognition of particular object types. In the system for recognizing ribs in X-ray images, by (Ballard et al./1978), objects are represented by a combination of a semantic net and procedures for recognizing instances of portions of the rib cage. The system by (Glicksman/1982) for the recognition of objects in aerial photographs includes references to specialized procedures within schemas that are embedded in a network representation of interobject relations. The system described in this section is of this second type:

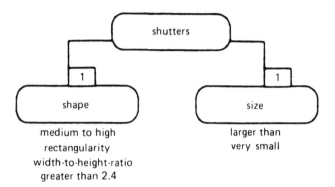

Figure 7.5   The score for shutter regions is determined by the weighted sum of two combination scores: one for shape and one for size. The combination scores are weighted sums of feature scores, and the feature scores are object-specific mappings from region statistics into a score value.

declarative information and procedural information are combined in the representation of objects.

In all the systems using specialized procedures, a procedure is applied only when there is some indication that it might succeed.   The task of object recognition has this circularity: indiscriminate application of specialized procedures wastes computational resources and leads to far too many false positives, however the knowledge to recognize the object is expressed in those procedures. The solution to this dilemma is to divide the recognition task into two parts. The first part is an inexpensive method for eliminating the most distracting false positives, leaving a set of possibilities greatly reduced in number and having some chance of success. This is typically followed by a second step that identifies instances of the target concept among the remaining candidates. Such a selection process usually involves computations that are relatively complex and costly. For example, in their system, Nagao and Matsuyama first identify regions that have characteristics of the objects they wish to identify in the image. Once the regions of correct characteristics have been labeled, more specialized and costly procedures are applied to a small area within the image. This is an application of the concept of *focus of attention* (Nagao and Matsuyama/ 1980).

In VISIONS, the steps from the interpretation strategy for the shutters schema illustrate how object-specific knowledge interacts with data from the image and hypotheses that describe partial interpretations. The initial

Figure 7.6 The steps of the *shutters* interpretation strategy apply increasingly abstract knowledge as more information about the objects becomes available. (a) The initial regions from a thresholding of the score value. (b) Regions are grouped by adjacency to form shutter-seed groups; only one group shown, for clarity. (c) and (d) The regions adjacent to each shutter-seed group are considered as a candidate group, and the subset of those regions matching a shutter model is selected. (e) and (f) A shutter-pair model is then applied to select pairs of shutters. (g) The shutter pairs are labeled in the image. (h) Later in the interpretation process, when the house-interpretation strategy produces the three-dimensional description of the house wall, shutter pairs are added to the house interpretation as features on the house wall.

(g)

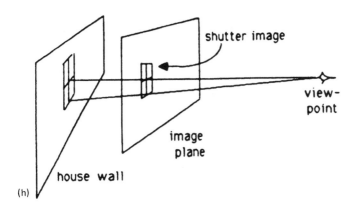

(h)

shutter hypotheses are formed by examining a scoring mechanism that combines region shape and size measures to select "shutter-like" regions in the image (Fig. 7.5). Details of that scoring process are given in Weymouth et al./1983 and Hanson and Riseman/1986. The selected regions serve as the starting point for the interpretation strategy that both searches the image for the best fit to a model and builds a description of the object (Fig. 7.6). The resulting partial interpretation is placed in STM as an object hypothesis.

During an interpretation, several schema instances will be active at the same time. Some of these need to communicate results and requests with other instances. This is done through links created in response to goals and hypotheses. When they are created, goals and hypotheses are deposited in a global blackboard. Their creation might cause the activation of additional schemas and, hence, of their interpretation strategies. Once active, these new interpretation strategies (with their schema instances) are linked to the interpretation strategy requesting the goal or creating the hypothesis; thus connected, the two interpretation strategies can exchange messages.

Interpretation proceeds until there are no additional contributions from the remaining active schema instances. As there can only be one scene schema active at a time, the final interpretation is taken to be the scene-interpretation network associated with the active-scene schema instance.

7.3.1 Two Example Interpretations

A time trace of the first example interpretation is shown in Fig. 7.7. The distance down from the top of the diagram shows the approximate passage of time; the basic time unit is the execution of one interpretation strategy step. Since the interpretations are carried out by a simulation of parallel activity, true concurrency does not occur; however, events at roughly the same position (vertically) are nearly concurrent. For example, at t_4, five instances are simultaneously active: two are suspended, the outdoor-scene and ground-plane instances, because they are waiting for a response to the goal requests; the other three (the sky, grass, and foliage schemas) continue to execute their interpretation strategies.

The hypotheses for sky, ground-plane, and foliage are returned to the outdoor-scene instance, and the second step in the outdoor-scene instance (at t_6) is the invocation of the house-scene schema. The house-scene schema instance is controlled by an interpretation strategy that first finds its parts; specifically, house and road. Again we see the branching out of parallel activation in the time line. The interpretation strategy associated with the road instance does not cause the activation of further schemas and eventually returns a road hypothesis (see t_{13} in Fig. 7.7).

The house, walls, and shutters schema instances interact through the hypotheses for the geometric structures. The creation of the house instance (t_7) starts the interpretation process for house. In this example, the interpretation strategy for house first selects the method based on finding the roof; thus, a request for roof activates the roof schema (approximately t_8), ultimately creating a roof hypothesis (see t_{11} in Fig. 7.7). In addition to, and parallel with, the activation of the roof schema, the house-interpretation strategy requests a goal for walls (also at approximately t_8). Through the actions of the interpretation strategies, each respective instance has issued a goal based on the expectations that specific objects might be present: house-scene for house, house for walls, and walls for shutters (t_9). The shutters-interpretation strategy responds to the shutters goal by creating

Goal Activated Interpretation

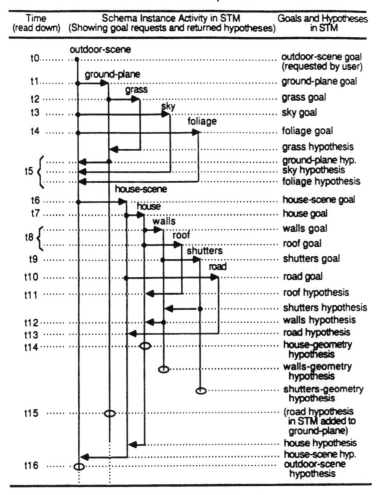

Time (read down)	Schema Instance Activity in STM (Showing goal requests and returned hypotheses)	Goals and Hypotheses in STM

Figure 7.7 The time relations of goal requests, schema activation, and hypothesis creation are shown (time increases downward). Right arrows indicate creation of a goal and the activation of a schema to respond to that goal; left arrows indicate the creation of a hypothesis structure in response to a goal. Additional events indicated by circles.

several shutter-pair hypotheses. These shutter hypotheses are returned to the walls instance which uses them to get the image area for the house walls. The walls instance then waits for the house-geometry hypothesis. The roof hypothesis enables the creation of the house-geometry hypothesis, created by the house instance (t_{14} in Fig. 7.7), which enables the positioning of the shutters in space.

With the creation of its hypothesis, the house-interpretation strategy terminates, returning the house hypothesis to the house-scene instance. The house-scene interpretation strategy constructs a hypothesis and returns it to the outdoor-scene instance. The interpretation strategy for outdoor scene adds this new hypothesis to the outdoor-scene hypothesis, and the interpretation is complete (t_{16} in Fig. 7.7).

The resulting outdoor-scene hypothesis is a network describing the image (see Fig. 7.8). Although the distinction exists primarily for illustrative purposes, the subclass arch (from B to A) is used from an object (A) that is a special case of another object (B), while the subpart arch (from A to B) is used from an object (A, the part) to an object composed of it and other objects (B, the whole).

In addition, we can construct an image of the scene corresponding to the original image by using the spatial and geometric information in the interpretation network. Such a synthetic image can be used to evaluate the extent and quality of the interpretation. Figure 7.9 shows the interpretation image for this interpretation. The details of the labeling of the house image are shown separately in Fig. 7.10.

The second example is an alternative interpretation of the same image. Here, interpretation is initiated by image events. As part of the schema for roof, a key event is the occurrence of a long horizontal line. This event causes the activation of the roof schema. When the roof schema creates a roof hypothesis and posts it to STM, the house-scene schema is activated through its key-event list (see t_1 in Fig. 7.11).

From this point, the interaction of the schemas is the same as in the goal-driven example, with two exceptions: the roof schema is not again activated, and the outdoor-scene schema is activated by key event. To communicate that the roof description already exists, the house-scene instance is instantiated (as part of the normal instantiation process) with a pointer to the key event that caused it to activate: the roof hypothesis. This is passed on to the house-schema instance as part of the goal description (at t_2). The house-interpretation strategy, receiving the roof description, proceeds without requesting the goal for roof. Thus, the roof schema is not activated by the house-schema interpretation strategy. The resulting interpretation (at t_3) is the same as in the previous example.

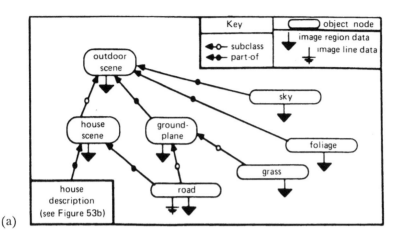

(a)

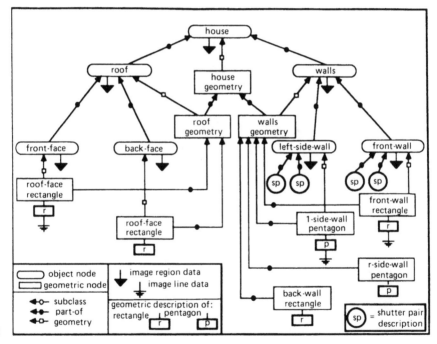

(b)

Figure 7.8 (a) The overall interpretation as produced by the interactions of the interpretation strategies of the active schema instances. (b) The portion of the interpretation network associated with the house hypothesis. The shutter-pair descriptions are not expanded (for clarity). Each shutter-pair node points to two shutters and has an associated rectangular surface description. The rectangle and pentagon surface descriptions are ordered sets of lines with three-dimensional end-points. Note that each shutter pair is associated with a particular wall.

	unlabeled		grass		sky
	road		foliage		house area

Figure 7.9   Projection of the interpretation network into the image plane.

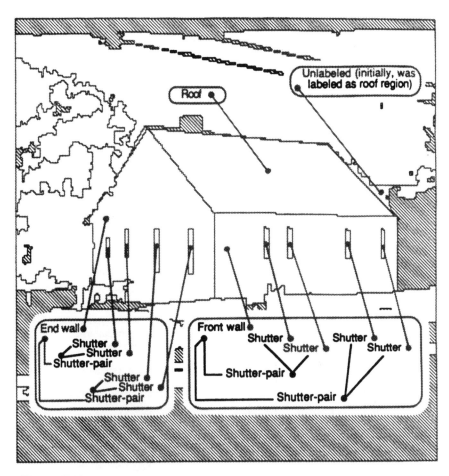

Figure 7.10 Details of the labeling of the house from Fig. 7.9.

Despite the fact that the two interpretations yield identical results, there are important differences in the course of the interpretation. The goal-driven interpretation finds evidence for the general scene before proceeding. Thus, we see (at t_6 in Fig. 7.7) that sky, ground, and foliage are sought and confirmed before we proceed. Furthermore, every action in this scenario is goal driven. The initial goal causes activation of a strategy that proceeds to find goals for each subordinate class of object; since the same general strategy is used throughout (with variations in method), the effect is one of goal-directed activation. Note that this is the result of the particular strategies used and not a necessary component

Data-Activated Interpretation

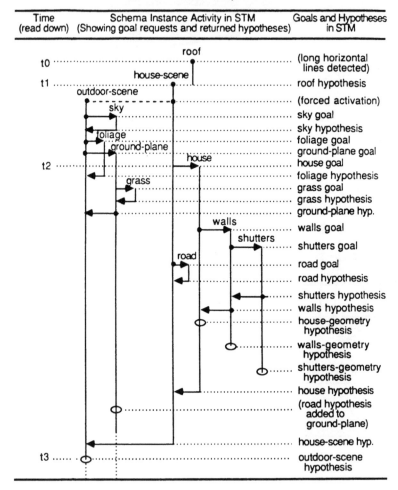

Time (read down)	Schema Instance Activity in STM (Showing goal requests and returned hypotheses)	Goals and Hypotheses in STM

Figure 7.11 The trace of hypothesis, goal, and schema instance creation for a data-activated interpretation of the example image. The activation of the roof schema causes the creation of a roof hypothesis; the creation of the roof hypothesis causes the activation of the house-scene schema. The activation of the house schema triggers the activation of the outdoor-scene schema.

of interpretation. The effect of the particular strategies is more clearly shown in the data-activated interpretation (7.11). In this interpretation, when the data-activated strategies are invoked, they also seek to carry out a goal-guided interpretation. It is because of this convergence of the two types of strategies that these two interpretations are similar. Clearly, the design of the strategies for interpretation has a great bearing on how the interpretation is carried out. The issues of how to design strategies and what constitutes a good strategy are subjects of current research.

Because the system exhibits the ability to place an overall hypothesis for each of these objects, such hypotheses could be used as a base for extending the interpretation using contextual clues. For example, Fig. 7.12 illustrates the steps of an interpretation strategy based on context. Similar interpretation strategies would enable the system to hypothesize the curbs

Figure 7.12 The interpretation strategy for wire assumes prior interpretation of sky. (a) The initial sky region. (b) The lines in the image from which the strategy selects long and parallel lines based on the context indicated by the sky hypothesis. (c) The final interpretation of the wires in the sky (each approximated by a long, narrow quadrilateral).

of the road, the chimney of the house, the windows (based on shutter pairs), and the placement of gutters on the leaves. With such additions, the interpretation could be extended to another level of detail.

7.4   AN EXAMPLE DOMAIN

As seen in the previous section, the process of performing experiments with artificial systems is one of incremental design. First, a model of an appropriate system is developed; it will have several levels of abstraction. Procedures are developed that form descriptions of entities at a given level of abstraction by selectively grouping less abstract elements. These procedures are experimentally tested. In the testing, generally, the model is found wanting and is modified. The modified design is again implemented and tested. One would then follow a cycle of design, implement, and test. In general, therefore, we would like the supporting architecture to be as flexible as possible. Blackboards provide the necessary flexibility for system development. Furthermore, the organization of knowledge into multiple levels of abstraction is an integral part of blackboard systems (Erman et al./ 1980, Nii/1986). By taking advantage of this natural organization of the problem space, for vision, we can identify those entities that are going to be used for inference and the types of inferences that are necessary. The inferences are the bases for the procedures in our design.

In this section, we look at the problem of image interpretation in the domain of inner-ear hair cells. Interpretation of hair-cell slides, in a laboratory setting, involves counting and mapping out the position of the cells. The laboratory technician not only must identify the presence of a cell, but also must examine it for defects and classify it based on those defects. The cells are counted by class and the results reported as a ratio of each class over fixed, preset portions of the tissue sample. The classification of a cell involves examining its three-dimensional structure. This is done by varying the focal plane of the microscope. From the point of view of designing a system for scene interpretation, this reexamination of the microscope slide is a reimaging of the scene from a sightly different viewpoint. This medical-scene interpretation is a labor-intensive inspection process requiring an extended training period.

In the hair-cell interpretation problem, we have identified five levels of abstraction, as shown in Fig. 7.13. The levels are: *gray image*, *edge image*, *cell*, *cell group*, and *cell map*. The incoming images are stored in image descriptions at the gray image level. Edge descriptions, extracted by an edge operator, linked, and segmented, are stored on the edge-image level. Cell hypotheses, stored on the cell level, describe cell position and cell contour and contain measures of cell shape. Cell-group hypotheses are

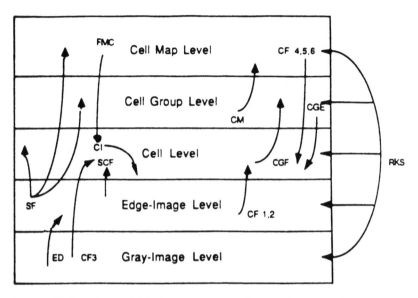

Figure 7.13 The multiple levels of abstraction within the blackboard provide a natural division of the processing necessary to perform the inspection task. Specialized processes, called knowledge sources, provide the mapping of information to new information within and between levels.

descriptions of local patterns of cells that fit the model of adjacent cells; they are fragments of the entire cell map. Cell-map hypotheses are descriptions of the cell mappings that account for all the known data.

Identifying the primitives for inference (i.e., the knowledge sources) facilitates the identification of the transformation that takes one primitive to another. As before, knowledge sources react to the creation of hypotheses on one level of abstraction and cause the creation of new hypotheses on that level or another level, higher or lower. Knowledge sources may group hypotheses and link them to form a new hypothesis, as, for example, forming a cell-group hypothesis from several cell hypotheses. Or they may just create a new hypothesis in reaction to information.

The use of knowledge sources reduces the overall problem to several smaller problems. In this use of the divide-and-conquer paradigm, the goal is to select knowledge sources so that their interactions are clearly identified and associated with hypothesis types. Knowledge sources interact only through the creation and detection of hypotheses. Thus, there needs to be a hypothesis type for each possible interaction. These hypothesis types correspond to different types of object entities. Then, each knowledge source is reacting to one type of object entity and adding knowledge to infer from that entity the existence of another entity.

Knowledge sources are designed so that they interact with hypotheses of only a few types. These hypothesis types are primarily identified by the level of abstraction at which they occur. In addition, at any level of abstraction, there are (potentially) large numbers of hypotheses. This creates an opportunity for parallelism. Knowledge sources interacting with unrelated levels of abstraction can be activated in parallel. Instances of the same knowledge source (i.e., copies of the knowledge source applied to different sets of input) can be activated in parallel when their input is disjoint.

Currently, we have identified 13 knowledge sources that are needed for this problem. They are identified in Fig. 7.13 and are:

ED. Edge Extraction. Operates on gray-level images and produces a set of edge segments.

SF. Scale Finder. Determines approximate image scale from the gray-level images. The image scale is used for control and to refine settings of default values in various hypothesis types.

CF1. Cell Finder. Groups edge segments into cell boundaries.

CF2. Cell Finder. Locates sterocilia from cell hypothesis and edge segments. The sterocilia is another part of the cell, usually in a different focal plane from the one in which the contour is clearly defined.

CF3. Cell Finder. Locates cell nuclei based on cell-contour hypothesis and image-intensity data. Again, may involve a different focal plane.

CF4. Cell Finder. Locates cell contours based on cell-group hypothesis and edge-segment data. May involve reactivation of ED knowledge source with different parameters.

CF5. Cell Finder. Locates sterocilia based on cell-group or cell-map hypotheses and edge-segment data. Again, the knowledge-source ED may have to be reinvoked.

CF6. Cell Finder. Locates cell nuclei based on cell-group or cell-map hypotheses and image data.

CGF. Cell-Group Finder. Constructs cell hypotheses for cells that are adjacent and whose position corresponds to the positions expected based on model information. Cell groups are also extended in response to CF4, CF5, and CF6 when those knowledge sources develop hypotheses for cells. The cell groups correspond to *islands of certainty* (Erman et al./1980).

CM. Cell Map. Generates a map hypothesis and updates map hypotheses based on cell-group and cell hypotheses. Also creates a default cell map (based on scale information from SF) to initiate control for cell-finder knowledge sources that require information from higher levels of abstraction.

SCF. Scarred Cell Finder. Looks for scar lines, which are a pattern of edge segments that indicates the presence of a dead cell (one of the classifications of cells).

FMC. Find Missing Cell. Creates missing-cell hypothesis. Based on the lack of evidence for a cell where one is expected because of cell-group or cell-map hypotheses. Missing-cell hypotheses are also used by the cell-map knowledge source, since missing cells account for particular types of deformities in the cell map.

RKS. Rank Knowledge Source. Determines which set of knowledge-source instances should be activated based on hypotheses scores and knowledge-source priorities. This is an example of a control knowledge source.

These knowledge sources interact through the hypotheses that they create and place on the blackboard. Since this is the only interaction, knowledge sources can be improved incrementally. When a knowledge source is removed, the effect on the system is that it no longer produces hypotheses. No other knowledge source will be *directly* affected. When a knowledge source is put in, it begins to produce hypotheses; if those hypotheses are responded to by other knowledge sources, then that new knowledge source becomes an active part of the system. In fact, human intervention can also be included, as long as it is styled after the type of interaction that a knowledge source makes. That is, a human-operator knowledge source would respond to particular hypotheses, be scheduled for activation, and possibly create additional hypotheses when activated.

The paths of interaction among the knowledge sources, by means of the hypotheses, set up several loops. For example, when a cell hypothesis is created near a cell group, the cell-group finder will create a new cell-group hypothesis extending the cell group. This new hypothesis could cause activation of a cell-finder knowledge source to find cells to further extend the cell-group hypothesis. Paths of control follow the creation of hypotheses, expanding from given hypotheses along lines that match the model and the data.

Control proceeds bottom-up initially (CF1, CF2, CF3, SF, CGF). With the instantiation of a cell group or a cell map, top-down control is also possible (SCF, CM, FMC). Cell groups are extended from initial hypotheses based on cell-group hypotheses or cell-map hypotheses (CF4, CF5, CF6, CGE).

Suppose now that the system is given two images at two different focal planes. These images are processed and edge descriptions are produced at the edge level. Edge operators usually produce edge fragments, such as those shown in Fig. 7.14. We have labeled various regions in each figure as corresponding to certain hair cells. Figure 7.14 illustrates different cell subparts appearing in the two planes of focus. It also illustrates both scarred cells and missing cells.

Bottom-up processes such as CF1 operate by abstracting data on lower levels on the blackboard into more structured descriptions. Examples of the CF1 data-abstraction process are shown in Fig. 7.15. Fig. 7.15 illustrates the kinds of hypotheses that are generated about cell boundaries at the cell level using edge fragments. One could equally well consider generation of true cell hypotheses based on sterocilia information.

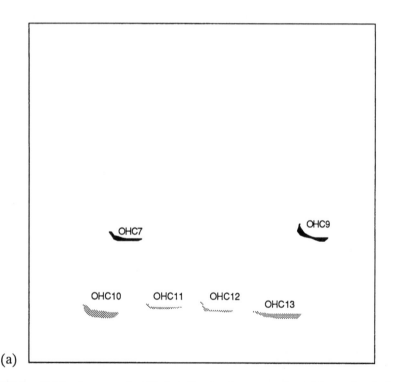

(a)

Figure 7.14   An example of hair-cell scene interpretation with two focal planes. Generally, one needs to resample the environment at different focal planes in order to locate present structures belonging to the three dimensional cell body.

There are two scars present in Fig. 7.14 (OHC6 and the gap between OHC8 and OHC9). The SCF knowledge source operates on cell-contour hypotheses and detects the position of scar lines within the area enclosed by the boundaries (OHC6). This is done for every cell hypothesis with a large enough degree of certainty. Because SCF has a goal, the area within the image to be searched is restricted. Now let us examine how the missing cell will be detected. Among the vast number of cell-group hypotheses, there are few that contain cell hypotheses with large certainty factors. The CGF knowledge source groups adjacent cell hypotheses in this case. Of course, this is only one of the group hypotheses. On the blackboard, we have the cell group shown in Fig. 7.16. A second high-scoring cell group is shown in Fig. 7.17.

The two cell groups CG hyp 1 and CG hyp 2 correspond to islands of certainty. The CGE knowledge source attempts to extend these groupings into larger descriptions based on size and location information of the cells in the cell group that it operates on. CGE extends each cell grouping by

(b)

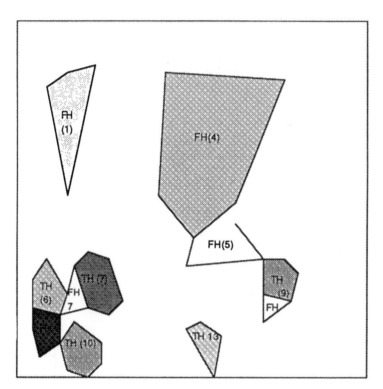

Figure 7.15   Multiple alternative hypotheses exist for different regions of scene. Using Fig. 7.14, a number of hypotheses may be locally consistent. However, it is the accumulation of evidence from local to global stages that is the final arbitrator.

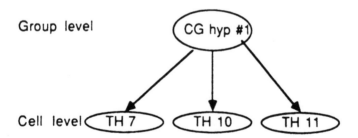

Figure 7.16   Continuing with the example of two focal planes, a cell group is created here. Starting from this cell group, top-down algorithms invoked on the image can extend the cell group to provide a more complete partial interpretation.

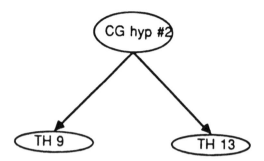

Figure 7.17   Another cell group. The rank knowledge source assigns a high score to this second cell group, reflecting the confidence of the system in the evidence present.

choosing the best cell hypothesis that is on the blackboard and is adjacent to the cell group. In this way, CG hyp 1 is extended to contain TH8, and CG hyp 2 is extended to contain TH12. It should be stated at this point that not all the cells need to be located for the cell map to be instantiated. CG hyp 1 and 2 are combined by the CM knowledge source. Based on this grouping, CM hypothesizes a cell map for the whole scene. When referring to cell groupings, we include the degenerate case of a single cell to also be a cell grouping, so that a cell map may be instantiated with minimal information. However, the more cells with higher scores that support a cell-map hypothesis, the higher the score will be for the corresponding cell map. In fact, we are allowing for a default cell map to also be produced. This default map uses only hypotheses about image scale.

Having generated cell maps such as CM hyp 1, the FMC knowledge source may be invoked, which locates old scars. FMC is invoked with the goal of finding an area in the image where the cell map expects a cell but no evidence has been found to support this expectation. Other top-down processes locating sterocilia, nuclei, and cell contours may also be invoked. These could be invoked when cell groups are instantiated also.

At any time, there are several possible extensions; how to determine which one to apply determines the blackboard system's control strategy. Our initial design is to adopt a straightforward strategy. It is essentially the one used in Hearsay (Lesser and Erman/1977) and other blackboard systems (Nii/1986). Each knowledge source, in addition to creating a hypothesis, makes a fitness measure on that hypothesis and includes the value of that measure as part of the hypothesis. When a knowledge source responds to a hypothesis, a record of the knowledge source and the

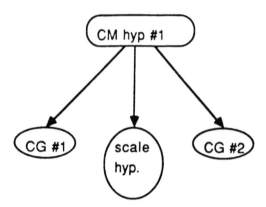

Figure 7.18 A cell-map hypothesis. The cell map is the most global symbolic object in the system. It is built, refined, and completed through combination of local evidence, thus capturing global structures in the abstraction hierarchy.

hypotheses to which it responded is put into a ready list and given a ranking by the *rks* knowledge source. Knowledge sources are activated from this list.

Control follows from this ranking mechanism. Those hypotheses with the highest fitness scores and those knowledge sources with the highest priority get activated first. If there is an interpretation that emerges from their activation, it will be one that is consistent with model and data. In practice, however, during the ongoing interpretation there are frequently no clear winners in the ranking. There will be several partial hypotheses that are not in conflict (that is, they deal with separate portions of the image), and there will be hypotheses that are in conflict. In general, there will be conflicting hypotheses with roughly the same fitness score. These are alternative plausible partial interpretations.

Our approach is to allow alternative plausible partial interpretations to proceed in parallel. We are designing a control mechanism to simulate the parallel activation of knowledge sources dealing with alternate hypotheses that are not directly dependent on one another. Although we have not developed the details of this portion of the architecture, we believe it will be a mechanism that will prove useful in dealing with error and uncertainty. In signal-understanding problems such as interpretation of image data, there are many alternative paths to the solution. Given a sufficiently rich branching of those paths and fitness measures that capture the notions of consistency with both data and model, we should be able to arrive at a consistent interpretation.

For development, we are also taking advantage of those extensions to the classical blackboard architecture offered by the GBB (generic blackboard) system (Corkill et al./1986). We are especially using the notion of dimensions in the blackboard and the concept of knowledge-source instance. In GBB, each blackboard can have any number of dimensions. Hypotheses are stored in an indexable data structure, so that it is easy to retrieve hypotheses that are near a given point in the dimensions of the blackboard. This is very useful in keeping track of the relationships between the hypothesis and the portion of the image that it explains. Also, hypotheses that are (potentially) in conflict can be retrieved based on the fact that they are dealing with the same portion of the image.

The concept of a knowledge-source instance is being used extensively in our design. As can be seen from the example shown in the next section, many of our knowledge sources operate in an area restricted in location with respect to a seed hypothesis. Furthermore, these knowledge sources do not interact with other hypotheses of the same type. Because of this, the knowledge source could be active in several places in the image at the same time. To capture this notion of parallelism, we have added the knowledge-source instance as a record of the progress of an active knowledge source. These records may eventually be used by control knowledge sources.

As indicated, our knowledge sources have access to an object model. The object model in this case is a very simple set of frame-based descriptions for each hypothesis type. The frame consists of a set of name-value pairs, the slots. Each slot includes a default value or default value function that is used to set the values of the hypothesis of that type on creation if no other value is available.

7.5   THE DEVELOPMENT OF KNOWLEDGE SOURCES

In this section, we have selected several knowledge sources for description. These are the ED, the CF1 (Amini et al./1989), and the CF4 (Amini et al./1988) knowledge sources. They will serve as illustrations of knowledge sources and provide some examples of the types of interactions with which knowledge sources must deal.

As a knowledge source is being developed and tested, our understanding of the problem increases. To illustrate the process, for the case of the cell-interpretation system, we discuss a possible reorganization of the suggested knowledge sources, discussed in the previous section. For example, we may see fit to add a cell-body knowledge source. This knowledge source would duplicate part of the CF1, CF2, and CF3 knowledge sources and would collect information from various focal planes, constructing hypotheses of cell bodies that were consistent with the data and the cell-body model.

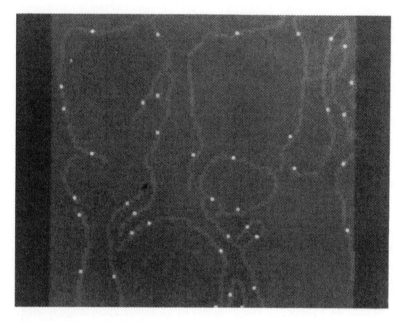

Figure 7.19   Edges are segmented at points of high curvature using a scale-space for curvatures based on the k-curvature function.

The features of the blackboard architecture with the extensions of blackboard dimensions, frame-based object models, and parallel activation of knowledge sources provide the necessary framework for integrating these knowledge sources. In these examples, we will see the usefulness of communicating through hypotheses, of being able to access hypotheses based on location, and the best-first control strategy, and of being able to experimentally alter the design without rebuilding the system.

The edge-extraction knowledge source (ED) selects and segments edge fragments from the gray-level image. The process of locating and linking edge elements is carried out by the Canny operator (Canny/1986); then, the resulting connected edge sets are segmented. If the initial edge values were to be linked and used for cell-boundary fragments, there would be problems due to the fact that those edge segments would often belong to more than one cell. Each initial edge-fragment hypothesis is formed by segmenting each linked edge at the zero-crossings of its local curvature. Curvature (Shahraray/1985) is measured using a scale-space description for curvature based on the k-curvature function (Rosenfeld and Johnston/1973), in which the curvature of the edge at every point is calculated given the coordinates

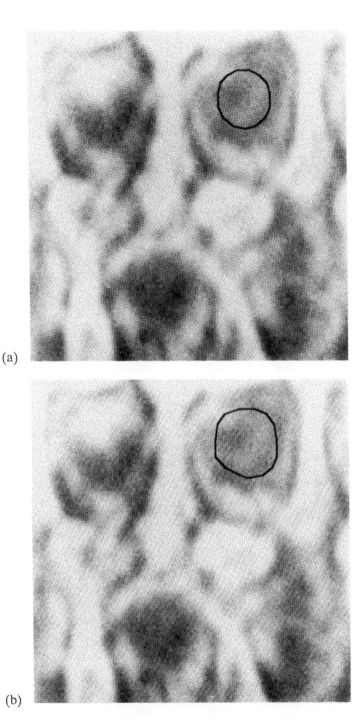

(a)

(b)

Figure 7.20 Illustrating a top-down process, Energy Minimizing Contours. The algorithm is given an initial contour. The Contour then attaches itself to outline of objects through an iterative process. The first image ((a), (b)) is the initial contour, a circular contour. In this case the contour expands rather than contracts, as is usually the case when localizing object boundaries.

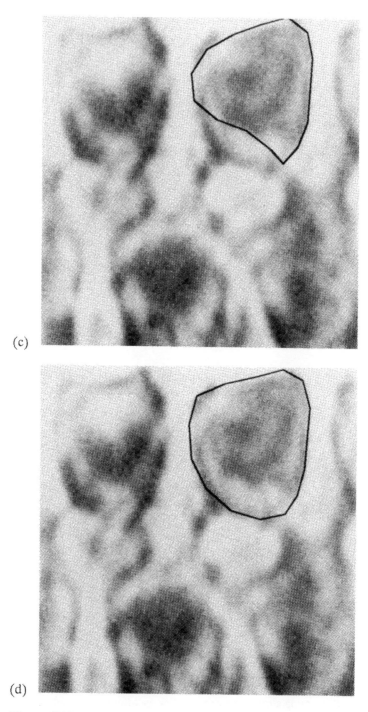

(c)

(d)

Figure 7.20 (Continued)

of the lattice through which it passes. Prior to the actual measurement stage, an autoregressive moving average (ARMA) filter is applied to the coordinate points of the edge curve to reduce the effects of quantization. The resulting curve fragments form the edge hypothesis used by the CF1 knowledge source (see Fig. 7.19).

Each edge hypothesis serves as a starting point for the cell-finder knowledge source CF1 (Sec. 7.5.1). This knowledge source takes the edge fragment and attempts to construct a cell boundary with that fragment as a starting point. The cell boundary is constructed using a best-first search controlled by constraints enforcing legal geometrical relations between fragments.

CF1 creates bottom-up groups of edge fragments that correspond to coherent closed shapes. CF4, on the other hand, is a top-down process for locating features of interest in images. In our system design, CF4, just as CF1, uses quite general methodologies for locating objects of interest in images. It uses model knowledge in the form of expectations and is based on energy-minimizing contours (Kass et al./1987, Amini et al./1988). A snake is an energy-minimizing spline guided by external constraint forces and influenced by image forces that pull it toward features such as edges, lines, and terminations. High-level control mechanisms can interact with the contour model by pushing it toward an appropriate local minimum. We do not provide the details of this algorithm here. The interested reader is referred to references given in the literature. To understand the basic idea, however, the algorithm is given an initial contour such as the one shown in Fig. 7.20. The contour then attaches itself to the outline of objects through an iterative process, as shown in the figure.

7.5.1   The CF1 Knowledge Source

In this section, we describe the CF1 knowledge source (Amini et al./1989), a scale- and orientation-invariant algorithm that hypothesizes cell positions in the image plane based on edge information. The methodology is one of searching among boundary fragments for a complete, closed boundary. It is unnecessary to find positions of all the cells in the image in a bottom-up fashion. Therefore, the goal of this algorithm is only to find enough cell contours (where the cell hypotheses can be made with high confidence) that the map scale (SF) can be confirmed and the approximate map position (CM) can be refined.

Following the application of the ED knowledge source, fragment hypotheses are placed in data structures, facilitating further processing on them. The major problem that needs to be circumvented is that fragments that do belong to a single cell are normally disconnected.

In circumventing this problem and in producing hypotheses about cells, for every edge segment in the image, a local search is made to find the most plausible cell boundary, part of which is the segment. The only criterion for labeling a series of edge segments a cell contour is that the centroids of the grouped segments form a closed polygon. Representing cells with polygons is, of course, not detailed, but does provide for a good description of their location without great loss of their boundary structure.

The process of polygon formation is based on search. Search is needed because it is not clear which segments should follow one another in producing a cell-contour hypothesis. A best-first search with no backtracking has proven useful for inferring 2-D object positions. An important property of CF1 is that because search is started for every edge segment in the image, no starting point is needed for CF1. However, multiple searches are needed because, a priori, there is no way to know which edge segments are produced by noise and which ones are not. Screening is provided by a "voting" mechanism—since there is a boundary hypothesis for each edge segment, each segment may be thought of as voting for the boundary hypothesis for which it is the starting segment. Once votes are gathered, global figures of merit are invoked to remove "bad" votes.

Cell hypotheses are formed for those boundary fragments that form closed contours. In determining the boundary fitness, in addition to measures of contour shape and shape orientation, one of the most intriguing ideas is to exploit redundant interpretations. Figure 7.21 illustrates the voting technique wherein each cell hypothesis that describes the same contour votes for the presence of that cell. Because many of the segments on a cell contour, when used as a hypothesis seed, result in the same cell description, their agreement turns out to be an excellent indication of the presence of a cell. This suggests separating the cell-finder knowledge source into two knowledge sources: a boundary knowledge source and a cell-body knowledge source.

In this new design, the cell-body knowledge source would initiate a cell-body hypothesis whenever a small number of cell-boundary hypotheses were instantiated in the same location. The initial cell body would have a low fitness score, reflecting little evidence. As the sterocilia and nuclei of the cell were instantiated—that is, as hypotheses for those cell parts were made in roughly the same cell location—if they were consistent with the model of the cell, a new cell hypothesis would be created extending the previous cell hypothesis with a higher fitness score. Additional boundary hypotheses in the same location could also be used to increase the fitness score. Such a change in design would be possible with only a change to the cell-finder knowledge sources because of the characteristics of the blackboard architecture.

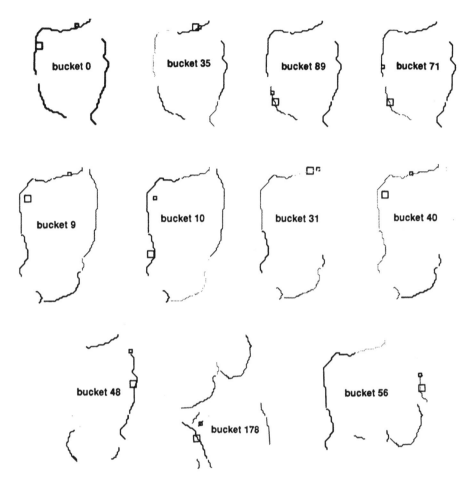

Figure 7.21 Each edge segment forms a seed for a boundary search. This seed starts a boundary-segment list, and a new segment is added to the list when it is ranked the best next segment in a local search.

In these examples, we see the potential for parallel activation of knowledge sources. The formation of each boundary, contour, and cell hypothesis draws only on local information. The formations of these hypotheses in distant parts of the image interact only at the higher levels of abstraction. If we allow the creation of hypotheses that are locally consistent at the lower levels of abstraction and only worry about global consistency at the higher levels of abstraction, then those hypotheses that are spatially separate can be pursued in parallel.

7.6 CONCLUSIONS

We propose the blackboard architecture (Nii/1986) for supporting processes of visual perception. Within this architecture, frames are used to represent hypotheses and object-model information, while knowledge sources represent the steps in inference under uncertainty. Each knowledge source uses object-model information to construct hypotheses consistent with both the data and the object models in response to other hypotheses or data. The flow of control moves through the creation of hypotheses and is channeled by the selection of knowledge sources for activation. In addition, parallel activation of noninteracting knowledge sources provides the potential for a speedup of computation.

REFERENCES

Anandan, P. (1984). "Computing Dense Displacement Fields With Confidence Measures in Scenes Containing Occlusions," Proceedings of Image Understanding Workshop, New Orleans, pp. 236–246.

Amini, A. A. and Weymouth, T. E. (1987). "Representation and Organization of Domain Knowledge in a Blackboard Architecture: A Case Study from Computer Vision," Proceedings of the 1987 Conference on Systems, Man, and Cybernetics.

Amini, A. A., Tehrani, S., and Weymouth, T. E. (1988). "Using Dynamic Programming for Minimizing the Energy of Active Contours in the Presence of Hard Constraints," Proceedings of the 2nd International Conference on Computer Vision, Tarpon Springs, Florida.

Amini, A. A., Weymouth, T. E., and Anderson, D. J. (1989). A parallel algorithm for determining 2D object positions using incomplete information about their boundaries, *Pattern Recognition, 22, no. 1:* .

Barrow, H. and Tenenbaum, J. (1976). *MSYS: A System for Reasoning About Scenes,* Technical Note 121, AI Center, Stanford Research Institute.

Ballard, D., Brown, C., and Feldman, J. (1978). An approach to knowledge-directed image analysis, *Computer Vision Systems* (A. Hanson and E. Riseman, eds.), Academic Press, New York.

Barnard, S. T. and Thompson, W. B. (1980). Disparity analysis of images, *IEEE Trans. on Pattern Analysis and Machine Intelligence, 2, no. 4:* 333–340.

Batchelor, B. G., Hill, D. A., and Hodgson, D. C. (eds.) (1985). *Automated Visual Inspection,* North Holland, UK.

Besl, P. J., Delp, E. J., and Jain, R. C. (1985). Automatic visual solder joint inspection, *IEEE J. of Robotics and Automation, RA-1, no. 1.*

Besl, P. J. and Jain, R. C. (1985). Three-dimensional object recognition, *ACM Computing Surveys, 17, no. 1:* 75–145.

Binford, T. (1982). Survey of model based image analysis systems, *International J. of Robotics Research, 1, no. 1:* 18–64.

Brady, M. (1982). Computational approaches to image understanding, *ACM Computing Surveys, Vol. 14:* 3–71.

Brooks, R. (1981). *Symbolic Reasoning Among 3-D Models and 2-D Images*, Technical Report STAN-CS-81-861, Dept. of Computer Science, Stanford University, Stanford, California.

Burns, J. B., Hanson, A. R., and Riseman, E. M. (1986). Extracting straight lines, *PAMI, 8, no. 4*: 425–455.

Canny, J. F. (1986). A computational approach to edge detection, *IEEE Trans. on Pattern Analysis and Machine Intelligence 8*: 679–698.

Corkill, D. D., Gallagher, K. Q., and Murray, K. E. (1986). "GBB: A generic blackboard system," *Proceedings of AAAI '86*, Philadelphia, Pennsylvania.

Erman, L., Hayes-Roth, F., Lesser, V., and Reddy, D. (1980). The Hearsay-II speech-understanding system: Integrating knowledge to resolve uncertainty, *Computing Surveys, 12, no. 2*: 213–253.

Glicksman, J. (1982). *A Cooperative Scheme for Image Understanding Using Multiple Sources of Information*, PhD Dissertation, University of British Columbia, Vancouver, BC.

Hanson, A. R. and Riseman, E. M. (1975). *The Design of a Semantically Directed Vision Processor (Revised and Updated)*, COINS Technical Report 75-C1, University of Massachusetts at Amherst.

Hanson, A. and Riseman, E. (1978). VISIONS: A computer system for interpreting scenes, *Computer Vision Systems* (A. Hanson and E. Riseman, eds.), Academic Press, New York, pp. 303–333.

Hanson, A. and Riseman, E. (1983). *A Summary of Image Understanding Research at the University of Massachusetts*, COINS Technical Report 83-35, University of Massachusetts at Amherst.

Hanson, A. R. and Riseman, E. M. (1986). A methodology for the development of general knowledge-based vision systems, *Vision, Brain, and Cooperative Computation*, (M. Arbib and A. Hanson, eds.) MIT Press, Cambridge, Massachusetts.

Haralick, R. M. and Shapiro, L. G. (1985). Image segmentation techniques, *Computer Vision, Graphics, and Image Processing, 29*: 100–132.

Horn, B. K. P. (1986). *Robot Vision*, MIT Press, Cambridge, Massachusetts.

Kass, M., Witkin, A., and Terzopolous, D. (1987). "Snakes: Active Contour Models," Proceedings of International Conference on Computer Vision, London, pp. 259–268.

Lesser, V. R. and Erman, L. D. (1977). A retrospective view of the Hearsay-II architecture, *IJCAI-77*: 790–800.

Lowe, D. G. and Binford, T. O. (1985). The recovery of three-dimensional structure from image curves, *IEEE Trans. on Pattern Analysis and Machine Intelligence, PAMI-7, no. 3*: 320–326.

Nagao, M. and Matsuyama, T. (1980). *A Structural Analysis of Complex Aerial Photographs*, Plenum Press, New York.

Nii, H. P. (1986). Blackboard systems (parts I and II), *The AI Magazine, 7, nos. 2 and 3*: 38–53; 82–106.

Ohta, Y. (1980). *A Region-Oriented Image-Analysis System by Computer*, PhD Dissertation, Dept. of Information Science, Kyoto University, Kyoto, Japan.

Rosenfeld, A. and Johnston, E. (1973). Angle detection on digital curves, *IEEE Trans. on Computers*, C-22, no. 9: 875–878.

Shahraray, B. (1985). *Measurement of Boundary Curvature of Quantized Shapes: A Method Based on Smoothing Spline Approximations*, PhD Dissertation, University of Michigan, Ann Arbor, Michigan.

Simon, H. A. (1981). *The Sciences of the Artificial*, 2d ed., MIT Press, Cambridge, Massachusetts.

Ullman, S. (1979). *The Interpretation of Visual Motion*, MIT Press, Cambridge, Massachusetts.

Weymouth, T. E., Griffith, J. S., Hanson, A. R., and Riseman, E. M. (1983). "Rule Based Strategies for Image Interpretation," Proceedings of AAAI '83, pp. 429–432; available in a longer version in Proceedings of the DARPA Image Understanding Workshop, Arlington, Virginia, pp. 193–202.

Weymouth, T. E. (1986). *Using Object Descriptions in a Schema Network for Machine Vision*, PhD Dissertation, available as COINS Technical Report 86-35, University of Massachusetts at Amherst.

Weymouth, T. E. and Amini, A. (1987). "A Framework for Knowledge-Based Computer Vision Using a Blackboard Architecture," Proceedings of SME Vision '87 Conference, Detroit, Michigan.

Weymouth, T. E. (1987a). "Knowledge-Based Spatial Reasoning: Using a Blackboard Architecture," presented at Workshop on Blackboard Systems, Seattle, Washington.

Weymouth, T. E. (1987b). "Incremental Inference: Spatial Reasoning Within a Blackboard Architecture," Proceedings of the 1987 Workshop on Spatial Reasoning and Multi-Sensor Fusion, St. Charles, Illinois.

Yost, W. and Nielsen, D. (1977). *Fundamentals of Hearing*, Holt, Rinehart, and Winston, New York.

8

Analysis of High-Resolution Aerial Images

Mohan M. Trivedi

University of Tennessee
Knoxville, Tennessee

8.1   INTRODUCTION

Images of earth acquired by aerial or satellite-based sensors are of great utility in a variety of applications. These include agriculture, land resource management, cartography, environmental sciences, oceanography, and military surveillance and reconnaissance. These images are typically acquired using high-resolution cameras, multispectral scanners, or radars. Trained photointerpreters can recognize features of interest by a careful analysis of these images. Due to the complex nature and vast amount of data involved, this process is typically quite tedious, labor-intensive, and expensive. Automating the image interpretation process has been an active area of research for several years. These efforts have resulted in some useful techniques for image enhancement and restoration, noise suppression, and registration. However, the development of automatic feature extraction and interpretation techniques has proven to be quite difficult.

Automated interpretation of high-resolution aerial images is a complex task consisting of a variety of functional modules. In Fig. 8.1, we have presented a typical flow of the image interpretation process. Image acquisition is the first step in the process. In the next step, images are processed to eliminate any geometrical distortions caused by the sensor system. If possible, during this step one would like to correct the images for various radiometric and atmospheric effects. These operations are generally known as preprocessing tasks. The next step corresponds to the

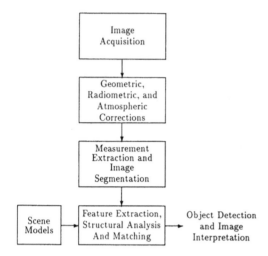

Figure 8.1 Computational tasks required in automatic interpretation of aerial images.

image segmentation module. In this, various operators are employed to extract image cues in spectral, spatial, or topographic domains. These measures are analyzed to partition the image into regions that share some homogeneity property. These segmented regions are sometimes referred to as blobs. It is hoped that these blobs correspond to the objects or their parts appearing in the scene. In the final step of processing, various features associated with the segmented blobs are extracted and compared against the object attributes stored in the knowledge base of the system. It may employ pattern-recognition procedures for matching spectral, spatial, and topographic features and graph matching or relaxation techniques to analyze relational structures. At the heart of the matching process is a control strategy to systematically and efficiently search for the solution of the matching problem.

In this chapter, we present a discussion of a number of issues related to the analysis of high-resolution aerial images. Specifically, we examine the aerial image formation process from the matter-energy interaction perspective and discuss approaches for segmentation and object detection.

8.2 ENERGY FLOW PROCESS IN PASSIVE REMOTE SENSING

Sensors mounted on an airborne platform record reflected or emitted radiant flux incident on the sensor aperture. In order to develop a better understanding of how the recorded intensity levels are related to the intrinsic properties of various objects appearing in the scene, it is instructive to follow the path of the illuminating beam striking the scene and the reflected beam incident on the sensor aperture. Figure 8.2 displays the four components of a remote sensing system: source, scene, sensor, and intervening medium. We shall use this figure to discuss the radiation transfer process underlying passive remote sensing in a manner similar to that described by Hugli and Frei (1983). In our presentation we consider that the images are acquired with a multispectral scanner type of sensor (Goetz et al., 1985). In such images the incident radiation associated with the same patch of scene (i.e., the "footprint") is split into several distinct wavelength bands and imagery is recorded simultaneously in these distinct bands. Thus, analysis of these images amounts to a case of multiple sensor information integration, with each sensor providing a measure of the incident flux in a particular wavelength band. The important feature of such imagery is that the images of different bands are in perfect spatial registration with one another.

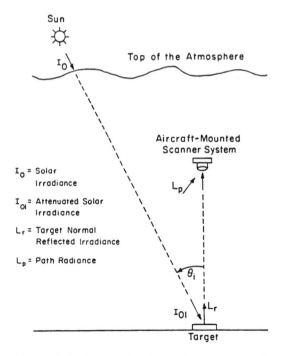

Figure 8.2   Energy flow in passive remote sensing.

Let $I_0(\lambda)$ represent the solar irradiance at the top of the atmosphere. This is attenuated while going through an atmospheric column having a transmission coefficient of $T_i(\lambda)$. The solar irradiance also becomes scattered in the atmosphere, giving rise to diffused skylight. The irradiance of the target surface orthogonal to the solar beam is

$$I_{01}(\lambda) = T_i(\lambda)I_0(\lambda)$$

and due to the foreshortening, the irradiance at the target surface by the solar beam is

$$I_{i1}(\lambda) = I_{01}(\lambda) \cos \Theta_i$$
$$= T_i(\lambda)I_0(\lambda) \cos \Theta_i$$

The irradiance of the target surface due to diffused spatially distributed skylight can be calculated using the radiance function $L_{i2}(\Theta_i, \Phi_i, \lambda)$ as follows:

$$I_{i2}(\lambda) = \int_{-\pi}^{\pi} \int_{0}^{\pi/2} L_{i2}(\Theta_i, \Phi_i, \lambda) \sin \Theta_i \cos \Theta_i \, d\Theta_i \, d\Phi_i$$

The total irradiance I_i on the target is the sum of irradiance due to direct solar beam and that due to diffused skylight, i.e.,

$$I_i(\lambda) = I_{i1}(\lambda) + I_{i2}(\lambda)$$

The radiance of the target illuminated by the above irradiance is

$$L_r(\lambda) = \int_{\text{hemisphere}} f_r(\Theta_i, \Phi_i; \Theta_r, \Phi_r; \lambda) \, dI_i$$

where the integration is performed over the hemisphere and $f_r(\Theta_i, \Phi_i, \Theta_r, \Phi_r; \lambda)$ is the bidirectional reflectance distribution function (BRDF) of the object surface (Nicodemus et al., 1977; Woodham and Gray, 1987; Slater, 1985). The BRDF specifies the reflectance properties of a surface. Both the incident beam geometry (i.e., Θ_i, Φ_i) and reflected beam geometry (i.e., Θ_r, Φ_r) are considered in the evaluation of the BRDF as shown in Fig. 8.3. It also depends on the wavelength λ. The BRDF of an object surface can be defined by

$$f_r(\theta_i, \phi_i; \theta_r, \phi_\gamma; \lambda) = \frac{dL_r(\lambda)}{dI(\lambda)}$$

where $dL_r(\lambda)$ is the reflected radiance and $dI(\lambda)$ the incident irradiance on the surface.

The radiance $L_r(\lambda)$ is attenuated and scattered on the path to the sensor. If $T_r(\lambda)$ denotes the transmission coefficient of the atmospheric column and

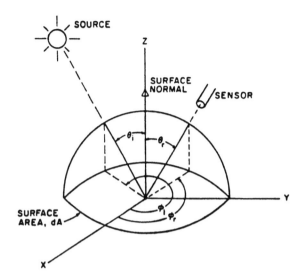

Figure 8.3   Geometrical relationships between the source, scene, and sensor.

$L_p(\lambda)$ denotes the path radiance due to scattering, then the radiant flux incident on the sensor aperture is

$$L_i(\lambda) = T_r(\lambda)L_r(\lambda) + L_p(\lambda)$$

The next step in the energy flow is to account for the sensor-related effects. This involves utilization of the electrooptical calibration information associated with the particular sensor type. Figure 8.4 shows such information for the sensor employed in our study. It shows the relationship between the amount of radiance incident on the sensor aperture and the digitized gray levels generated for a near-infrared channel of the multispectral scanner.

If we assume such linear conversion between the incident radiation and recorded gray levels in a multispectral scanner $g(\lambda)$, the gray levels recorded in the band corresponding to in wavelength λ will be given as

$$g(\lambda) = m(\lambda)L_i(\lambda) + b(\lambda)$$

where $m(\lambda)$ and $b(\lambda)$ are the scanner calibration parameters for wavelength λ.

The above equations show the interrelationships between the recorded gray level intensity values in different channels of a multispectral scanner and various parameters influencing the four components (source, scene, sensor, and medium) of a remote sensing system. The incorporation of

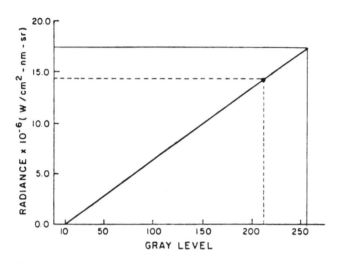

Figure 8.4 Radiometric calibration information showing conversion of incident radiance to gray levels. The plot represents the linear conversion properties governing a near-infrared channel of the multispectral scanner utilized in our study.

Figure 8.5 Incorporation of atmospheric correction models in multispectral image analysis.

appropriate models describing the nature of these components can provide estimates of the parameters appearing in the above equations.

Recently, we have demonstrated successfully the incorporation of atmospheric models in analyzing high-resolution multispectral images (Trivedi et al., 1987). We utilized models developed by the Air Force Geophysical Laboratory for correcting atmospheric absorption and scattering effects (Kneizys et al., 1983). An example of applying such correction is illustrated in Fig. 8.5. Note that the corrected image offers better contrast and more discriminatory information than the original image.

8.3   EXTRACTION OF IMAGE CUES

We have identified the image understanding task as one involving the matching of image-derived information with the scene-domain models associated with the various objects expected to appear in the scene. It is therefore important to develop image operators capable of extracting characteristics of the objects appearing in the image. Signals recorded by the remote sensing cameras (or scanners) can be analyzed to derive object characterization in spectral, spatial, temporal, and polarization domains (Wyatt, 1978). Due to the difficulties associated with the accurate sensing of polarization effects, this domain is not utilized widely. Temporal domain involves the analysis of multiple frames of images to derive motion cues, or seasonal variations.

8.3.1   Operators for Spectral Analysis

Several land cover features, both man-made and natural, have distinctive spectral characterization (Colwell, 1983). Examples are vegetation, water, snow, concrete, asphalt, and soils. If the recorded intensity values can be related directly to the intrinsic spectral properties of the object surfaces, then reliable spectral feature analysis is possible. Establishment of this correspondence is not easy. Images are 2-D representations of the 3-D scene, and therefore, such representations are ambiguous. For an isotropic (ideally diffusing) surface that is illuminated by a point source with illumination intensity I, the image intensity L can be calculated as

$$L = IR \cos \theta$$

where R represents the reflectivity of the surface and θ is the angle of incidence (Horn, 1977). This relationship shows that there are innumerable possible combinations of reflectance, illumination, and orientation that can give the same image intensity value. In the remote sensing of natural and cultural features, the problem becomes much more complex because of the

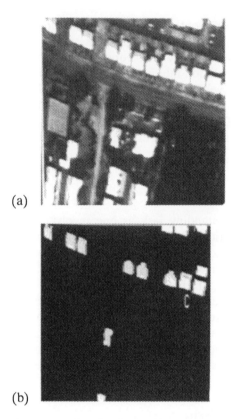

(a)

(b)

Figure 8.6 Results obtained by spectral analysis of the test image. (a) Image in the near-infrared channel. (b) Detection of small buildings. (c) Detection of carports. (d) Detection of large buildings. (e) Detection of vegetative cover.

effects of the atmospheric medium, shadows and mutual illumination, and the electrooptical properties of the sensors. The recorded intensity values are affected by all of the above factors. Clearly, for accurate interpretation of the scene composition, one would like to decouple the effects on recorded image intensities due to the intrinsic properties of the objects from other effects. Such a framework for developing a vision system is based on the hypothesis that the final interpretation task becomes more manageable if various characteristics, such as surface reflectance, illumination, range, surface orientation, etc., are examined separately. Zucker et al. (1975) Barrow and Tenenbaum (1978), and Gregory (1970) have noted that the

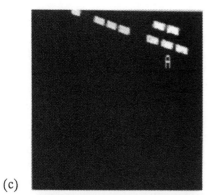

(c)

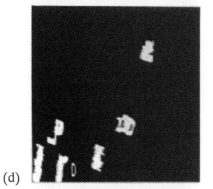

(d)

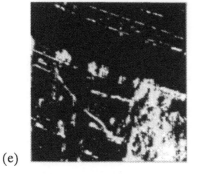

(e)

human vision system can perceive variations in incident illumination, color or reflectance, and orientation independently.

To solve the above problem, one needs appropriate models that describe the cause-effect relationship between the parameters involved in the system. The system consists of the illuminating source, scene, sensor, and the intervening medium between the source, scene, and sensor. In a passive system for reflective optical wavelengths, the source is the sun and its properties are well known. It remains to investigate the properties of the scene, sensor, and atmospheric column and their interactions. This requires reflectance/radiance models for the scene constituents, an electrooptical model for the sensor, and a model for describing the atmospheric effects. The modeling and analysis of these variables will be an important factor in building robust vision systems that operate on scenes taken under different conditions at different times.

Let us consider an example of the use of spectral analysis. Fig. 8.6 shows a high-resolution scene of a housing area acquired by a multispectral scanner. Eleven channels record intensity values for reflected scene radiation in the visible and near-infrared wavelengths and also the emitted radiation in a thermal infrared band. The processing shows the utility of spectral analysis in identifying carports, buildings, and vegetative areas in the scene. The roofing material used for the carports is different from the material used for the buildings. The analysis procedure identifies a cluster in measurement space associated with a specific object (Trivedi, 1987). Regions are then formed and subjected to simple size and shape constraints to eliminate false objects. Fig. 8.6 shows typical results.

8.3.2   Operators for Spatial Analysis

Spatial domain operators are required to extract intrinsic object characteristics such as size, shape, and texture. In a straightforward implementation, spatial analysis can be performed after completing the spectral domain analysis. Images are first analyzed to identify regions of homogeneous spectral properties and then these spectrally homogeneous regions are analyzed to extract their spatial domain characteristics. Simple shape measures such as area, principal axis, height, width, perimeter, and aspect ratio can provide important features for many applications. Extraction of shape information can be accomplished by utilizing moments or Fourier shape descriptors (Mitchell and Grogan, 1984; Dudani et al., 1977). Such descriptors are of special utility in applications involving high-resolution images of targets.

Another important spatial domain characterization quite frequently utilized in aerial image analysis is that of texture. Haralick has presented a survey of texture analysis approaches for image analysis (Haralick, 1981).

In our past research we have investigated a texture analysis operator based on the gray-level cooccurrence matrices (Harlow et al., 1986b; Conners et al., 1984, Trivedi et al., 1984). These matrices describe the second-order statistics of gray levels associated with a pair of pixels in a region. The matrices are of the $N \times N$ dimension, where N is the number of gray levels. Elements in the matrix correspond to the estimated probabilities of transition from gray level i to gray level j for a prescribed distance and orientation of the pixel pairs considered and the size and shape of the region examined. Typically, these matrices are further processed to extract a set of quantitative measures. GLC-based texture analysis has been demonstrated to be successful in analyzing complex textured images of urban areas as well as for local object detection tasks such as airplane, building, or military target detection.

Recently, we have utilized an operator for analyzing the topographical variations for object detection tasks (Trivedi et al., 1990). Topographic analysis is somewhat similar to texture analysis in which image intensity variation profiles in a selected target area are analyzed. These operators provide a robust and reliable characterization of image features into classes such as peaks, valleys, ridges, convex, concave, and planar shapes.

8.4 A FRAMEWORK FOR LOW-LEVEL SEGMENTATION OF AERIAL IMAGES

Two common approaches to segmentation are edge or boundary formation and region growing (Nevatia, 1982; Riseman and Arbib, 1977). Of these, the edge detection approach is based on recognizing properties of dissimilarities in an image, whereas region growing utilizes similarity properties. Edge detection techniques are quite sensitive to noise. Real-world images are always corrupted by noise; therefore, successful segmentation based on edge detection usually depends on higher-level semantic knowledge (Nevatia, 1975; Tang and Huang, 1980). Region growing techniques offer better noise immunity and therefore do not require as much reliance on semantic knowledge as edge detection techniques.

Considering the complex structure details appearing in an aerial image, a hierarchical segmentation framework provides several benefits (Pavlidis, 1979; and Harlow et al., 1986a). Such a formulation enables us to relate the hierarchical natures of processing and image detail (Rosenthal and Bajcsy, 1978). At lower processing levels, coarser classes appearing in the image are considered; at higher levels, finer class details are examined. For example, in the segmentation of an urban scene, classes such as commercial, residential, or agricultural would be considered at a lower level than classes such as buildings, houses, cars, etc.

The flowchart shown in Fig. 8.7 allows further explanation of the above concepts. At processing level i, measures are extracted from each of the level i regions. These regions are partitioned into c_i classes using a cluster formation algorithm. The c_i classes are selected based on the homogeneity of their perceptual appearance. Naturally, the above partitioning would directly depend on the ability of the measures employed in characterizing perceptual properties. Each region is subjected to a homogeneity test. If it is found to be nonhomogeneous (such would be the case if the region was composed of a mixture of two or more of the c_i classes), then it is split into smaller regions. These regions are processed at the next higher level to find groupings of perceptually homogeneous regions in c_{i+1} classes using the basic clustering procedure.

The concept of clustering measurement vectors into perceptually homogeneous groups has been employed by several investigators in the past (Conners et al., 1984; Trivedi et al., 1984; Bryant, 1979; Goldberg and Shlien, 1978; Nagy and Tolaba, 1972). The main advantage of this approach lies in the fact that one does not require training samples for the development of unsupervised classification of the image data. This is particularly important in the analysis of remotely sensed data where the availability of

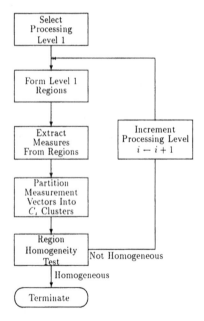

Figure 8.7   Flowchart of the steps for the low-level segmentation procedure.

reliable training samples is at best very limited. As mentioned earlier, it is important to use measures that characterize low-level perceptual properties in order for this approach to be successful. Some of the earlier studies using this approach in remote sensing had to face the practical problem of high computational burden and therefore performed a rather limited cluster analysis. With the dramatic growth in computer performance, a study by Coleman and Andrews (1979) showed that analyzing large aerial images using the k-means clustering algorithm was, in practice, quite feasible.

Two important observations about the above studies should be made: First, in all of them "hard" cluster analysis was used. That is, each element was assigned to one and only one of the c classes under examination. Second, only a single-level segmentation of the images was performed. In order to overcome these limitations, one can utilize a fuzzy cluster analysis approach (Trivedi and Bezdek, 1986). In this instead of assigning an image element to only one of the classes, we derive a membership vector that assigns a membership value, a number between 0 and 1, for each class considered. Membership values can be analyzed to identify homogeneous image elements and mixture elements. Mixture samples can later be split and the clustering process repeated at another resolution level. The specific cluster analysis algorithm that has been utilized successfully in several image analysis studies is the fuzzy c-means algorithm. This algorithm is based on the minimization of the fuzzy equivalent of the within group sum-of-squared-error objective function.

8.5 OBJECT DETECTION USING MULTIRESOLUTION ANALYSIS

One of the main tasks performed by a computer vision system is to identify the presence of a unique object in the scene. The term unique object is used in a generic manner to specify an entity with well-defined physical characteristics. Examples of specific object detection tasks are many. While analyzing aerial images of urban scenes, unique objects such as individual houses or roads may have to be detected (Harlow et al., 1986a), or in interpreting reconnaissance images, objects such as airplanes, tanks, or bridges may have to be detected (Trivedi and Harlow, 1985; Sevigny et al., 1983; Lutton and Mitchell, 1980). The development of techniques for performing such object detection tasks depends on the ability to utilize prior knowledge of various object characteristics and how these properties manifest themselves in a particular type of imagery. The object detection systems are required to perform within the accuracy and speed tolerances for a given application. Also, in many such applications these systems operate with minimal supervision.

In this section, we focus on the problem of object detection in multispectral aerial images. The conceptual formulation of our approach is similar to the one described by Tenenbaum (1973). That study represents one of the earliest works on object detection. The application domain considered in that approach is the analysis of a typical office scene. Tenenbaum recommended the utilization of multiple sensor data and recognized the importance of the system's ability to utilize knowledge about the material composition of objects and perceptual cues in their detection.

Object detection methods can be viewed as having two steps. In the first step, appropriate measurements are extracted from an image using low-level image processing techniques. These measurements are then utilized to derive a preliminary segmentation of the image into regions. In the second step, higher-level descriptors of the segmented regions are compared with the object properties stored in the scene-specific knowledge base to derive the final object location map. Utilization of scene-dependent knowledge at the latter stage is justified mainly due to the fact that processing higher-level descriptors requires significantly more computation than simple low-level operators, and also, the inferencing methods used for the verification of object presence are applied to only selected areas in the image with well specified semantic relationships easing the difficult combinatorial explosion problem. In the above formulation, the final results will, however, be quite sensitive to errors arising in the segmentation of the first step (Havens and Mackworth, 1983).

8.5.1 Segmentation Utilizing Qualitative Knowledge of the Spectral Response

The quality of segmentation results can be improved by utilizing knowledge of object properties in the above first step of object detection system. It is hoped that this will produce better-quality results, offsetting any loss of generality of the segmentation procedure. Other investigators have also recommended such use of scene-specific information in segmentation (Barrow and Tenenbaum, 1981). It will be desirable to utilize easily computable operators to drive the segmentation process. The use of intensities recorded in different spectral bands of a multispectral scanner for segmentation has been shown to be effective in several studies (Cannon et al., 1986; Trivedi et al., 1982; Swain and Davis, 1978). In our study, we incorporate knowledge of the spectral properties of the objects to be detected to improve the performance. The specification of such spectral property knowledge need not be given in absolute terms but rather *relative* terms, i.e., a certain object is brighter than the background in a particular channel and darker in some other channel. This can be illustrated using a number of specific cases encountered in a typical remote sensing application. For example, buildings are typically warmer than the background classes; therefore, regions associated with buildings in a multispectral image should have relatively higher intensity values in the thermal infrared band. Vegetation is highly reflective in the near-infrared bands; thus, regions with vegetative cover should have relatively higher intensity values in these bands. Water has very low reflectance in the visible as well as the near-infrared bands; therefore, regions having waterbodies should have relatively low intensity values in these bands. Similar rules can be developed for other object classes such as roads (concrete and asphalt) or roofing material.

The above is an important feature of the approach because in most aerial scenes, it is impossible to accurately extract absolute spectral characteristics in terms of reflectance and emittance from an image input. Also, relative specification of object properties is a more robust characterization that remains invariant over altitude, source/viewing angle, and environment factors like atmosphere affecting the image.

8.5.2   Implementing the Object Detection Approach

A practical object detection system has to satisfy requirements associated with accuracy, speed, robustness, and the extent of allowed supervision. Although one may desire a system that is very accurate, fast, robust and requires no supervision, in practice, a compromise has to be made with the often conflicting requirements. In our implementation, we address these requirements. The accuracy of the approach is mainly due to proper spectral characterization of the objects and the segmentation procedure. Speed of performance is achieved by utilizing a hierarchical approach based on processing multiresolution copies of an image. The robustness of the approach is derived by using relative spectral characteristics rather than absolute ones. And finally, supervision of the approach is minimized by employing a fuzzy-cluster-analysis-based segmentor and the general knowledge of the spectral properties of the objects.

The object detection process starts with creating and storing multiple resolution copies of the multispectral images. An important factor for implementation of a segmentation procedure is the nature of the data structure employed. Our framework employs a pyramid data structure (PDS) (Tanimoto and Pavlidis, 1975; Kelly, 1971). This structure allows one to represent the original image at various levels that differ in resolution (Fig. 8.8). We shall compute the level $k - 1$ version of the image using the level k image and the following averaging operation:

$$I_{k-1}(i,j) = \frac{[I_k(2i, 2j) + I_k(2i - 1, 2j) + I_k(2i - 1, 2j - 1) + I_k(2i, 2j - 1)]}{4}$$

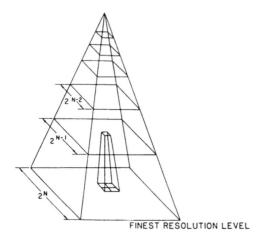

FINEST RESOLUTION LEVEL

Figure 8.8   Pyramid data structure utilized in the multiresolution image analysis.

where I_l represents the matrix storing the level l version image.

Several investigators have examined and used PDS in image analysis (Rosenthal and Bajcsy, 1978; Tanimoto and Pavlidis, 1975; Kelly, 1971; Riseman and Hanson, 1975; Rosenfeld, 1984; Burt, 1984; Pietikainen and Rosenfeld, 1981). The main advantage of PDS is the savings in computation

Figure 8.9   Satellite image (near-infrared channel) used in river detection experiment.

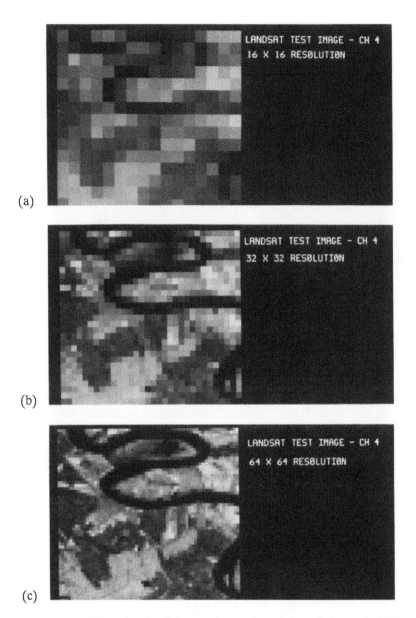

Figure 8.10   Three levels of the test image (near-infrared channel) displayed at (a) 16 × 16, (b) 32 × 32, and (c) 64 × 64 resolutions.

Figure 8.11   Results of river detection experiment.

time that can be realized. This is accomplished by using results of the analysis performed on a coarser resolution, a smaller version of the image to generate a "plan" to guide the analysis performed on the higher-resolution, larger-version image. Storage requirements for the complete pyramid are only 30% higher than for the original image (Pavlidis, 1977). Rosenthal and Bajcsy (1978) made another important observation about PDS that is of particular significance to the segmentation framework proposed. Their study showed that PDS allows one to link the conceptual hierarchy to the visual hierarchy of the image. This allows savings in search time and processing operations for the object detection type of image analysis tasks. Thus, for hierarchical segmentation PDS seems very appropriate.

We emphasize, however, that the PDS approach also has limitations. In particular, the effect of averaging may destroy some essential image details. Also, the same object will not often possess identical characterizations at two different levels. This is particularly true when one employs texture operators in the analysis. The spectral characterization of an object, on the other hand, may differ less at different levels. It is for this reason that multispectral properties are used in the experimental verification phase of our study.

The analysis begins with the segmentation of the lowest resolution level image. The segmentation is performed using a cluster-analysis-based approach (Trivedi and Bezdek, 1986). This technique employs a fuzzy clustering algorithm to generate a locally optimum partition of

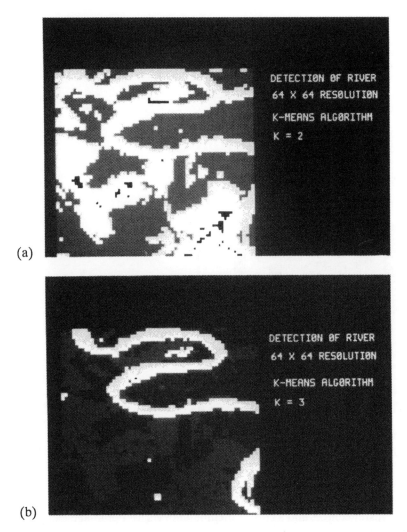

(a)

(b)

Figure 8.12 Results of applying *k*-means cluster analysis algorithm for detection of river on the 64 × 64 resolution image directly. Results for (a) two clusters and (b) three clusters.

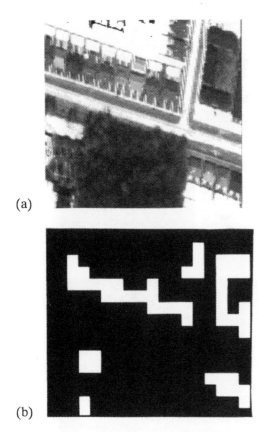

(a)

(b)

Figure 8.13 Automatic recognition of buildings appearing in an aerial image. The vision system utilizes information about the multispectral properties of the objects

the measurement vectors extracted from an image. Segmentation results include cluster centers and a cluster membership vector per image element. The components of this vector signify to what extent an image element corresponds to each one of the k clusters. The cluster centers denote prototypes of k classes in the data set.

A decision regarding which image elements need to be analyzed further is now made using prior knowledge of spectral characteristics of the objects. Examination of the cluster centers can provide guidance about which cluster center would correspond to the regions containing objects. Based on such an evaluation, only a part of the image is considered for the analysis of the next higher-level resolution image in the pyramid. The image elements of this subimage are again subjected to the above clustering-based

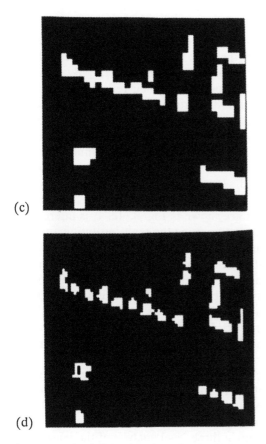

(c)

(d)

in accurately delineating them in the images. (a) The input image; (b–d) the detected buildings at 16 × 16, 32 × 32, and 64 × 64 resolutions.

segmentation and evaluation. The process is repeated until analysis with the finest resolution copy of the image is performed.

8.5.3 Experimental Verification of the Approach

In order to evaluate the utility of the above object detection approach, several experiments involving satellite and aerial images of varying spatial resolutions were conducted. Results of an experiment dealing with the detection of a river (or water bodies) is described with the help of Figs. 8.9–8.11. Figure 8.9 shows the near-infrared channel satellite (Landsat) image of the Mississippi river and its surrounding area. Three channels, visible and two near-infrared, were used in the experiment involving 16 × 16,

32 × 32, and 64 × 64 resolution copies of a section of the image. This test image appearing at three different levels of the pyramid is displayed in Fig. 8.10. Analysis begins with the 16 × 16 copy of the image and the two-class segmentation results are displayed in Fig. 8.11a. By utilizing the general knowledge about water bodies, that their reflectance in near-infrared bands is very low (Colwell, 1983), bright regions in the image were considered to correspond to the river. The gray area represents background and black indicates elements having membership values corresponding to a mixture of object and background class. The analysis continues with the 32 × 32 version image where only selected image elements, those considered to contain a river, are segmented. Results of this analysis are shown in Fig. 8.11b. Finally, the 64 × 64 version image is analyzed and the final results of the river detection experiment are shown in Fig. 8.11c. It is interesting to compare the results of river detection using the multiresolution analysis with those acquired by performing only a single-channel cluster analysis. In Fig. 8.12, we present these results. A k-means cluster analysis algorithm was utilized to find two clusters as shown in part (a). These results are obviously inaccurate and useless for the river detection task. If we attempt to identify three clusters, then the performance is improved as shown in part (b). This is still not as accurate as the final results of the multiresolution approach and, of course, single resolution analysis requires examination of a much larger number of image elements.

The robustness of the above object detection approach is evaluated by applying it to much higher-resolution aerial images. The goal was to detect objects such as buildings, carports, roads, and vegetative cover in high-resolution multispectral images. Figure 8.13a shows the thermal infrared channel image of the scene. Again, a three-level analysis of the image was utilized along with the general knowledge of the spectral properties, that buildings are brighter than the background in the thermal infrared band to detect them. Final results of the analysis are presented in Fig. 8.13b. Note that all the buildings are detected accurately. The approach was also successful in detecting other objects such as carports, roads, and vegetative areas in the scene.

8.6   CONCLUDING REMARKS

The synoptic coverage provided by aerial images is of great utility in a wide variety of applications. Progress in sensor technology has resulted in the development of quite sophisticated aerial image acquisition systems having high spectral and spatial resolutions. Imagery acquired by such sensors cannot be analyzed by the simple pixel-classification approaches. In this chapter, we discussed various issues related to the analysis of high-resolution aerial images. Specifically, we presented the matter-energy interaction process governing the image acquisition phase. We also discussed approaches for extracting information from spectral and spatial domains. Finally, we discussed segmentation and multiresolution object detection procedures. Experimental results have shown the robustness of these procedures in successful object detection.

ACKNOWLEDGMENTS

Our research in aerial image understanding has been supported by a number of funding agencies over the past several years. We gratefully acknowledge the following sources of support: Army Research Office Grant DAAG23-82-K-0189; Defense Mapping Agency Grant DMA800-85-C-0011; U. S. Army Grant DACA-39-84-C00017; and U. S. Army Grant DAAK-70-87-P2569. Interactions with Drs. Richard W. Conners and Charles A. Harlow were useful. Graduate students have contributed to this research and the assistance of Joe Besselman, ChuXin Chen, and Mark Farley is acknowledged. Janet Smith helped in the preparation of the manuscript.

REFERENCES

H. G. Barrow and J. M. Tenenbaum (1978), "Recovering intrinsic scene characteristics from images, *Computer Vision Systems* (A. R. Hanson and E. M. Riseman, eds.), Academic Press, New York, pp. 3–26.

H. Barrow and J. Tenenbaum (1981), Computational vision, *Proc. IEEE*, *69*: 572–595.

J. Bryant (1979), On the clustering of multidimensional pictorial data, *Pattern Recog.*, *11*: 115–125.

P. J. Burt (1984), The pyramid as a structure for efficient computation, in *Multiresolution Image Processing and Analysis*, Springer-Verlag, New York, pp. 6–35.

R. L. Cannon, J. V. Dave, J. C. Bezdek, and M. M. Trivedi (1986), Segmentation of a thematic mapper image using the fuzzy C-means clustering algorithm, *IEEE Trans. Geosci. and Remote Sensing*, GE-24: 400–408.

G. B. Coleman and H. C. Andrews (1979), Image segmentation by clustering, *Proc. IEEE*, *67*: 773–785.

R. Colwell (1983), *Manual of Remote Sensing*, 2nd ed., American Society of Photogrammetry, Falls Church, Va.

R. W. Conners, M. M. Trivedi, and C. A. Harlow (1984), Segmentation of a high resolution urban scene using texture operators, *Comp. Vision, Graph. and Image Proc.*, *25*: 272–310.

A. F. H. Goetz, J. B. Wellman, and W. L. Barnes (1985), Optical remote sensing of the earth, *Proc. IEEE*, *73*: 950–969.

M. Goldberg and S. Shlien (1978), A clustering scheme for multispectral images, *IEEE Trans. Syst., Man, and Cyber., SMC-8*: 86–92.

R. L. Gregory (1970), *The Intelligent Eye*, McGraw-Hill, New York.

R. M. Haralick (1981), Statistical and structural approaches to texture, *Proc. IEEE, 67*: 786–809.

C. A. Harlow, M. M. Trivedi, R. W. Conners, and D. Phillips (1986a), Scene analysis of high resolution aerial scenes, *Opt. Eng., 25*: 347–355.

C. A. Harlow, M. M. Trivedi, and R. W. Conners (1986b), Use of texture operators in image segmentation, *Opt. Eng., 25*: 1200–1206.

W. Havens and A. Mackworth (1983), Representing knowledge of the visual world, *Comp., 16*(10): 90–96.

B. Horn (1977), Understanding image intensities, *Artif. Intell., 8*: 201–231.

H. Hugli and W. Frei (1983), Understanding anisotropic reflectance in mountainous terrain, *Photogrammetric Eng. and Remote Sensing, 49*: 671–683.

M. D. Kelly (1971), Edge detection by computer using planning, *Machine Intell., 6*.

F. X. Kneizys, E. P. Shettle, W. O. Gallery, J. J. H. Chetwynd, L. W. Abreu, J. E. A. Shelby, S. A. Clough, and R. W. Fenn (1983), Atmospheric transmittance/radiance: computer code LOWTRAN 6, Tech. Rept. AFGL-TR-8-3-0187, AFGL, Hanscom AFB, Ma.

S. M. Lutton and O. R. Mitchell (1980), "Adaptive Segmentation of Unique Objects," *Proceedings of the 5th International Conference on Pattern Recognition*, Miami, Florida, pp. 548–550.

G. Nagy and J. Tolaba (1972), Nonsupervised crop classification through airborne multispectral observations, *IBM J. Res. Develop., 16*: pp. 138–153.

R. Nevatia (1976), Locating object boundaries in textured environment, *IEEE Trans. Comp., 25*: 1170–1175.

R. Nevatia (1982), *Machine Perception*, Prentice-Hall, Englewood Cliffs, N.J.

F. E. Nicodemus, J. C. Richmond, J. J. Hsie, I. W. Ginsburg, and T. Limperis (1977), Geometrical considerations and nomenclature for reflectance, *NBS Monograph 160*, National Bureau of Standards, Washington, D.C.

T. Pavlidis (1977), *Structural Pattern Recognition*, Springer-Verlag, New York.

T. Pavlidis (1979), Hierarchies in structural pattern recognition, *Proc. IEEE, 67*: 737–744.

M. Pietikainen and A. Rosenfeld (1981), Image segmentation by texture using pyramid node linking, *IEEE Trans. Syst., Man, and Cyber. SMC-11*: 822–825.

E. M. Riseman and A. R. Hanson (1975), Design of a semantically directed vision processor, Tech. Note 75C-1, Dept. of Comp. and Inform. Sci., Univ. of Massachusetts, Amherst.

E. M. Riseman and M. A. Arbib (1977), Computational techniques in visual segmentation of static scenes, *Comp. Graph. and Image Proc., 6*: 221–276.

A. Rosenfeld (1984), *Multiresolution Image Processing and Analysis*, Springer-Verlag, New York.

D. Rosenthal and R. Bajcsy (1978), "Conceptual and Visual Focussing in the Recognition Process as Induced by Queries," *Proceedings of 4th International Conference on Pattern Recognition*, pp. 417–420.

L. G. Sevigny, H. Jensen, M. Bohner, E. Ostevold, S. Grinaker, and J. Dehne (1983), Discrimination and classification of vehicles in natural scenes from thermal imagery, *Comp. Vision, Graph., and Image Proc., 24*: 229–243.

P. N. Slater (1985), Radiometric considerations in remote sensing, *Proc. IEEE, 73*: 997–1001.

P. H. Swain and S. M. Davis (1978), *Remote Sensing: The Quantitative Approach*, McGraw-Hill, New York.

J. M. Tenenbaum (1973), On locating objects by their distinguishing features in multisensory images, *Comp. Graph. and Image Proc., 2*: 308–320.

G. Y. Tang and T. S. Huang (1980), Using the creation machine to locate airplanes on aerial photos, *Pattern Recog., 12*: 431–442.

S. Tanimoto and T. Pavlidis (1975), A hierarchical data structure for picture processing, *Comp. Graph. and Image Proc. 4*: 104–119.

M. M. Trivedi, C. L. Wyatt, and D. R. Anderson (1982), A multispectral approach to remote detection of deer, *Photogrammetric Eng. and Remote Sensing, 48*: 1879–1889.

M. M. Trivedi, C. A. Harlow, R. W. Conners, and S. Goh (1984), Object detection based on gray level cooccurrence, *Comp. Vision, Graph. and Image Proc., 28*: 199–219.

M. M. Trivedi and C. A. Harlow (1985), Identification of unique objects in high resolution aerial images, *Opt. Eng. 24*: 502–506.

M. M. Trivedi and J. C. Bezdek (1986), Low-level segmentation of aerial images with fuzzy clustering, *IEEE Trans. Syst., Man, and Cyber., SMC-16*: 589–598.

M. M. Trivedi (1987), Analysis of aerial images using fuzzy clustering, *The Analysis of Fuzzy Information*, Vol. 3 (J. C. Bezdek, ed.), CRC Press, Boca Raton, Florida.

M. M. Trivedi, S. C. Zhou, and M. L. Farley (1987), "Corrections for Atmospheric Effects in Multispectral Images," *Proceedings of the Infrared Image Processing and Enhancement Conference*, Orlando, Florida.

M. M. Trivedi, C. Chen, and D. H. Cress (1990), Object detection by step-wise analysis of spectral, spatial, and topographic features, *Comp. Vision, Graph., and Image Proc.*.

R. J. Woodham and M. H. Gray (1987), An analytic method for radiometric correction of satellite multispectral scanner data, *IEEE Trans. Geosci. and Remote Sensing, GE-25*: 258–271.

C. L. Wyatt (1978), *Radiometric Calibration: Theory and Methods*, Academic Press, New York.

S. Zucker, A. Rosenfeld, and L. S. Davis (1975), "General Purpose Models: Expectations About the Unexpected," Vol. 716, *Proceedings 4th International Conference on Artificial Intelligence*.

9

Image Formation and Characterization for Three-Dimensional Vision

Dragana Brzakovic

University of Tennessee
Knoxville, Tennessee

Rafael C. Gonzalez

Perceptics Corporation
Knoxville, Tennessee

9.1 INTRODUCTION

Complex machine-vision tasks, such as autonomous vehicle guidance, automation of industrial processes, and object inspection, require the ability to deal with inherently three-dimensional environments. Understanding the structure of a 3-D environment requires assessing the position of objects and identifying the objects. In robot guidance, for example, this understanding provides for successful obstacle avoidance and the acquisition of parts from unknown environments. 3-D scene understanding is also essential for the inspection of surface finishes and the detection of missing or incorrectly placed parts. In principle, using 3-D vision systems in industrial processes leads to production cost savings, increased quality control, consistent performance, and 100% inspection, and, more importantly, provides for the automation of dangerous tasks in hazardous environments.

This work was supported in part by Martin Marietta Aerospace, Orlando, Florida.

Two types of vision system, *active* and *passive*, are used to obtain information about the 3-D environment. Active systems obtain images of a 3-D scene by projecting energy (e.g., light or ultrasound) onto a scene. The acquired images contain explicit distance between points in a 3-D scene and a known point such as the robot's location. Theoretically, emissions of various types of energy can be used to generate depth images. Examples of possible sources of energy are radio waves and ultrasound. However, the methods associated with these energies do not yield sufficient resolution for most applications of interest. Lasers, on the other hand, have found wide a range of applications in active 3-D vision systems. Examples are pulsed-mode (Lewis and Johnston/1977) and modulated continuous range finders (Svetkoff et al./1984). Lasers have also been used extensively in structured-light methods that are described in detail later in this chapter.

Passive vision systems infer 3-D information about scenes from images acquired in the visible region of the electromagnetic energy spectrum. These systems infer information about the depth of a scene by analyzing images in a manner similar to that used in human vision. This suggests that a fruitful research direction is to study functional similarities between camera-based and human vision systems. Research in the psychology of visual perception has provided valuable insight in the development of computer vision systems for 3-D image interpretation. In particular, Gibson's studies (Gibson/1960) have stimulated the development of computer-based theories of 3-D scene perception. Gibson noted 23 distinct cues that play important roles in determining a scene's structure. Although the usefulness of some of these cues is still controversial, research in the psychology of visual perception has established that binocular vision, shading, and texture are the most important cues. For a recent study on the importance and incorporation of these cues, see (Cavanagh/1987).

While interpretation of depth images acquired by active systems is an easier task, the scope applicability of passive systems is wider. Active systems have found application primarily in industry and are used in tasks such as servicing machine tools (Shneier et al./1986). In contrast to active 3-D vision systems, passive 3-D vision systems are applicable to many types of environments and do not require a controlled environment. Moreover, passive systems can be used in tasks that require that the vision system not be detected; e.g., military surveillance. Passive 3-D vision systems have been used extensively for interpreting aerial images (Case/1981, Konecny and Pape/1981), autonomous vehicle guidance (Moravec/1981, Hannah/1980), and robotics (Baylou et al./1984).

The first part of this chapter (Secs. 9.2 and 9.3) deals with issues in image formation and characterization for 3-D vision, including methods of data acquisition, and data segmentation and examples of particular applications.

The second part of the chapter (Sec. 9.4) presents a detailed discussion of a 3-D vision system that infers depth from texture and binocular views. We also discuss the application of this system to problems in obstacle avoidance for autonomous navigation.

9.2 CAMERA MODELS

The first step in designing any camera-based 3-D vision system is to model the image-formation process. Most frequently, cameras are modeled by pinhole lenses. The image-formation process is modeled by a *perspective* geometric projection, wherein the projection of a point P (in a 3-D scene) is defined by the intersection of a projection plane (i.e., the image plane) with a projecting ray emanating from the center of the projection and passing through P. The other type of planar geometric projection is the *orthographic projection*, which is applied to modeling image formation wherein the distance between the projection plane and the center of the projection is infinite. Perspective projection is a more general model of image formation and includes orthographic projection as well, while orthographic projection is limited to modeling image-formation processes, such as aerial photography, where the size of the objects in the 3-D scene is small compared to the distance between the objects and the image plane.

A perspective-projection model of particular importance in 3-D vision applications is one defined in an *image-based coordinate system* (Fig. 9.1). In this system, the coordinate axes are chosen such that the image plane, I, coincides with the $z = 0$ plane, the origin of the coordinate system coincides with the center of I, and the center of projection is located at $(0, 0, \lambda)$; therefore, a point $P(X, Y, Z)$ in a 3-D scene projects onto I as $p(x', y', 0)$,

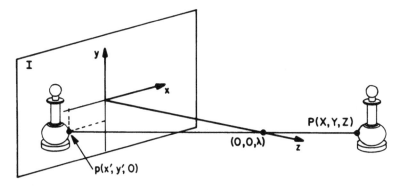

Figure 9.1   Perspective projection in the image-based coordinate system.

where

$$x' = \frac{\lambda X}{\lambda - Z},\tag{9.1}$$

$$y' = \frac{\lambda Y}{\lambda - Z},\tag{9.2}$$

and λ is the distance between the projection plane and the center of projection. When λ becomes very large (i.e., $\lambda \to \infty$), this perspective-projection model reduces to orthographic projection. From Eqs. (9.1) and (9.2), it then follows that, for orthographic projection, $x' = X$ and $y' = Y$.

Camera-based methods usually rely on knowing the set of parameters that specify the relationships between points in a 3-D space and their projections. The most frequently required parameter is the position of the center of projection; i.e., the focal length. Additional parameters are the camera offsets and angles of pan and tilt. The procedure for determining such parameters is referred to as *camera calibration*. Calibration procedures generally obtain camera parameters by using projections of points whose 3-D coordinates are known and solving the resulting projection equations that model the image-formation process. When the calibration parameters need to be known accurately, the differences between the projection models and the corresponding physical camera systems can be taken into account by performing distortion-calibration procedures; see (Moravec/1979) for details. The mathematical aspects of calibration procedures are described by Fu et al./1987, and many practical aspects are discussed by (Yakimovsky and Cunningham/1978 and Tsai/1987).

9.3   ACQUISITION AND INTERPRETATION OF 3-D DATA

While the technology of acquiring images differs for various systems, the corresponding methods share the common problem of coping with many-to-one mappings between the scenes and the acquired images. Usually, a scene is interpreted by combining signal-processing and artificial-intelligence techniques. In particular, knowledge about objects likely to be encountered in a scene greatly enhances the vision system's performance. Vision systems that perform a priori known tasks in controlled environments generally function accurately and at high speed; e.g., in assembly-line inspection. On the other hand, when the structure of a scene is poorly known (e.g., robot guidance in an unknown environment), the performance and speed of present-day 3-D vision systems are often inadequate.

The remainder of this section describes the most commonly used 3-D vision methods. It is important to note that some of these methods lead to depth measurements based on a monocular view of a scene. By contrast, other methods lead to depth measurements based on multiple views of a scene. There are many difficulties associated with generating a 3-D scene description from a single 2-D image. Information available in a single image, even with some a priori knowledge about the scene, is generally not sufficient to obtain complete 3-D information. Some methods, such as *shape-from-skewed-symmetry methods*, employ basic rules of projective geometry to perform recognition of specific 3-D objects (Kanade/1981, Mulgaonkar et al./1986). These methods do not use a priori knowledge about the objects in the scene, other than a class of surfaces that constitute the scene and some possible spatial relationships between the objects. More robust 3-D vision systems employ cues other than perspective geometry in inferring a 3-D scene description. Two particularly interesting classes of methods employ shading and texture information in 3-D scene understanding.

Also, some methods (e.g., binocular vision) give the actual 3-D coordinates of individual points, while others (e.g., methods of shading and texture) give only the position of points relative to each other. The methods that yield relative depth usually describe objects in terms of their surfaces and the orientations of these surfaces relative to the viewer (i.e., the camera). Most frequently, the orientation of the surface $S : z = f(x,y)$ is described in terms of the surface normal, $\mathbf{N}$. The surface normal is most commonly given in the *gradient space* described by

$$p = \frac{\partial f(x,y)}{\partial x}, \tag{9.3}$$

and

$$q = \frac{\partial f(x,y)}{\partial y}, \tag{9.4}$$

so that $\mathbf{N} = (p,q,-1)$, as shown in Fig. 9.2. An alternative description, used in methods that use textural information, is to describe the surface normal in terms of the surface slant, σ, and the surface tilt, τ. In an image-based coordinate system, σ is the angle between the surface, S, and the image, I, and τ is the angle between the projection of the surface normal onto I and the x-axis in I (see Fig. 9.2). Note that σ and τ are related to the parameters of the gradient space via the formulas

$$\sigma = \tan^{-1} \sqrt{p^2 + q^2} \tag{9.5}$$

and

$$\tau = \tan^{-1} \frac{q}{p}. \tag{9.6}$$

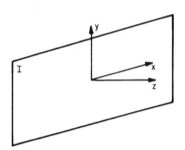 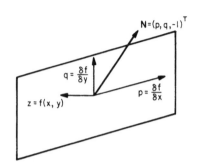

Figure 9.2   Parameters of the gradient space and their relationship to surface tilt and slant.

9.3.1   Structured-Light Systems

Structured-light methods obtain the coordinates of points in a scene by measuring the coordinates of their projections onto the image plane and by using the information about the scene illumination. These methods operate only on those points in the scene that are illuminated by, for example, a stripe of light. This stripe may be produced by a laser or by projecting a beam of light through a slit. The applicability of using projection through a slit is more restrictive than using a laser, because the image of a projected stripe must be obtained in a darkened room. On the other hand, lasers can be used primarily in environments where appropriate safety precautions are exercised.

Various structured-light systems have been described by (Shirai/1972, Popplestone et al./1975, Nitzan et al./1977), and, more recently, by (Oshima and Shirai/1983). A typical structured-light system contains a light projector and a TV camera that picks up the reflected light (see Fig. 9.3). A mirror is rotated so that the light plane covers many points in the field of view. Acquisition of depth images in this manner is generally very slow. The process may be accelerated by projecting multiple beams of light. This approach is faced with the problem of determining the correspondence between the projected stripes of light and their images, since the positions of the projected stripes (relative to each other) may change when projected onto the scene. The problem of establishing this correspondence, also known as *indexing*, has been overcome recently in an elegant way by projecting a color grid and identifying the elements of the grid based on their color (Boyer and Kak/1987). This method has been successfully applied to 3-D scene understanding wherein the color content of the scene is predominantly neutral.

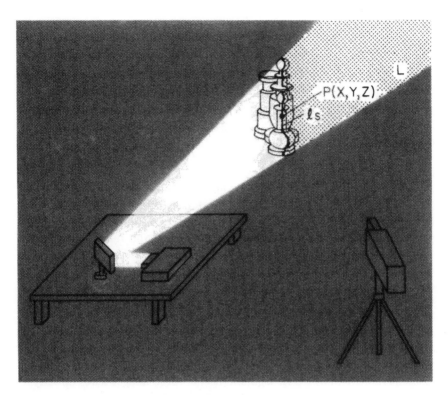

Figure 9.3 A structured-light system.

Structured-light methods use the equation of a known light plane, L, to determine the coordinates of a point $P(X,Y,Z)$ on the illuminated segment, l_s (Fig. 9.3). The equation of the projecting ray [determined by P and the center of projection $c(0,0,\lambda)$] is

$$\frac{x-x'}{-x'} = \frac{y-y'}{-y'} = \frac{z}{\lambda}. \tag{9.7}$$

Since point P lies in the plane L with equation,

$$Ax + By + Cz + 1 = 0, \tag{9.8}$$

it follows from Eqs. (9.7) and (9.8) that the coordinates of P are

$$X = \frac{-x' - C\lambda}{Ax' + By' - C\lambda}, \tag{9.9}$$

$$Y = \frac{-y' - C\lambda}{Ax' + By' - C\lambda}, \tag{9.10}$$

and

$$Z = \frac{\lambda(Ax' + By' + 1)}{Ax' + By' - C\lambda}. \tag{9.11}$$

Based on Eqs. (9.9) through (9.11), determination of the coordinates of P requires knowledge of the parameters A, B, and C. Therefore, structured-light systems must be calibrated to obtain these parameters. Details of calibration procedures are described by (Mensbach/1986).

The resulting depth images usually need further processing; in particular, it is necessary to segment them into individual objects which, in turn, requires grouping of data points into geometrical structures such as planes (Henderson/1983), cylindrical surfaces (Hebert and Ponce/1982), or junctions (Sugihara/1979). (Different aspects of surface representation for the purpose of object detection are analyzed by [Besl and Jain/1986].) In the cases where multiple stripes or patterns are projected onto a scene, the images can be treated by methods that analyze scenes whose surfaces are characterized by known texture patterns. Therefore, methods (such as those described in Sec. 9.3.4) that use textural information to deduce scene layout can be applied to the images obtained by projecting particular illumination patterns. Furthermore, these images are easy to segment, since the discontinuities in the image pattern correspond to changes in surface orientation or abrupt distance changes between the objects in the scene and the image plane. In many cases, segmentation of these images leads directly to object recognition; for example, see (Tomita and Kanade/1984).

9.3.2 Laser-Ranging Systems

Temporal and spatial coherence of laser beams has led to laser usage in active range finders known as *time-of-flight* range finders. Two types of systems, pulsed-laser systems and continuous-beam lasers, are presently in use.

Pulsed-laser systems obtain range measurements by measuring the time between emitting a pulse of light and its reflection from the reflecting surface. Given distance, D, between the source and the reflecting surface and assuming that pulse travels at speed of light, c, by measuring time of flight, T, one obtains

$$D = \frac{cT}{2}. \tag{9.12}$$

Generally, pulsed-laser systems can be used for long range, up to 20 km; however, they are frequently slow and provide poor precision (Monchaud/1986).

Continuous-beam lasers measure the phase shift between the emitted and returned beams. Referring to Fig. 9.4, and assuming that the beam of laser light of wavelength λ is split into two beams that travel distances L and D, respectively, it is apparent that the reflected beam travels distance $D' = L + 2D$ when it arrives at the phase measurement. Consequently, the phase shift ϕ is related to traveled distances by

$$D' = L + \frac{\phi}{360}\lambda. \tag{9.13}$$

From Eq. (9.13), it follows that the unique solution exists only when $\phi < 360°$ or equivalently, $2D < \lambda$. Due to the very small wavelength of lasers, in practical application laser light is usually modulated by using a waveform of much higher wavelength; for details, see (Fu et al./1987). The basic advantage of continuous-beam lasers over pulsed-light techniques is that they provide both intensity and range information. However, they require considerably higher power. Special characteristics of particular laser-range systems are detailed by (Fu et al./1987, Besl and Jain/1985 and Monchaud/1986).

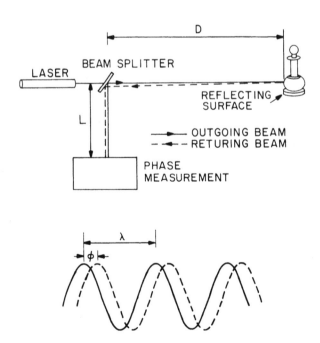

Figure 9.4   Range measurement by phase shift.

9.3.3   3-D Object Shape from Shading

The importance of shading visible surfaces in creating realistic images is well understood in art and computer graphics. Shading reveals the three-dimensionality of objects, the omission of which would result in ambiguous shape information. Shading is determined by the reflectance laws of physics; it is a function of illumination, the reflecting characteristics of the surface, and the material. In computer vision, the reflectance laws of physics are fundamental to the development of *shape-from-shading methods*. These laws are used to deduce the normal of any surface in the scene and to implicitly deduce the scene's relative depth from image intensities. The basic assumption underlying these methods is that the imaging device produces an image in which intensities, $i(x,y)$, are directly proportional to the reflectance function ϕ. The reflectance function of a perfectly diffuse surface, known as a Lambertian surface, is modeled as

$$\phi_l = \rho \cos(\eta), \tag{9.14}$$

where ρ is the *reflectance factor*, which accounts for the reflectance properties of the surface's material, and η is the incident angle of illumination (Fig. 9.5). The presence of specular reflection is accounted for by incorporating the emergent angle e into the reflectance function, which, in this case, becomes

$$\phi_s = \frac{\rho \cos(\eta)}{\cos(e)}. \tag{9.15}$$

In the gradient space, the quantities $\cos i$ and $\cos e$ are given by the following normalized dot products:

$$\cos \eta = \frac{\mathbf{N} \cdot \mathbf{S}}{\|\mathbf{N}\|\|\mathbf{S}\|} \tag{9.16}$$

and

$$\cos e = \frac{\mathbf{N} \cdot \mathbf{v}}{\|\mathbf{N}\|\|\mathbf{v}\|}, \tag{9.17}$$

respectively, where $\mathbf{N} = (p,q,-1)$, $\mathbf{S} = (p_s,q_s,-1)$ is a vector pointing in the direction of the light source, and $\mathbf{v} = (0,0,1)$ is a vector pointing in the direction of the viewer. In most cases, the reflectance function incorporates both ϕ_l and ϕ_s and the phase angle (see Fig. 9.5); for complete models of the reflectance function, see (Bui-Tuong/1975).

The problem of obtaining a 3-D scene description from shading is usually simplified by considering scenes in which objects are small compared to the viewing distance. This simplification arises from using an orthographic projection to model the image-formation process. In orthographic projection,

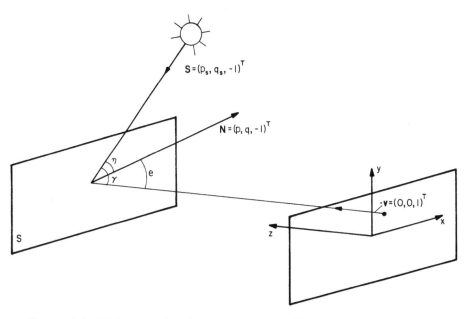

Figure 9.5 Reflectance-function nomenclature. Symbol γ denotes the phase angle.

the viewing direction is constant, so the phase angle is also constant for all surface elements. In this case, Eqs. (9.14) and (9.15) become

$$\phi_l = \frac{\rho(1 + pp_s + qq_s)}{\sqrt{1 + p^2 + q^2}\sqrt{1 + p_s^2 + q_s^2}}, \tag{9.18}$$

and

$$\phi_s = \frac{\rho(1 + pp_s + qq_s)}{\sqrt{1 + p_s^2 + q_s^2}}, \tag{9.19}$$

respectively. The right sides of Eqs. (9.18) and (9.19) are called the *reflectance maps* and are usually denoted by $R(p,q)$. A reflectance map can be represented graphically in the gradient space by contours of constant brightness. The contours can be obtained either experimentally, by directly measuring the reflectance of the object under consideration, or theoretically, if the surface reflectance is known as a function of angles η and e; for details, see (Horn/1975). The equation in which ϕ is replaced by image intensities; i.e., an equation of the form

$$i(x,y) = R(p,q), \tag{9.20}$$

is called the *image irradiance equation.*

In general, the surface normal cannot be determined from the intensities of an image of the surface, because one needs to solve a nonlinear first-order partial-differential equation (PDE) with two unknowns of the type shown in Eqs. (9.18) and (9.19). The additional equation needed to make this problem well-posed is usually obtained by imposing a constraint on the geometrical properties of the surface. Horn/1975 has imposed a surface smoothness constraint to solve the shape-from-shading problem by the modified characteristic strip-expansion method, wherein the surface-smoothness equation and the PDE [i.e., Eq. (9.18) or (9.19)] are replaced by five ordinary differential equations for $x, y, z, p,$ and q. (The class of image-irradiance equations for which this method works has been investigated by [Bruss/1981].) Ikeuchi and Horn/1981 have reformulated the original problem addressed by the shape-from-shading method as a problem of finding a surface orientation that minimizes the integral of the brightness error by employing calculus of variations to obtain the normal of a surface from its image intensities (Ikeuchi and Horn/1981, Horn and Brooks/ 1986). An alternative approach to obtaining information about a 3-D scene from image intensities has been proposed by (Pentland/1984), who has inferred various properties of scenes by considering image regions instead of individual image points and by estimating the average value of measured image properties.

Shape-from-shading methods require precise knowledge about light sources and are limited to a controlled environment. Nevertheless, these methods are still applicable to cases when this knowledge is not available, although significant errors may occur in such cases. For example, (Ikeuchi and Horn/1981) report that an error of 7.5° in estimating the source direction may lead to a 20% error in the measurement of surface orientation.

9.3.4   3-D Object Shape from Texture

In many cases, objects in an image are perceived as regions having distinct texture properties, rather than as clearly defined regions of contrasting brightness. The term *texture* usually refers to a large number of alike patterns, called *texels* (*texture elements*), densely and evenly arranged over a field of view. Mathematically, the modeling of texture relies on the theory of regular figures (Zucker/1976), statistics (Haralick/1979), and structural-pattern recognition (Lu and Fu/1978). Texture features serve as a basis for classifying parts of an image and are essential in many image-processing tasks. Consequently, understanding texture features has attracted considerable attention in recent years; see (Rosenfeld/1986) for references.

The hypothesis that texture is the primary cue of visual space perception by humans (Gibson/1960) has led to a number of computational theories relating texture to scene description. The common theme of these studies, collectively known as *shape-from-texture methods*, is that surface orientation (and, implicitly, relative depth) may be obtained by measuring changes in the shape, size, and spacing of texels. Shape-from-texture methods are important for surface analysis and, in particular, for determining the relative distances from obstacles for autonomous vehicle guidance in unknown environments. An example of such an application is described in a greater detail in Sec. 9.4.

Shape-from-texture methods describe a scene in terms of planar surfaces and their orientations relative to the image plane. Surface orientation is most frequently described by the surface tilt, τ, and the surface slant, σ. Curved surfaces are approximated locally by planes. (An example of a 3-D scene description from texture is shown in Fig. 9.6.) Information available in a single image, without a priori knowledge about the scene, is not sufficient to measure accurately both τ and σ. Usually, shape-from-texture methods can yield fairly accurate measurements of the surface tilt (by statistically averaging a number of measurements), but yield less accurate measurements of surface slant (Kender/1980 and Witkin/1981).

 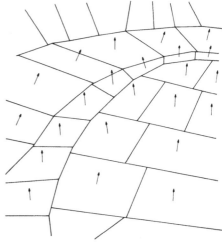

Figure 9.6 An example of a 3-D scene description obtained from texture; surfaces' orientations are represented by surfaces' normals.

Observable changes in texture in an image, I, that relate to orientation of S are termed *scaling* and *foreshortening*. Scaling describes the fact that texels in I appear smaller as the surface recedes from the image plane; i.e., the viewer. Foreshortening describes the compression of texture elements in I in the direction of τ; for this reason, this direction is usually called the *direction-of-texture gradient*. The choice of the model of image-formation process determines whether measurements of scaling or foreshortening (or both) are necessary to determine the parameters σ and τ.

Models using orthographic projection rely on foreshortening. Witkin/ 1981 designed a maximum-likelihood estimator of surface orientation by assuming that changes in orientation of texel edges arise from foreshortening and measuring these changes. Models using perspective projection obtain a 3-D scene description by using either knowledge-based or model-free methods. Kender/1980 designed a knowledge-based method that uses the Normalized Texture Property Map. This map encodes observed effects of surface orientation on a particular texture property. The strategy underlying Kender's method is similar to that of methods that deduce 3-D information from shading.

Model-free methods capitalize on the fact that texture regularity remains preserved (in an image) along the lines parallel to the surface S (Stevens/ 1980, Brzakovic and Tou/1984). These lines are called the *characteristic lines*. Note that the characteristic lines are orthogonal to the direction-of-texture gradient. The angle that a characteristic line in an image forms with the x-axis, measured counterclockwise, is the *characteristic angle* (for an illustration of these terms, see Fig. 9.7). Typically, measuring τ requires the determination of the characteristic angle or, equivalently, the direction in which texture regularity is preserved. Once the characteristic angle is determined, surface slant can be estimated by measuring changes of texture elements along the direction-of-texture gradient. The dominant orientations of the patterns can be measured using standard texture-analysis methods, such as power-spectrum analysis (Bajscy/1973), co-occurrence matrices (Haralick/1979), or estimation of anisotropy (Kass and Witkin/ 1987).

9.3.5   Measurements of Depth from Binocular Views

Studies on the psychology of binocular vision, such as that of (Julesz/1970), have motivated research in computer vision dealing with the development of methods for image understanding using two views of a scene. Such methods belong to the area known as *stereo vision*. Note that the origin of the word stereo is $\sigma\tau\epsilon\rho\epsilon o\varsigma$, which means "solid" in Greek; thus, the term stereo is not necessarily limited to two images, but may imply 3-

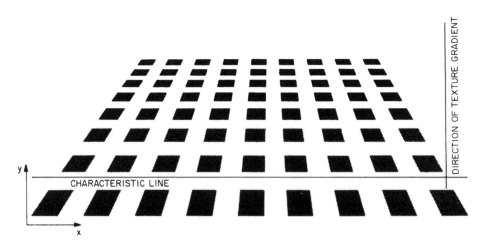

Figure 9.7   Shape-from-texture nomenclature.

D vision in general. This distinction has been made by authors such as (Haddow et al./1985), who refer to two views of a scene as stereoscopic binocular vision rather than just stereo vision. Theoretically, two views of a scene provide exact 3-D coordinates of individual points for static scenes; therefore, methods that rely on two views possess an intrinsic theoretical advantage over methods that rely on a single view. 3-D vision systems that employ binocular views have found a wide range of applications from analysis of aerial imaging (Case/1981) to robotic vision (Gennery/1980).

The objective of binocular-vision systems is to determine the coordinates of a point P in a 3-D scene from its projections p_1 and p_2 onto image planes I_1 and I_2, respectively (see Fig. 9.8). Typically, this is achieved by computational methods that include four major steps: (1) camera modeling, (2) feature acquisition, (3) image matching, and (4) determination of 3-D coordinates. The basic ideas underlying each of these steps are described below; details and a more complete list of references can be found in (Barnard and Fishler/1982).

The key to achieving binocular vision is to establish that p_1 and p_2 are the projections of the *same* point P onto I_1 and I_2, respectively. Establishing this fact is referred to as solving the *correspondence problem* or *image matching*. The difference between the positions of the two points p_1 and p_2, both originating from point P, is known as *parallax* or *disparity*. Parallax is a function not only of the location of P, but also of the characteristics of the two cameras that generate I_1 and I_2. Cameras are generally modeled as pinhole lenses. Thus, the image-formation process is modeled using the

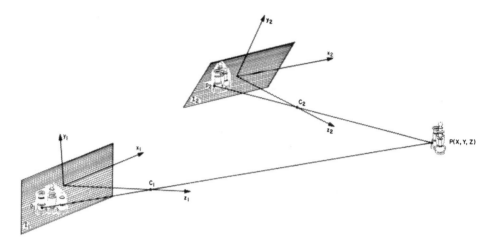

Figure 9.8   Example of an image-plane arrangement for binocular vision.

methods of perspective projection. More refined camera models usually correct for various distortions by using compensation methods (Moravec/ 1979).

The coordinates of P are expressed in terms of its projections p_1 and p_2 by introducing two image-based coordinate systems (see Fig. 9.8), related by

$$(x_2, y_2, z_2) = (x_1, y_1, z_1) \cdot R + (x_T, y_T, z_T), \qquad (9.21)$$

where

$$R = \begin{bmatrix} \alpha_{11} & \alpha_{12} & \alpha_{13} \\ \alpha_{21} & \alpha_{22} & \alpha_{23} \\ \alpha_{31} & \alpha_{32} & \alpha_{33} \end{bmatrix} \qquad (9.22)$$

is a rotation matrix, whose elements are the direction cosines of the x_2-, y_2-, and z_2-axes relative to the x_1-, y_1-, and z_1-axes, respectively. The subscript T in Eq. (9.21) denotes translation parameters. The coordinates of P may be expressed either in an image-based coordinate system or in a global coordinate system whose position relative to the origins of (x_1, y_1, z_1) and (x_2, y_2, z_2) is known.

Specific camera arrangements can simplify the correspondence problem; however, this problem is difficult to solve in general. In particular, occluded boundaries, optical illusions, and shadows pose major difficulties and may cause the correspondence problem to be ill-posed (Goshtasby/1984). The most commonly used camera arrangement involves two image planes that are parallel, in which case the matrix R becomes a unitary matrix; i.e.,

$$R = \begin{bmatrix} 1 & 0 & 0 \\ 0 & 1 & 0 \\ 0 & 0 & 1 \end{bmatrix}. \tag{9.23}$$

Furthermore, if $y_T = 0$ and $z_T = 0$, a point $p_1 = (x_1',y_1')$ in I_1 has its correspondent $p_2 = (x_2',y_2')$ on the line $y_1' = y_2'$ in I_2. The camera arrangement described above has been the subject of studies by (Marr and Poggio/1979, Grimson/1980, and Goshtasby/1984). Other camera arrangements are possible. A particularly interesting camera arrangement involves a single camera moving along the z-axis; this arrangement is suitable for robotics applications (Itoh et al./1984, O'Brien and Jain/1984, Alvertos et al./1986). Sec. 9.4.2 describes in greater detail the advantages of this camera arrangement.

As previously mentioned, the second computational step toward achieving the objective of binocular vision is feature acquisition. The function of this step is to increase the likelihood of successful matches by considering only image points with unique properties. Usually, these points are nontexture pixels with high neighborhood variances (Barnard and Fishler/ 1982). This selection process is accomplished by specially designed operators that identify pixels that have high variance (Moravec/1981, Hannah/ 1980). An approach to feature acquisition involves edge detection, a popular implementation of which is based on choosing zero-crossings as feature points (Marr and Poggio/1979, Grimson/1980). A recent study by (Clark and Lawrence/1986), however, indicates that the use of the differences in spatial-frequency contents of images leads to more accurate results than does the use of zero-crossings.

The objective of the image-matching computational step is to establish the correspondence between a selected point p_1 in I_1 and a point in a particular region of I_2 (as determined by the camera arrangement) that has characteristics similar to p_1. Two general methodologies, *correlative* and *combinatorial*, are employed in image matching. The correlative methodologies use area correlation to establish when two points p_1 and p_2 originate from the same point P; area correlation is suitable for textured images in which the disparity between I_1 and I_2 is very small (Lucas and Kanade/ 1981, Moravec/1979). By contrast, combinatorial methodologies use distinct properties of selected points to find the correspondence between I_1 and I_2. The most commonly used properties are: sign and magnitude of intensity change (Marr and Poggio/1979, Grimson/1980), edges (Baker/1980, Baker and Binford/1981), and chain envelopes about edge contours (Bellugi et al./1985). Some combinatorial methods capitalize on a priori knowledge about the scene; e.g., knowledge about object shape (Baylou et al./1984) or probability distribution of particular features (Gennery/1981).

The number of erroneous matches can be reduced by adding a third camera (Bellugi et al./1985, Ito and Ishii/1986) or by excluding shadows and occluding points from the matching procedure (Goshtasby/1984). The accuracy in establishing the correspondence between p_1 and p_2 can be increased by considering multiple views and statistically averaging matches between different pairs of images (Moravec/1979).

When the camera parameters are known, determination of the 3-D co-ordinates of a point is straightforward; see (Yakimovsky and Cunningham/ 1978) for details on camera calibration. As an example, we consider parallel image planes and $y_T = 0$ and $z_T = 0$; then,

$$X = \frac{x_T x_1'}{x_2' + \lambda - x_1'}, \tag{9.24}$$

$$Y = \frac{y_1' X}{x_1'}, \tag{9.25}$$

and

$$Z = \lambda(1 - \frac{X}{x_1'}). \tag{9.26}$$

When the cameras are mounted on a dynamic system and their positions are not exactly known, initial control points (Hannah/1980) are used to simultaneously calculate the coordinates of P and the camera parameters.

9.3.6   Incorporation of Binocular Vision with Shading and Texture

Early attempts to obtain 3-D scene descriptions from images were confined to rigidly constrained environments and could not accommodate informa-tion about more than one cue at a time; e.g., texture, shading. More recent research focuses on incorporating multiple sources of information regarding the scene layout. In general, the availability of two or more cues simplifies the problem of scene understanding and increases the accuracy of vision results. A class of such methods is aimed at dynamic scene under-standing and determining motion within the scene; for example, see (Jenkin and Tsotsos/1986, Mitiche/1984). The following material briefly describes two other types of methods that incorporate binocular vision with shading and texture.

The first method is the *photometric stereo method* (Woodham/1981, Colleman and Jain/1982). This method uses information from two or more images generated by varying the incident illumination and preserving the same viewpoint. Photometric stereo is best suited for determining the position and orientation of objects in a controlled environment. By preserving the same imaging geometry, this method overcomes the difficulties associated with the correspondence problem; i.e., the pixel $p_i(x_0,y_0)$ in image I_i and the pixel $p_j(x_0,y_0)$ in image I_j are unambiguously known to correspond to the same point $P(X,Y,Z)$ in the scene under consideration. Thus, the objective of the photometric stereo method is to measure the unit surface normal $N = (N_x, N_y, N_z)$ using image intensities $i_j(x_0,y_0) \, j = 1, 2, \ldots$.

Given three known sources of illumination S_1, S_2, S_3 and the corresponding intensity values i_1, i_2, i_3, the three unknown components of the surface normal are determined by using the irradiance equations

$$i_j(x,y) = \rho\cos(N, S_i), \qquad j = 1, 2, 3, \tag{9.27}$$

where the parameter ρ is the reflectance factor as described in Sec. 9.3.3. By rewriting these equations in matrix form, the surface normal can be expressed as

$$N = \frac{1}{\rho}[M]^{-1}V, \tag{9.28}$$

where M denotes the source matrix, whose rows contain the components of S_1, S_2, and S_3, and V is the vector whose elements are intensities i_j. Equation (9.28) shows that by using the photometric stereo method, the surface normal can be measured directly from the image intensities.

The second method discussed in this section is the *texture stereo method*. This method obtains accurate measurements of the orientations of various texture surfaces in a scene by taking two views of the scene, obtained by rotating the camera between the two exposures. The two image-based coordinate systems are related by the rotational matrix R (Eq. [9.21]). The various surface orientations are then inferred solely from the geometry of the perspective projection underlying the model of the image-formation process. Just like the photometric stereo method, the texture stereo method avoids the need for the complicated and computationally expensive matching procedures required to solve the correspondence problem; for details, see (Brzakovic and White/1986).

The texture stereo method capitalizes on the fact that the characteristic lines in both images are parallel to the surface S (see Sec. 9.3.4). Therefore, two families of characteristic lines uniquely determine the components N_x, N_y, N_z of the surface normal N, giving

$$N_x = \sin\Omega_1(\alpha_{13}\cos\Omega_2 + \alpha_{23}\sin\Omega_2), \tag{9.29}$$

$$N_y = -\cos\Omega_1(\alpha_{13}\cos\Omega_2 + \alpha_{23}\sin\Omega_2), \tag{9.30}$$

and

$$N_z = \cos \Omega_1 (\alpha_{21} \cos \Omega_2 + \alpha_{22} \sin \Omega_2)$$
$$- \sin \Omega_1 (\alpha_{11} \cos \Omega_2 + \alpha_{21} \sin \Omega_2); \quad (9.31)$$

where Ω_1 and Ω_2 are the characteristic angles and α_{ij} are the direction cosines that relate the two image-based coordinate systems. From Eqs. (9.29)–(9.31), it follows that the components of the surface normal can be determined by measuring the characteristic angles in textured regions in the two images. In turn, the characteristic angles can be measured by comparing the Fourier power spectrum of the texture regions of interest (Brzakovic and White/1986).

9.4   A 3-D VISION SYSTEM FOR OBSTACLE AVOIDANCE

This section describes a 3-D vision system that combines binocular vision with the shape-from-texture method to determine the distances to objects in a scene. Specifically, the system is used to determine the distances to obstacles placed on planar textured surfaces, such as floors and walls. These distances are important in tasks such as guidance of autonomous vehicles in unknown environments. The task of obstacle avoidance in other environments has been considered by, among others, (Olin et al./1987 and Trendl and Kriegman/1987). The 3-D vision system described here analyzes two images of a scene acquired by moving the camera toward the scene between the two exposures. The analysis is performed in two steps. First, the orientations of planar textured surfaces are estimated using textural information in one of the acquired images. This step provides information on the relative positions of obstacles located on textured surfaces. In the second step, the exact position of the obstacle nearest to the camera is determined using binocular-vision methods. Perspective projection is used in both steps to model the image-formation process and the calculations are carried out in the image-based coordinate system described in Sec. 9.2. The remainder of this chapter details the methods used in the two stages. This discussion pertains only to the vision system and does not address issues related to decision-making and automatic control.

9.4.1   Texture Vision Subsystem

The texture vision subsystem (TVS) provides information about the relative positioning of obstacles on a textured background. Therefore, TVS simplifies the task of the binocular-vision subsystem, which determines only the exact 3-D position of the nearest obstacle. The other important role of

TVS is to segment textured images so that the edges (e.g., between the floor and walls) can be used for vehicle guidance. Also, TVS detects changes in surface orientation, such as the one shown in Fig. 9.9(b), that are of importance in vehicle guidance. The output of TVS is a segmented image with labeled subregions that are to be processed by the binocular-vision subsystem (Fig. 9.10[b]). These subregions correspond to the obstacles closest to the camera.

TVS uses a single view of a scene to obtain the description of the scene in terms of planar surfaces and their orientations relative to the viewer. Since no a priori knowledge about the scene is assumed, both surface tilt and surface slant cannot be estimated (Sec. 9.3.4). Instead, TVS measures surface tilt and then roughly estimates the surface slant.

Two fundamental assumptions are made regarding texture patterns. The first pertains to the fact that the individual texture elements lie fully on the surface. The second assumption pertains to the regularity of a frontal view of texture. Texture is assumed to be globally regular and statistically homogeneous; however, variations in the size and/or arrangement of the texture elements are allowed. Under these assumptions, the essence of our approach to texture analysis involves the notions of characteristic lines and characteristic angles (see Sec. 9.3.4 and Fig. 9.7). Geometrically, the characteristic lines are determined by the intersection of the surface plane and the image plane, and the characteristic angle Ω is related to the surface tilt τ by

$$\tau = \Omega \pm \frac{\pi}{2}. \tag{9.32}$$

Another useful texture descriptor is the *characteristic period*, defined to be the ratio of the length of a characteristic line segment and the number of texture elements lying along that line segment.

Based on the notions of the characteristic angle and the characteristic periods, TVS infers possible scene configurations in two steps. First, textured subregions that correspond to planar surfaces in a scene are determined. These subregions are referred to as *uniform regions*. Each uniform region is mapped into a possible surface orientation. The second step determines boundaries that delineate different textured regions. The two steps are preceded by a separation between textured and nontextured regions in an image. This separation, in turn, employs the variance of intensity and assigns pixels that have high variances to textured regions by using standard thresholding methods.

(A) Determination of Uniform Regions
Due to the fact that the individual texture elements are subject to perspective projection and, consequently, the textures in an image lose the

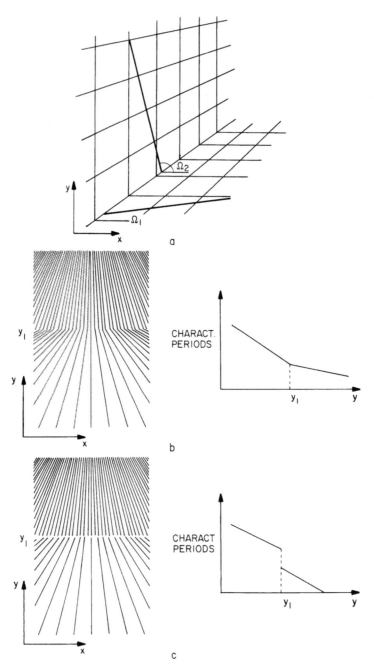

Figure 9.9 Examples of three possible orientations of the surface planes relative to the image plane and corresponding changes in textural properties: (a) two surfaces that are tilted differently relative to the image, (b) two surfaces that have the same tilt but different slants, and (c) two parallel surfaces.

homogeneity intrinsic to frontal views, our approach to extracting uniform regions in an image involves identification of abrupt changes in texture properties at their boundaries. The following two types of changes are of special interest: (1) Changes arising from the presence of a physical edge corresponding to a corner between two planes that are differently tilted relative to the image plane; the corresponding image regions are characterized by different characteristic angles (Fig. 9.9[a]); and (2) changes that originate from the presence of a physical edge corresponding to the intersection of two planar surfaces that have the same tilt and are slanted differently relative to the image plane. The corresponding image subregions have the same characteristic angle, but the projection of the physical edge is characterized by discontinuity in the rate of change of the characteristic periods (Fig. 9.9[b]). In addition, changes in image texture may arise from abrupt changes in the surface distance from the camera or changes in structural properties of texture patterns. The corresponding boundary in an image is characterized by discontinuity in the characteristic periods (Fig. 9.9[c]). These types of boundaries are not of practical importance in vehicle guidance in the described environment.

The computational procedure for identifying uniform regions is developed in the framework of a split-and-merge algorithm; for a general discussion on this subject, see (Gonzalez and Wintz/1987). The essence of the computational procedure is merging the neighboring textured subregions into uniform regions if they have the same characteristic angles. In this work, region merging and the measurement of the characteristic angle is performed by using the energy map of the 2-D Fourier power spectrum. This map is obtained by calculating the energy associated with a particular angle ϕ as

$$P(\phi) = \sum_{r=1}^{\frac{W}{2}} P(r, \phi), \qquad (9.33)$$

where W is the size of the image subregion under consideration, and

$$P(r, \phi) = [F_R^2(r, \phi) + F_I^2(r, \phi)]^{\frac{1}{2}}. \qquad (9.34)$$

F_R and F_I are the real and imaginary parts of the Fourier transform expressed in polar coordinates. The extrema in the energy map are related to texture directionality and allow differentiation between directional and nondirectional textures; for more details, see (Bajscy/1973 and Gonzalez and Wintz/1987). These extrema also are related to the surface orientation in the case of arbitrary views of textured surfaces (Bajscy and Lieberman/1976). Generally, texture preserves regularity along the characteristic lines and changes the most in the orthogonal direction; i.e., in the direction-of-texture gradient. As a consequence, texture directionality extrema in $P(\phi)$ close to the characteristic angle are favored due to the retention of parallelism along the characteristic lines, while the extrema in the direction of the texture gradient are the least favored, since the texture pattern changes the most in that direction (Brzakovic and White/1986).

Region merging is performed based on pattern vectors whose elements are the angles at which maxima in energy maps of the corresponding regions occur. Based on the neighboring criterion and the elements of the pattern vectors, the regions are merged into uniform regions; for details, see (Brzakovic and Tou/1984, Brzakovic and White/1986). Image subregions that have different pattern vectors from their neighbors are labeled as *ambiguity zones* and are used later in determining edges between uniform regions. An example of the results obtained using this approach is shown in Fig. 9.10.

Determination of the characteristic angle from the energy map is a difficult task both theoretically and computationally. In cases of nondirectional textures, it is possible to determine the characteristic angle precisely by analyzing the energy map and identifying the most prominent direction of energy. Determination of the characteristic angle leads directly to determination of the surface tilt. Instead of attempting to determine surface slant, TVS measures the density of texel edges along the characteristic lines. This, in turn, allows determination of the direction of surface slant (Fig. 9.11). By comparing the positions of obstacles placed on textured surfaces of known orientations, TVS determines the closest obstacle to the camera and labels the corresponding image subregion for the binocular-vision subsystem to process.

(B) Estimation of Boundaries Between Textured Regions

Formation of the uniform regions and ambiguity zones where boundaries lie reduces the problem of boundary determination between unknown textured regions into boundary determination between textured regions of known properties. The constraints arising from the possible scene layout determine the three general boundary types. The boundary types are:

Type 1. A linear boundary of the form $y = mx + n$ lying in the ambiguity zone between two uniform regions that have the same characteristic angles, Ω (Fig. 9.9[b]). In this case, the boundary is parallel to the characteristic line and its position is determined by the position of discontinuity in the characteristic periods (Fig. 9.9[b]).

Type 2. A linear boundary lying in the ambiguity zone between two uniform regions that have different characteristic angles (Fig. 9.9[a]). In this case, the values of the coefficients m and n are not known

Figure 9.10 An example of the uniform regions and ambiguity zones obtained by using energy maps. Each of the uniform regions is assigned a particular intensity, while the ambiguity zones are black.

a priori and are estimated by using linear recursive filtering. The filter estimates these coefficients from measurements performed in the known textured regions. Details of this method are described by (Brzakovic and Liakopoulos/1986).

Type 3. A boundary of an arbitrary shape lying in the ambiguity zone between two regions that have the same tilt and slant but correspond to structurally different textures or lie on planes that are placed at different distances from the camera (Fig. 9.9[c]). In this case, boundaries are approximated by piecewise polynomial functions, and the boundary estimation is performed by a linear recursive filter (Brzakovic and Liakopoulos/1986).

By deducing the boundary type from the measurements of the surface tilt and surface slant, TVS determines the boundaries between textured

Figure 9.11   The estimated surface slant (indicated by arrows showing the direction of change) for the image shown in Fig. 9.10.

regions. An example of the results obtained using this approach is shown in Fig. 9.12(a). The obtained boundaries can be used in autonomous vehicle guidance, as described by (Dickinson et al./1987).

9.4.2   Binocular Vision Subsystem

The objective of the binocular vision subsystem (BVS) is to obtain 3-D co-ordinates of pixels labeled by TVS as belonging to the closest obstacle. The task of BVS is simplified in that it has to consider only image regions that do not contain textured patterns. Establishing correspondence between individual points in textured regions is a very difficult problem; therefore, in textured images, correspondence is usually established between image sub-regions rather than between individual points (Lucas and Kanade/1981).

As described in Sec. 9.3.5, difficulties associated with the correspondence problem can be simplified by considering particular camera arrangements. In this work, we consider a camera arrangement consisting of a single camera that acquires two images of a scene by moving toward the scene between two exposures. The camera arrangement and examples of acquired images are shown in Fig. 9.13. The relationship between two coordinate systems (Eq. [9.21]) involves the matrix R described by Eq. (9.23) and $x_T = 0, y_T = 0, z_T = \delta$. By applying Eq. (9.3) to the projections, p_1 and p_2, of a point $P(X, Y, Z)$, it follows that

$$\frac{X}{x_1} = \frac{Y}{y_1} = \frac{\lambda + \delta - Z}{\lambda} \tag{9.35}$$

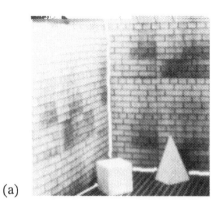

(a)

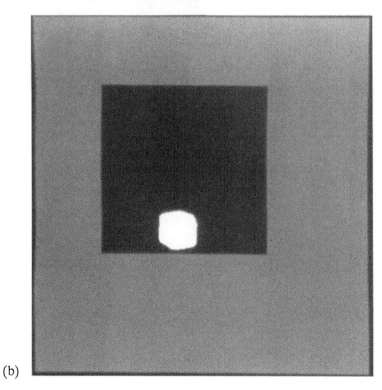

(b)

Figure 9.12 Outputs of TVS: (a) segmented texture regions; (b) extracted obstacle nearest to the viewer.

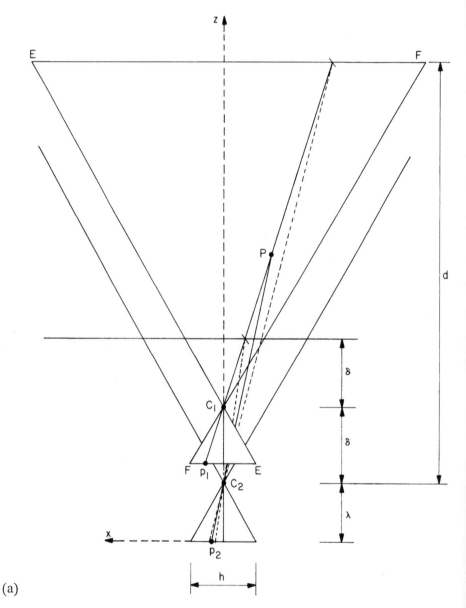

(a)

Figure 9.13 Camera arrangement wherein cameras are displaced along the z-axis: (a) geometrical arrangement; (b) examples of acquired images.

(b)

and

$$\frac{X}{x_2} = \frac{Y}{y_2} = \frac{\lambda - Z}{\lambda}, \tag{9.36}$$

where subscript 1 denotes projections onto the front image (i.e., the image acquired by positioning camera closer to the scene, and 2 denotes the projection onto the back image. From Eqs. (9.35) and (9.36), it further follows that

$$\frac{y_2}{x_2} = \frac{y_1}{x_1}. \tag{9.37}$$

Therefore, the correspondence to any point $p_1(x_1, y_1)$ in I_1 lies on the line determined by Eq. (9.37) in I_2. Moreover, point p_2 lies between the coordinate origin (in the local image-based coordinate system) and (x_1, y_1). The length of the line segment to be considered in establishing the correspondence can be further reduced by taking into consideration camera resolution. An example of such considerations, wherein the correspondence of a point p_1 in image I_1 is on a line segment of length $1/4h$ in I_2 (where h is as shown in Fig. 9.13) is described by (Alvertos et al./1986).

The advantages of the above camera arrangement relative to other arrangements lie in: (1) reducing of the number of points to be considered in establishing correspondence between the two images, (2) reducing the number of ambiguous matches when establishing correspondence of p_1 with points in I_2, and (3) rapid decrease of errors that arise in computing 3-D coordinates of a point as the camera displacement δ increases. These advantages are described in a greater detail by (Alvertos et al./1986). Note that in order to take advantage of this camera arrangement, it is necessary to establish correspondence starting with the front image. Consequently, the front image is analyzed by TVS.

The second step in using binocular-vision methods is feature extraction. In this work, we chose intensity to be the appropriate feature for image matching. We assume that image intensity i is related to the scene reflectance ϕ by

$$i = K\phi \cos^4 \alpha, \tag{9.38}$$

where K is the parameter describing characteristics of the particular camera, and α is the projection angle of a point P onto the image plane (for details, see Horn/1986). The relationship between intensities i_1 and i_2 of points p_1 and p_2, respectively, is

$$\frac{i_1}{i_2} = \frac{\cos^4 \alpha_1}{\cos^4 \alpha_2}. \tag{9.39}$$

From geometrical considerations, it follows that

$$\cos^4 \alpha_j = \frac{\lambda^4}{(x_j^2 + y_j^2 + \lambda^2)^2} \qquad j = 1, 2. \tag{9.40}$$

From Eq. (9.40), it also follows that the ratio of intensities i_1 and i_2 is related to the coordinates of p_1 and p_2 in their image-based coordinate systems.

The image-matching algorithm used in this chapter involves, as the first step, determination of the appropriate segment on the line described by Eq. (9.37) that contains possible correspondences for a selected point $p_1(x_1,y_1)$ in I_1. Then, the matching algorithm selects a point $p_2(x_2,y_2)$ on this segment that minimizes a functional, Q, where

$$Q = \left| \frac{i_1}{i_2} - \frac{(x_2^2 + y_2^2 + \lambda^2)^2}{(x_1^2 + y_1^2 + \lambda^2)^2} \right|^2 \tag{9.41}$$

as the appropriate correspondence.

Upon establishing correspondence between p_1 and p_2, the 3-D coordinates of $P(X, Y, Z)$ are

$$X = \frac{x_1 \delta x_2}{\lambda(x_2 - x_1)}, \tag{9.42}$$

$$Y = \frac{y_1 y_2 \delta}{\lambda(y_2 - y_1)}, \tag{9.43}$$

and

$$Z = \lambda + \frac{x_1 \delta}{x_1 - x_2}. \tag{9.44}$$

Figure 9.14 Closest points on the object of interest determined by BVS.

As a result, BVS obtains a 3-D description of objects of interest and determines the minimum distance between the obstacles and the camera. The primary objective of BVS is to determine the points on objects of interest that are closest to the viewer. Examples of such points are shown in Fig. 9.14. The resulting information about the obstacles can then be used by decision-making and automatic-control models for appropriate guidance of an autonomous vehicle.

REFERENCES

Alvertos, N., Brzakovic, D., and Gonzalez, R. C. (1986). "Correspondence in Pairs of Images Acquired by Camera Displacement in Depth," Proceedings of SPIE Conference on Advances in Intelligent Robotics Systems, Cambridge, Massachusetts, pp. 131–136.

Bajscy, R. and Lieberman, L. (1976). Texture gradient as a depth cue, *Computer Graphics and Image Processing*, 5: 52–67.

Bajscy, R. (1973). "Computer Description of Textured Surfaces," Proceedings of 1973 International Joint Conference on Artificial Intelligence, Stanford, California.

Baker, H. H. (1980). "Edge Based Stereo Correlation," Proceedings of Image Understanding Workshop, College Park, Maryland, pp. 168–175.

Baker, H. H. and Binford, T. O. (1981). "Depth from Edge and Intensity Based Stereo," Proceedings of 7th International Joint Conference on Artificial Intelligence, Vancouver, BC, pp. 631–636.

Barnard, S. T. and Fishler, M. A. (1982). Computational stereo, *ACM Computing Surveys, 14, no. 4*: 553–572.

Baylou, P., El Hadj A. B., Monsion, M., Bouvet, C., and Bousseau, G. (1984). Detection and three-dimensional localization by stereoscopic visual sensor and its application to a robot for picking asparagus, *Pattern Recognition, 17, no. 4:* 377–384.

Bellugi, D., Dammagio, G., and Schiavoni, A. (1985). "RAMAN: A 3-D Stereoscopic Vision System," Proceedings of 4th Scandinavian Conference on Image Processing, Trondheim, Norway, pp. 445–453.

Besl, P. and Jain, R. (1985). "Range Image Understanding," Proceedings of Conference on Computer Vision and Pattern Recognition, San Francisco, pp. 430–449.

Besl, P. and Jain, R. (1986). Invariant surface characteristics for 3D object recognition in range images, *Computer Vision, Graphics, and Image Processing, 33:* 33–80.

Boyer, K. L. and Kak, A. C. (1987). Color-encoded structured light for rapid active ranging, *IEEE Trans. on Pattern Analysis and Machine Intelligence, PAMI-9, no. 1:* 14–28.

Bruss, A. R. (1980). *The Image Irradiance Equation: Its Solution and Application,* PhD Dissertation, MIT, Cambridge, Massachusetts.

Brzakovic, D. and Liakopoulos, A. (1986). Estimation theory based segmentation of texture images, *Advances in Image Processing and Pattern Recognition* (V. Cappellini and R. Marconi, eds.) North-Holland, Amsterdam, pp. 234–238.

Brzakovic, D. and Tou, J. T. (1984). "Image Understanding via Texture Analysis," Proceedings of 1st Conference on Artificial Intelligence Applications, Denver, Colorado, pp. 585–590.

Brzakovic, D. and White, B. (1986). "Surface Orientation Using Texture Gradient Derived from Two Views of a Scene," Proceedings of 8th International Conference on Pattern Recognition, Paris, pp. 1045–1047.

Bui-Tuong, P. (1975). Illumination for computer-generated pictures, *Communications of the ACM, 18, no. 6:* 311–317.

Case, J. G. (1981). Automation in photogrammetry, *Photogram. Eng. Remote Sensing, 47, no. 3:* 335–341.

Cavanagh, P. (1987). Reconstructing the third dimension: Interactions between color, texture, motion, binocular disparity and shape, *Computer Vision, Graphics, and Image Processing, 37:* 171–195.

Clark, J. J. and Lawrence, P. D. (1986). A theoretical basis for diffrequency stereo, *Computer Vision, Graphics, and Image Processing, 35:* 1–19.

Colleman, E. N. and Jain R. (1982). Obtaining 3-dimensional shape of textured and specular surfaces using four-source photometry, *Computer Graphics and Image Processing, 18:* 309–328.

Dickinson, S. J., Moigne, J. J., Waltzman, R., and Davis, L. S. (1987). "Expert Vision System For Autonomous Land Vehicle Road Following," *SPIE Proceedings, 786.*

Fu, K. S., Gonzalez, R. C., and Lee, C. S. G. (1987). *Robotics: Control, Sensing, Vision and Intelligence,* McGraw-Hill, New York.

Gennery, D. G. (1980). "Object Detection and Measurement Using Stereo Vision," Proceedings of Image Understanding Workshop, College Park, Maryland, pp. 161–167.

Gennery, D. B. (1981). "A Feature-Based Scene Matcher," Proceedings of 7th International Joint Conference on Artificial Intelligence, Vancouver, BC, pp. 667–673.

Gibson, J. J. (1960). *Visual World*, The Riverside, Cambridge, Massachusetts.

Gonzalez, R. C. and Wintz, P. (1987). *Digital Image Processing*, 2d ed., Addison-Wesley, Reading, Massachusetts.

Goshtasby, A. (1984). "A Refined Technique for Stereo Depth Perception," Proceedings of IEEE Workshop on Computer Vision, Annapolis, Maryland, pp. 3–7.

Grimson, W. E. L. (1980). "Aspects of a Computational Theory of Human Stereo Vision," Proceedings of Image Understanding Workshop, College Park, Maryland, pp. 128–149.

Haddow, E. R., Boyce, J. F., and Lloyd, S. A. (1985). "A New Binocular Stereo Algorithm," Proceedings of 4th Scandinavian Conference on Image Processing, Trondheim, Norway, pp. 175–182.

Hannah, M. T. (1980). "Bootstrap Stereo," Proceedings of Image Understanding Workshop, College Park, Maryland, pp. 201–208.

Haralick, R. M. (1979). Statistical and structural approaches to texture, *Proc. of the IEEE, 67, no. 5.*

Hebert, M. and Ponce, J. (1982). "A New Method for Segmenting 3-D Scenes into Primitives," Proceedings of the International Joint Conference on Pattern Recognition.

Henderson, T. C. (1983). Efficient 3-D object representations for industrial vision systems, *IEEE Trans. on Pattern Analysis and Machine Intelligence, PAMI-5, no. 6.*: 609–617.

Horn, B. K. P. (1975). Obtaining shape from shading information, *Psychology of Computer Vision*, (P. H. Winston, ed.), McGraw-Hill, New York.

Horn, B. K. P. (1986). *Robot Vision*, MIT Press, Cambridge, Massachusetts.

Horn, B. K. P. and Brooks, M. J. (1986). The variational approach to shape from shading, *Computer Vision, Graphics, and Image Processing, 33*: 174–208.

Ikeuchi, K. and Horn, B. K. P. (1981). Numerical shape from shading and occluding boundaries, *Artificial Intelligence, 17*: 141–184.

Ito, M. and Ishii, A. (1986). Three-view stereo analysis, *IEEE Trans. on Pattern Analysis and Machine Intelligence, PAMI-8, no. 6*: 524–532.

Itoh, H., Miyauchi, A., and Ozawa, A. (1984). "Distance Measuring Method Using Only Simple Vision Constructed for Moving Robots," Proceedings of 7th International Conference on Pattern Recognition, Montreal, pp. 192–195.

Jenkin, M. and Tsotsos, J. K. (1986). Applying temporal constraints to dynamic stereo problem, *Computer Vision, Graphics, and Image Processing, 33*: 16–32.

Julesz, B. (1970). *The Cyclopean Eye*, Academic Press, New York.

Kanade, T., (1981). Recovery of three-dimensional shape of an object from a single view, *Artificial Intelligence, 17.*

Kass, M. and Witkin, A. (1987). Analyzing oriented patterns, *Computer Vision, Graphics, and Image Processing, 37*: 362–385.

Kender, J. R. (1980). *Shape from Texture*, PhD Dissertation, Carnegie-Mellon University, Pittsburgh, Pennsylvania.

Konecny, C. and Pape, D. (1981). Correlation techniques and devices, *Photogram. Eng. Remote Sensing, 47, no. 3*: 323–333.

Lewis, R. A. and Johnston, A. R., (1977). "A Scanning Laser Range-Finder for Robotic Vehicle," Proceedings of International Joint Conference on Artificial Intelligence.

Lu, S. Y. and Fu, K. S. (1978). A syntactic approach to texture analysis, *Computer Graphics and Image Processing, 7*: 303–330.

Lucas, B. D. and Kanade, T. (1981). "An Iterative Image Registration Techniques with an Application to Stereo Vision," Proceedings of 7th International Joint Conference on Artificial Intelligence, Vancouver, BC, pp. 674–679.

Marr, D. and Poggio, T. (1979). A computational theory of human stereo vision, *Proc. of Royal Society of London B-204*: 301–328.

Mensbach, P. (1986). Calibration of a camera and light source by fitting to a physical model, *Computer Vision, Graphics, and Image Processing, 35*: 200–219.

Mitiche, A. (1984). "On Combining Stereopsis and Kineopsis for Space Perception," Proceedings of 1st Conference on Artificial Intelligence Applications, Denver, Colorado, pp. 156–160.

Monchaud, S. (1986). "Contribution to Range Finding Techniques for Third Generation Robots," Proceedings of Intelligent Autonomous Systems Conference, Amsterdam, Netherlands, pp. 459–469.

Moravec, H. P. (1979). "Visual Mapping by a Robot Rover," Proceedings of 6th International Joint Conference on Artificial Intelligence, Tokyo, Japan, pp. 598–600.

Moravec, H. P. (1981). "Rover Visual Obstacle Avoidance," Proceedings of 7th International Joint Conference on Artificial Intelligence, Vancouver, BC, pp. 785–790.

Mulgaonkar, P. G., Shapiro, L. G., and Haralick, R. M. (1986). Shape from perspective: A rule-based approach, *Computer Vision, Graphics, and Image Processing, 36*: 298–320.

Nitzan, D., Brain, A. E., and Duda, R. O. (1977). The measurement and use of registered reflectance and range data in scene analysis, *Proc. IEEE, 65*: 206–220.

O'Brien, N. and Jain, R. (1984). "Axial Motion Stereo," Proceedings of IEEE Workshop on Computer Vision, Annapolis, Maryland, pp. 88–92.

Olin, K. E., Vilnrotter, F. M., Darly, M. J., and Reiser, K. (1987). "Developments in Knowledge-Based Vision for Obstacle Detection and Avoidance," Proceedings of Image Understanding Workshop, Los Angeles, pp. 78–83.

Oshima, M. and Shirai, Y. (1983). Object recognition using three-dimensional information, *IEEE Trans. on Pattern Analysis and Machine Intelligence, PAMI-5, no. 4*: 353–361.

Pentland, A. P. (1984). Local shading analysis, *IEEE Trans. on Pattern Analysis, and Machine Intelligence, PAMI-6, no. 2*: 170–187.

Popplestone, R. J., Brown, C. M., Ambler, A. P., and Crawford, G. F. (1975). "Forming Models of Plane-and-Cylinder Faceted Bodies from Light Stripes," Proceedings of 4th International Joint Conference on Artificial Intelligence, Tbilisi, USSR.

Rosenfeld, A. (1986). Image processing: 1986, *Computer Vision, Graphics, and Image Processing, 38, no. 2.*

Shirai, Y. (1972). Recognition of polyhedrons with a range finder, *Pattern Recognition, 4*: 243–250.

Shneier, M. O., Lumia, R., and Kent, E. W. (1986). Model-based strategies for high-level robot vision, *Computer Vision, Graphics, and Image Processing, 33*: 293–306.

Stevens, K. A. (1980). *Surface Perception by Local Analysis of Texture and Contour,* Report AI-512, Artificial Intelligence Lab, MIT, Cambridge, Massachusetts.

Sugihara, K. (1979). Range-data analysis guided by junction dictionary, *Artificial Intelligence, 12*: 41–69.

Svetkoff, D. J., Leonard, P. F., Sampson, R. E., and Jain, R. (1984). Techniques for real-time 3-D feature extraction using range information, *SPIE Robot Vision, 336.*

Tomita, F. and Kanade, T. (1984). "A 3-D Vision System: Generating and Matching a Shape Descriptions in Range Images," Proceedings of 1st Conference on Artificial Intelligence Applications, Denver, Colorado, pp. 156–160.

Trendl, E. and Kriegman, D. J. (1987). "Vision and Visual Exploration for the Stanford Mobile Robot," Proceedings of Image Understanding Workshop, Los Angeles, pp. 407–416.

Tsai, R. Y. (1987). A versatile camera calibration technique for high-accuracy 3D machine vision methodology using off-the-shelf TV cameras and lenses, *IEEE J. of Robotics and Automation, RA-3, no. 4*: 323–344.

Witkin, A. P. (1981). Recovery of surface shape and orientation from texture, *Artificial Intelligence, 17*: 17–47.

Woodham, R. J. (1981). Analyzing images of curved surfaces, *Artificial Intelligence, 17, nos. 1–3*: 117–140.

Yakimovsky, Y. and Cunningham, R. (1978). A system for extracting three-dimensional measurements from a stereo pair of TV cameras, *Computer Graphics and Image Processing, 7*: 195–210.

Zucker, S. (1976). Toward a model of texture, *Computer Graphics and Image Processing, 5, no. 2*: 190–202.

10

Enhancement of Fingerprints Using Digital and Optical Techniques

Thomas F. Krile and John F. Walkup

Texas Tech University
Lubbock, Texas

10.1 INTRODUCTION

Few areas in criminalistics offer as much potential for the fruitful application of high technology as does the enhancement of degraded fingerprints. In this chapter, we review some fundamental properties of fingerprints and then consider how digital and optical image-processing techniques can be used effectively to enhance degraded or smudged fingerprints. The digital processing techniques offer advantages in flexibility (e.g., both linear and nonlinear operations) and accuracy, whereas the optical processing techniques are inherently very fast and highly parallel. The successes achieved with these enhancement techniques indicate that more research is needed to fully develop the promise of applying image processing in both fingerprint processing and other areas of criminalistics.

10.2 FINGERPRINT FUNDAMENTALS

The enhancement of fingerprints for the purpose of characterization is an attractive area for applying either digital or optical processing techniques. The reader is referred to other references for some of the rich history of the science of fingerprinting (FBI/1963, Moenssens/1971, Eleccion/1973,

Olsen/1978). Of necessity, we will deal mainly with a very basic outline of fingerprint fundamentals, especially highlighting those aspects most related to the development of digital and optical image-processing modalities.

As is well known, fingerprints uniquely identify an individual. Due to the binary nature of the ridge-pattern images, they constitute a particularly interesting image type. Figure 10.1 shows a number of general fingerprint patterns (FBI/1963). An initial observation is the fact that though the ridge-to-ridge spacings are essentially invariant spatially, the ridge orientations do change spatially, thus indicating the value of looking at use of the Fourier transform (i.e., spatial-frequency domain) for characterization purposes.

The individual characteristics that uniquely identify a fingerprint are called minutiae, which are essentially the various types of ridge-pattern deviations, as shown in Fig. 10.2. Thus, the basic ridge patterns, shown in Fig. 10.1, plus the minutiae and their locations on the fingerprint pattern, uniquely characterize a fingerprint. As shown in Fig. 10.2, the three basic types of minutiae are bifurcations, ridge endings, and short independent ridges. All ridge characteristics can be described by combining the basic types. Figures 10.3(a) and 10.3(b) (Moenssens/1971) show examples of lifted and inked fingerprints, respectively, with identifiable minutiae indicated. Note the blurring and noise effects in the lifted fingerprint; these are the types of degradations we wish to remove with enhancement and restoration techniques.

When one goes from a latent fingerprint to a lifted fingerprint, whether some sort of dusting technique or more recently developed technique such as argon laser fluorescence (Menzel/1980) is used, the print obtained is commonly degraded when compared with a carefully inked fingerprint, as shown in Fig. 10.3 To remove the degradations and increase the probability of accurately classifying the fingerprint, some combination of image restoration and enhancement operations may be necessary. The restoration operations might compensate for blurring due to smudging of the prints or camera defocusing, for example. They also might function to suppress image noise sources, such as film-grain noise, which are often signal-dependent (Kasturi and Walkup/1986).

Enhancement operations, on the other hand, may actually distort an image, but are employed to make it easier to extract information from an image. Since fingerprint ridges constitute edges, edge-enhancement algorithms are particularly attractive candidates for application to fingerprints.

With the above discussion as background, we will now move to detailed descriptions with examples of both digital and optical image-processing modalities, with specific application to fingerprints.

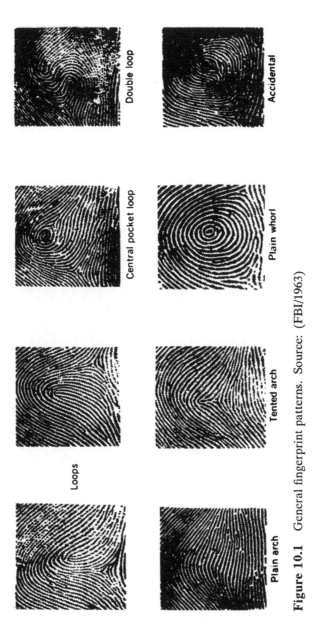

Figure 10.1 General fingerprint patterns. Source: (FBI/1963)

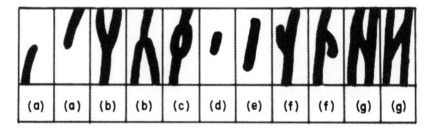

Figure 10.2 Seven basic types of fingerprint minutiae: (a) ridge termination, (b) fork or bifurcation, (c) enclosure or lake, (d) island, (e) short independent ridge, (f) spur or hook, and (g) crossover.

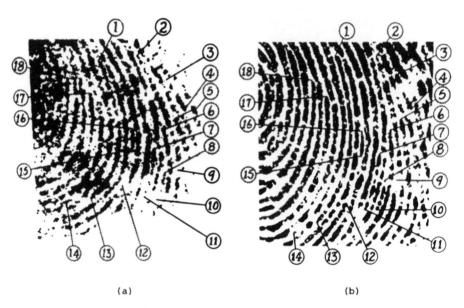

Figure 10.3 Comparison between: (a) lifted fingerprint, and (b) inked finger-print, indicating various minutiae. Source: (Moenssens/1971)

10.3 IMAGE-PROCESSING MODALITIES

Basic studies of fingerprint-enhancement techniques have used two different image-processing modalities: digital and optical. In the digital case, the fingerprint is acquired by a TV camera (vidicon or CCD) either directly from the latent print or from a photograph or slide transparency of the print.

The image is then digitized and sorted in a computer for subsequent application of digital-enhancement algorithms. In the optical case, a photographic transparency of the fingerprint is used as the input to the optical processing system, and the fingerprint is enhanced in real time by special filters usually recorded on film. Digital processing is more powerful than optical processing in that both linear and nonlinear enhancement operations can be performed readily, whereas only linear operations are usually available optically. However, optical processing is much faster than digital processing and, as will be shown, even the restricted class of linear enhancement operations can produce useful results. In the following two sections, digital and optical processing techniques that are especially suited for fingerprint enhancement will be examined in detail and a number of illustrative examples shown.

10.3.1 Digital Fingerprint Enhancement

In digital-image processing (Rosenfeld and Kak/1976, Andrews and Hunt/ 1977, Pratt/1978, Ekstrom/1984, Baxes/1984, Niblack/1986, Gonzales and Wintz/1987), an image is normally formed as a large square array of picture elements (pixels), usually 256×256 or 512×512, although even larger arrays are increasingly available. Each pixel value has a particular brightness, or gray-level, value assigned to it (e.g., usually one of the 256 possible levels for a typical 8-bit display device). The pixel values, along with their associated positions (addresses) are entered as numbers into a computer and are manipulated by special computer algorithms to produce a new, enhanced digital image that is then output to a display for human evaluation.

In the remainder of this section, a number of different enhancement algorithms that are particularly applicable to fingerprint images will be discussed (Chiralo and Berdon/1978, Lehar and Stevens/1984). They can be divided into two classes: point processing and neighborhood processing. In the case of point processing, individual pixel values are manipulated independently of other pixels. In neighborhood processing, groups of pixels are operated on either in the space domain or in the spatial-frequency domain. The operations performed can be either linear or nonlinear (as contrasted with the optical case). Applications of point processing will be discussed first.

(A) Point-Processing Techniques

Several types of point processing, such as contrast manipulations and histogram equalization, are performed by operating on the histogram of an image. A histogram is a function that shows the distribution of pixel gray values in an image. As such, when normalized, it can be viewed as a discrete probability-density function of the image intensities. Latent

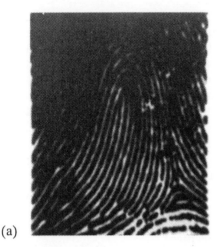

(a)

Figure 10.4 Histogram processing: (a) original image, (b) histogram of original image, (c) stretched histogram, (d) resulting enhanced image, and (e) image after contrast reversal. Source: (Krile et al./1987)

fingerprints often have a limited dynamic range; i.e., their intensity values occupy only a small portion of the possible gray-level range, resulting in narrow histograms.

One common method of fingerprint enhancement is to increase the dynamic range of the image so as to fill the entire gray-level range available. If an image's minimum-intensity gray-level is I_{min} and its maximum value is I_{max}, then a linear contrast-stretch operator is defined by:

$$I_{out} = 255(I_{in} - I_{min})/(I_{max} - I_{min}) \tag{10.1}$$

where I_{in} is the original (input) image pixel gray-level, I_{out} is the enhanced (output) image pixel gray-level, and the assumed gray-level range is 0 to 255; i.e., 8 bits of gray-level resolution.

Figure 10.4 illustrates an example of this enhancement operation. Figure 10.4(a) shows an original latent fingerprint, and Fig. 10.4(b) displays the print's histogram. Notice that the initial histogram does not occupy the entire dynamic range available, and most of the values are in the dark end of the gray-level range; i.e., toward zero. Note also that the histogram is bimodal, a characteristic of fingerprint images. Figure 10.4(c) shows the histogram after applying the linear contrast-stretch operator, and Fig. 10.4(d) shows the resulting enhanced image. It can be seen that the contrast has been much improved, and details that were not seen before are now visible. It should be pointed out that effective image-enhancement techniques do

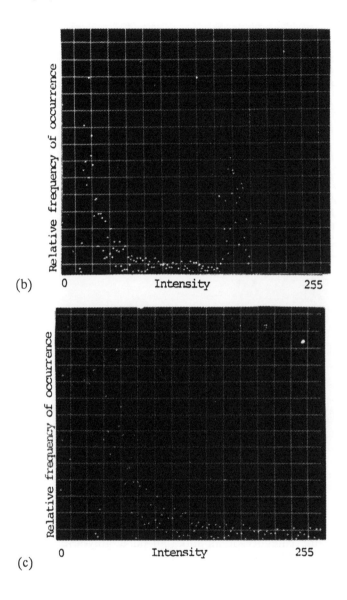

(b)

(c)

not make something out of nothing, but merely take information that is already in the input image and put it into the detectable range of the human visual system. There is, however, always the danger of generating artifacts, which we will see in some of the more complex enhancement algorithms to come.

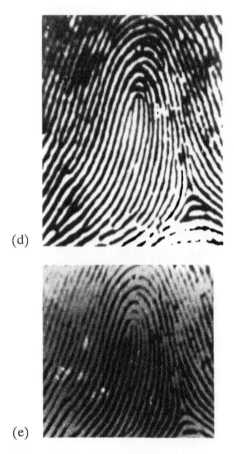

(d)

(e)

Figure 10.4   (Continued)

There are a number of other contrast-enhancement operators available belonging to this same category of input/output-point gray-level trans-formations. Some of the most useful for binary-line images, such as fingerprints, are thresholding (all gray-levels above a certain level are out-put as white, the rest are set to black), level slicing (all gray-levels within a certain narrow range are set to white, the rest are set to black), and con-trast reversal. An example of the last operation is shown in Fig. 10.4(e), where the gray-level values have been reversed. This is particularly effec-tive because small changes in gray-levels in the bright portions of an image are more difficult for humans to detect than are small changes in the dark portions, because of the visual system's inherently logarithmic response to light intensity. Thus, more information about an image can be obtained if both a positive and a negative version are examined.

A second family of point-enhancement operations is histogram manip-ulation. In this case, the histogram is considered as a probability-density function, and transformations are used to change the original image density to a different probability-density function for enhancement. A common ex-ample is histogram equalization, where the original histogram is changed to a discrete, quantized version of a uniform probability density. The idea is to redistribute input-image gray-levels that are close together (and, hence, not visually distinguishable) to gray-levels that are farther apart in the en-hanced image, the net result being quite similar to contrast stretching. A number of other histogram transformations are also available, such as log-arithmic or hyperbolic transforms that are designed to match the response of the human visual system.

A third type of point-image enhancement is pseudocoloring. In this technique, each gray-level value is assigned a different color for display purposes. This is very effective, because humans can distinguish only about 30 different gray-levels at any point in an image, whereas we can distinguish several hundred color hues. However, there is much psychovisual research yet to be done to determine optimal techniques for assigning colors to gray-levels to obtain the best visual interpretation.

The point-processing operations described above can be performed very quickly with even a microcomputer-based image-processing system; yet, they are quite effective in enhancing the information contained in an image without introducing artifacts. These are the techniques that one would normally try first when confronted with an image-enhancement task.

(B)   Neighborhood-Processing Techniques

In those situations where one is trying to change the information content of the image (e.g., to suppress the effects of additive noise or other degradations while highlighting the object's features of interest), more complex neighborhood processing is used. Here, a pixel value in the enhanced image is determined by a function operating on several pixel values in the neighborhood of the pixel in the original image. This function can be either linear or nonlinear in the digital image-processing case, but is usually constrained to the linear case for real-time optical image processors, although recent advances in nonlinear optical materials may change that situation. In any case, only nonlinear neighborhood processing will be discussed under the digital processing category in this chapter, and linear processing will be discussed in the optical image-processing section. In particular, three nonlinear techniques that are especially appropriate for fingerprint applications will be treated under the digital category: homomorphic filtering, adaptive binarization, and adaptive wedge filtering.

If the degradation of a fingerprint can be modeled as being due to a multiplicative term $d(x,y)$, then a technique known as homomorphic filtering can be applied (Barsallo/1985). In this case, we assume we have

$$g(x,y) = d(x,y)f(x,y) \qquad (10.2)$$

where $f(x,y)$ is the desired fingerprint, $g(x,y)$ is the degraded latent print, and $d(x,y)$ is the multiplicative degradation function. This may be the situation when the print is laid down with unequal pressures over the print area, resulting in some regions with good contrast and others with very low contrast. Usually, the degradation $d(x,y)$ will have slowly changing spatial variations and, thus, can be considered as a low-spatial-frequency function. In the spatial-frequency domain, Eq. (10.2) becomes

$$G(f_x,f_y) = D(f_x,f_y) * F(f_x,f_y) \qquad (10.3)$$

where f_x,f_y are spatial frequency variables, $*$ is the convolution operator, and capital letters refer to the Fourier transforms of functions with the equivalent lower-case letters; e.g.,

$$G(f_x,f_y) = \int\!\!\!\int_{-\infty}^{\infty} g(x,y)\exp[-j2\pi(f_x x + f_y y]dxdy \qquad (10.4)$$

Since the degradation term $D(f_x,f_y)$ is convolved with the desired fingerprint spectrum $F(f_x,f_y)$, the usual linear filtering techniques cannot be applied to separate the two in the spatial-frequency domain. However, if the natural logarithm of the degraded image is taken, we have:

$$\ln[g(x,y)] = \ln[d(x,y)] + \ln[f(x,y)] \qquad (10.5)$$

and in the spatial-frequency domain:

$$\mathcal{F}\{\ln[g(x,y)]\} = \mathcal{F}\{\ln[d(x,y)]\} + \mathcal{F}\{\ln[f(x,y)]\} \qquad (10.6)$$

where $\mathcal{F}\{\cdot\}$ is the two-dimensional Fourier-transform operator as defined in Eq. (10.4). Now, since the spatial-frequency content of the degradation term has primarily low spatial frequencies and the desired fingerprint has its important information at relatively high spatial frequencies, a simple high-pass filter can be used to remove the degradation term. After inverse filtering and then exponentiating the result, an enhanced fingerprint image is obtained.

An example to illustrate homomorphic filtering is shown in Fig. 10.5. In Fig.10.5(a), a fingerprint has been artificially smudged by multiplying a good print with an inverse Gaussian function $d(x,y)$. This results in progressively greater loss of information toward the middle of the fingerprint. This degraded print was then logged and Fourier transformed

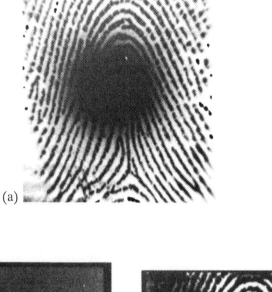

(a)

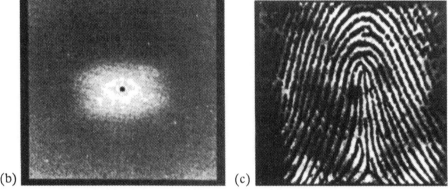

(b)   (c)

Figure 10.5   Homomorphic filtering: (a) degraded fingerprint, (b) filtered Fourier spectrum, and (c) enhanced image. Source: (Krile et al./1987).

to give the spatial-frequency result shown in Fig. 10.5(b). The black spot at the center of the figure represents a high-pass filter that is blocking out the low-frequency degradation terms at the spatial-frequency origin. The filter is a Butterworth high-pass filter designed to minimize ringing artifacts (i.e., Gibb's phenomenon) back in the spatial domain. After inverse transformation of Fig. 10.5(b) and exponentiation, the enhanced fingerprint is obtained as shown in Fig. 10.5(c), with the center portion

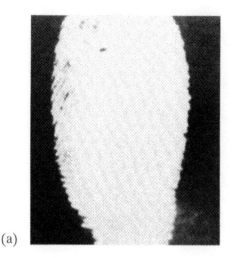

(a)

Figure 10.6   Adaptive binarization: (a) degraded fingerprint, (b) result of one-pass binarization, and (c) result of two-pass binarization. Source: (Krile et al./1987)

of the original fingerprint restored.   At this stage, the point contrast-enhancement techniques discussed earlier are usually applied to improve the results even further, if possible.

It should be noted that this technique works well for fingerprints because the important ridge and minutiae information lies in roughly circular bands far from the spatial-frequency origin as compared to the degradation information. This can be seen in Fig. 10.5(b) and is further developed in the section on optical processing.

Since fingerprint information consists of the presence or absence of ridges and minutiae and, thus, is binary by nature, enhancement techniques that result in a binary image are especially attractive.   However, simple thresholding to produce a binary image will not usually be successful, because most latent fingerprints have a spatially varying background intensity. To compensate for the local background variations, and at the same time produce a binary enhanced image, an adaptive-binarization technique has been successively applied (Andrews/1976, Barsallo/1985). An $M \times M$ window (e.g., 3×3 or 5×5 pixels) is placed over each pixel in the latent print image, and the average gray-level within that window is computed.   Then, the center pixel in the window is thresholded to 0 (black) or 255 (white), depending on whether the value of the center pixel is less than or greater than the local average, respectively. This technique is sensitive to the size, M, of the window. If M is small, say 3 pixels, there is less smoothing of the image,

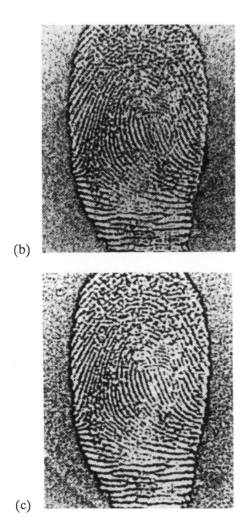

(b)

(c)

so that details of the print ridges and minutiae are sharper, but there are also a lot of outlier points (i.e., salt-and-pepper noise). As M increases, say to 10 or 20, the noise decreases but more smoothing occurs, so the minutiae are less distinct. Thus, some experimentation is necessary for a given print to find the M that results in the best print detail with the least noise.

Figure 10.6(a) shows a fingerprint whose contrast in the center has been artificially reduced by application of a multiplicative Gaussian smudge. Figure 10.6(b) shows the result of one pass of the adaptive-binarization algorithm using a 5×5-pixel window. One can see a number of outlier points that, in Fig. 10.6(c), have been reduced by making another pass through the algorithm using the same window size as before. This multiple-pass technique seems to reduce the noise, without at the same time blurring the minutiae.

Another characteristic of fingerprints that can be exploited for enhancement purposes (besides their binary nature, which was used in the adaptive-binarization algorithm) is that they consist of almost linear (i.e., straight-line) features on a local scale. Such features can be enhanced by a new algorithm we call Fourier short-space adaptive filtering (Tarng/1987). The procedure consists of several steps: (1) divide the image into small subimages, (2) Fourier transform the subimages, (3) modify each subimage spectrum with an adaptive wedge filter suitable for linear features, (4) inverse Fourier transform the filtered subimages, and (5) reconstruct the spatial-domain-enhanced image. Each of these steps will be discussed in more detail below.

The subimage size determined in step (1) is a tradeoff between having a window large enough to represent the features of interest adequately and yet small enough that the features have only a small amount of curvature within the subimage. For fingerprint images of 256×256 pixels, 32×32-pixel subimages prove to be a good compromise. Once the subimages are chosen, each has its average intensity value subtracted from all its pixels. The average value is restored later in the algorithm to be the same for all subimages, thus reducing the effects of variations in background illumination. To segment the image, operate on the subimages, and then reconstruct the enhanced image without introducing artifacts due to the subimage edges, a window-and-overlap technique is used. Each 32×32 subimage overlaps its nearest neighbors by 50%, so a 256×256-pixel image has $15 \times 15 = 225$ subimages. Also, each subimage is multiplied by a 2-D Hamming window function $W(i,j)$, where

$$W(i,j) = W(i)W(j); i,j = 0,\ldots,N-1 \qquad (10.7)$$

Here, $N \times N$ is the window (i.e., subimage) size and

$$W(n) = 0.54 + 0.46\cos\{n - [(N-1)/2]\}(2\pi/N));$$
$$n = 0,\ldots,N-1$$

Step (2) consists of applying the Fast Fourier Transform (FFT) algorithm to each of the subimages individually to get their respective spatial-frequency spectra. At this point, one could employ preprocessing filters, such as high-pass, low-pass, or band-pass, to reduce the effects of particular types of noise. In our case, however, we are interested primarily in enhancing linear feature information, so we proceed immediately to step

(3). Here, a special adaptive wedge filter is used to enhance the linear features of interest and suppress most of the unwanted noise. The filter will be described in detail after this discussion of the overall algorithm.

Step (4) entails setting the DC, or average, value of each filtered subimage from step (3) to the same value and then preforming the inverse FFT on each subimage. Finally, reconstruction is performed in step (5) by again weighting each processed subimage with a Hamming window, replacing the subimages in their original positions, and summing the results. If the overall processed image has low contrast, it can then be improved by a standard contrast stretch or histogram manipulation, as discussed earlier. A flowchart of the overall Fourier short-space adaptive algorithm is shown in Fig. 10.7.

The heart of this algorithm is the adaptive wedge filter, which will now be discussed. The main function of this filter is to pass the information related to linear image features and to suppress other spectral components. The wedge filter with angular orientation ϕ, shown in Fig. 10.8, is well suited to this task when its passband region (crosshatched area) is oriented perpendicular to the linear features in the original subimage. Since the linear features in the subimage will not be perfectly parallel to each other, they can have some degree of variation after processing, depending on the width (i.e., angular bandwidth θ) of the filter. In the example to follow, we will typically use filters oriented in eight different directions, with a 50% overlap between each wedge filter and its neighbor (i.e., an angular bandwidth of 45 deg). The number of wedge filters depends, of course, on the angular resolution desired.

A flowchart of the adaptive-wedge-filtering algorithm is shown in Fig. 10.9. First, the family of wedge filters chosen is applied to the Fourier spectrum of a subimage, and the wedge filter order is ranked according to the total amount of spectral energy contained in each filter's passband area. This determines which spatial direction contains the most significant linear features. At this point, a threshold can be employed to eliminate those wedges that contain only a very small amount of spectral energy, which is most probably due to noise in the original image. The wedge filters to be used are then selected from the reduced filter list in decreasing spectral energy order and according to how many spatial directions we want to enhance. In the case of fingerprints, usually only one direction (i.e., that with the highest spectral energy) is chosen. Now, the chosen filters are combined to form the adaptive wedge filter. Finally, the filter coefficients are modified to reduce artifacts in the enhanced image by attenuating the unwanted spectral components to a small nonzero value, rather than eliminating them entirely. Also, a Hamming-type window function can be applied to the wedge filters to smooth their sharp edges and, thus,

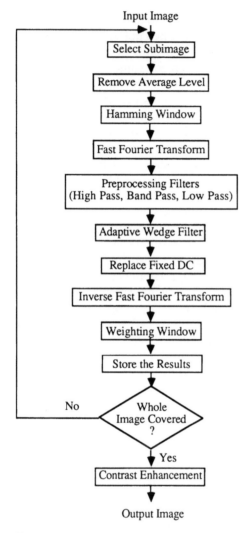

Input Image

Figure 10.7   Flowchart of Fourier short space adaptive filtering algorithm.

further reduce artifacts (Peli/1987). This adaptive-wedge-filter algorithm is applied separately to each subimage, after which the enhanced image is reconstructed as explained earlier.

As an example of the application of this algorithm to fingerprints, we refer to Fig. 10.10. Figure 10.10(a) shows the original image of a slightly degraded fingerprint. The Fourier spectra of the 225 32 × 32-pixel subimages are shown in Fig. 10.10(b). Since most of the spectral

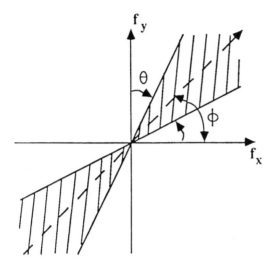

Figure 10.8   Wedge filter.

Input Spectrum of Subimage

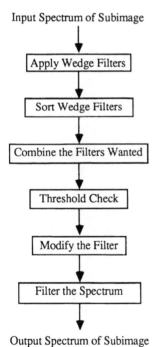

Output Spectrum of Subimage

Figure 10.9   Flowchart of adaptive-wedge-filtering algorithm.

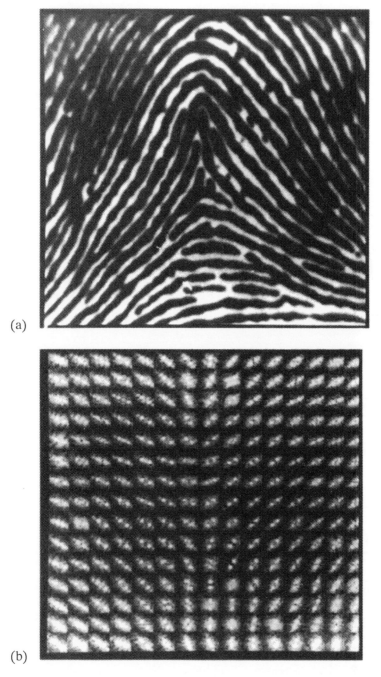

(a)

(b)

Figure 10.10 Example of adaptive wedge filtering: (a) degraded original image, (b) Fourier transforms of subimages from (a), (c) spectra of (b) after adaptive wedge filtering, and (d) final enhanced image. Source: (Krile et al./1987)

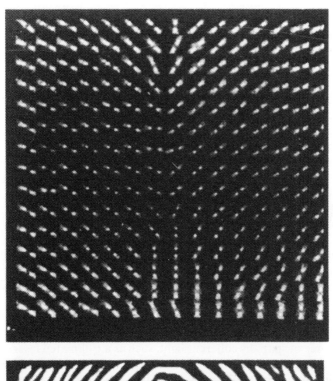

(c)

(d)

energies are located within one wedge band in each subimage, only one wedge filter (that with the highest spectral energy content) was used per subimage filter. The Fourier spectra of the subimages after filtering are shown in Fig. 10.10(c), and the final enhanced image (after binarization) is shown in Fig. 10.10(d). Note that in this case some gaps between ridges have been filled. This is an example of artifacts that can be introduced with neighborhood processing, even when the more common edge-effect artifacts (e.g., ringing) have been carefully eliminated.

A number of point- and neighborhood-type image-processing techniques that are especially applicable to fingerprint enhancement and that are usually implemented on a digital image-processing system have been discussed above. In the section to follow, fingerprint-enhancement techniques using the optical image-processing modality will be discussed.

10.3.2   Optical Fingerprint Enhancement

A second attractive processing modality, especially for linear operations, is optical image processing (Goodman/1968, Goodman/1977, Considine and Gonsalves/1978, Lee/1981, Stark/1982, Horner/1987, Krile et al./1987). The key advantages here are processing speed and the fact that all portions of the fingerprint images are processed in parallel. The basic principle of Fourier (frequency) plane processing is indicated in Fig. 10.11 (Goodman/ 1978). A source of spatially coherent light, such as a collimated laser beam, illuminates an input transparency $f(x,y)$; e.g., a fingerprint (positive or negative), located in plane $P1$. Assuming that the input plane lies in the front focal plane of the positive (converging) lens having focal length f, lens $L1$ forms the Fourier transform, $F(u,v)$, of the input image $f(x,y)$ in

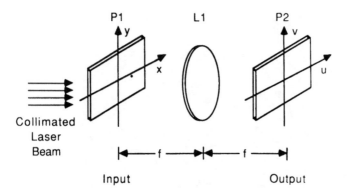

Figure 10.11   A coherent Fourier transform system. Source: (Goodman/1968)

its back focal plane $P2$, where

$$F(u,v)) = \int\!\!\!\int_{-\infty}^{\infty} f(x,y)\exp(-j2\pi(xu+yv)]dxdy \qquad (10.9)$$

where $u = x_f/\lambda f$ and $v = y_f/\lambda f$ are scaled (by the product λf) spatial frequencies (cycles/meter); λ is the wavelength of the coherent source; f, as noted above, is the focal length of the Fourier transforming lens $L1$; and x_f, y_f are the distances measured in meters along the u and v axes, respectively, in plane $P2$.

Given the Fourier transforming system of Fig. 10.11, one can construct the coherent image-processing system of Fig. 10.12 (Goodman/1968). The difference is that now plane $P2$ contains a filter $H(u,v)$ that modifies the spectral content of the input image. Here, both lens $L2$ and lens $L3$ perform direct Fourier transforms. By reversing the directions of the x' and y' axes (as shown) relative to the directions of the x and y (input-plane) axes, one can view the second direct Fourier transform as an inverse Fourier transform, so that the output-plane *intensity* (modulus squared of complex field) pattern is given by

$$\begin{aligned} g(x',y') &= |\mathcal{F}^{-1}[F(u,v)H(u,v)]|^2 \\ &= |f(x,y)*h(x,y)|^2 \end{aligned} \qquad (10.10)$$

where $\mathcal{F}^{-1}$ indicates the inverse Fourier transform and $*$ indicates a convolution operation. Equation (10.10) indicates that the $P3$ plane output image is indeed a linearly filtered (and hopefully enhanced) version of the $P1$-plane input image.

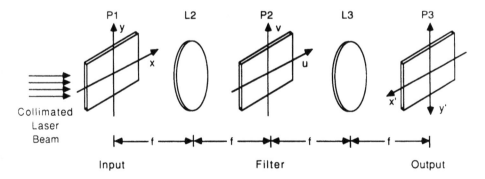

Figure 10.12 A coherent image-processing system. Source: (Goodman/1968)

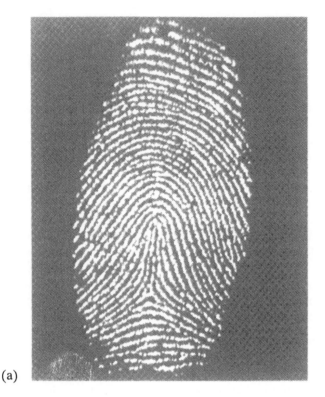

(a)

Figure 10.13 Tented-arch characteristics: (a) fingerprint; (b) optical Fourier spectrum. Source: (Krile et al./1987)

It is worth noting that by using essentially real-time spatial light modulators in plane $P1$, data can be rapidly input into the system, thus increasing its throughput dramatically. A drawback at this time is the fact that we are essentially restricted to linear processing operations. It should be noted, however, that such linear image-processing systems can be small and inexpensive and clearly can be very fast, so these are powerful alternatives to digital processing techniques, as will now be demonstrated using examples.

Figures 10.13 and 10.14 demonstrate two basic fingerprint patterns— Figs. 10.13(a) and 10.14(a)—and their corresponding Fourier spectra— Figs. 10.13(b) and 10.14(b)—as obtained using the optical processing system of Fig. 10.11. Although each print has a unique spatial-frequency spectral signature, all the basic print types of Fig. 10.1 have certain features in common based on their structures. Since the predominant ridge patterns are periodic, with small variations in period, the frequency spectra are

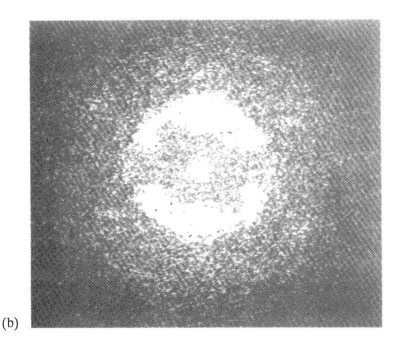

(b)

characterized by circular bands of spatial frequencies whose extents follow from the basic ridge orientations in the print. To illustrate, Fig. 10.13(b) indicates a ring with gaps on the sides due to the fact that very few of the ridges in Fig. 10.13(a) are oriented exactly vertically. In contrast, the twinned loop of Fig. 10.14(a) is characterized by the fact that the vast majority of the ridges run vertically, so that Fig. 10.14(b) has most of the spatial-frequency spectral energy on the sides.

It should be noted here, while discussing the spectral-domain characterization of fingerprints, that each of the fine detail minutiae of Fig. 10.2 have unique Fourier spectra that provide signatures for them and whose spectral energy lies in the bands seen in Figs. 10.13(b) and 10.14(b). These have also been studied in our laboratory (Olimb/1986), and the interested reader should consult the reference cited.

A final comment on Figs. 10.13 and 10.14 is that, due to the essentially binary nature of most fingerprints, harmonic energy is frequently visible in the Fourier spectra (i.e., rings at larger radii), as is noise, some of which may be due to the presence of dust or dirt particles, and is associated with the coherence of the source (i.e., laser speckle).

We now consider the use of coherent optical filtering for the purpose of enhancing degraded fingerprints. Here again, as in the digital processing

(a)

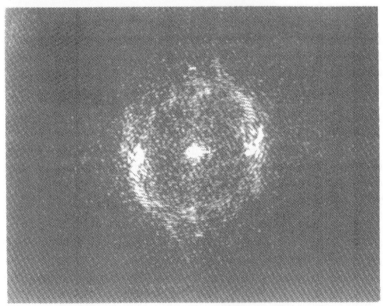

(b)

Figure 10.14 Twinned loop characteristics: (a) fingerprint; (b) Fourier spectrum. Source: (Krile et al./1987)

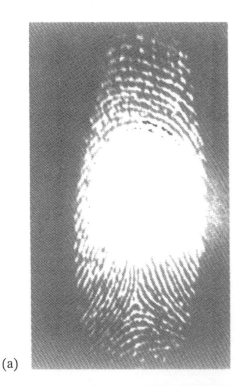

(a)

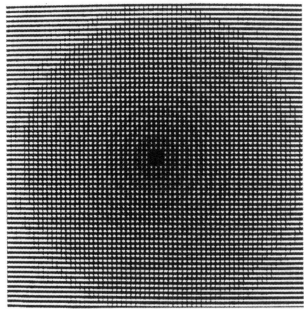

(b)

Figure 10.15 Example of enhancement by optical spatial filtering: (a) degraded fingerprint, (b) computer-generated Laplacian frequency plane filter, (c) Laplacian-enhanced image, and (d) high-pass-filtered enhanced image. Source: (Krile et al./ 1987)

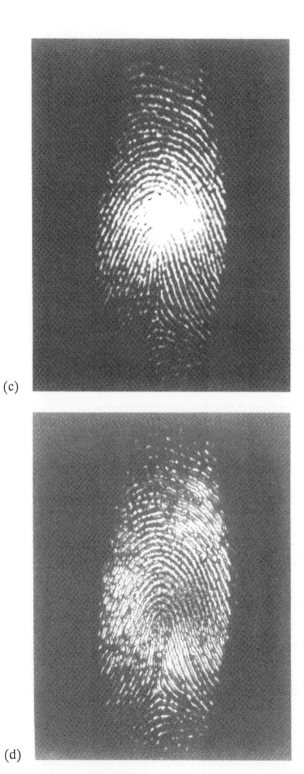

(c)

(d)

Figure 10.15 (Continued)

modality, a variety of processing techniques are available, ranging from simple high-pass filtering to contrast reversal to the use of computer-generated filter masks for Laplacian edge enhancement.

As shown in Fig. 10.15(a), a degraded or smudged print may be characterized by a low-frequency blur or smudge that obliterates much of the central detail. An attractive approach here is the use of a Laplacian filter that performs a two-dimensional linear-differentiation operation on the input, an operation characterized by

$$\nabla^2 f(x,y) = \frac{\partial^2 f(x,y)}{\partial x^2} + \frac{\partial^2 f(x,y)}{\partial y^2} \qquad (10.11)$$

where $f(x,y)$ is the 2-D input fingerprint. This filter must clearly bring out the high-spatial-frequency terms over the low-frequency terms and, thus, accentuate edges while suppressing the slowly varying smudge function.

Using the well-known Burckhardt method for computer-generated holograms (Burckhardt/1970), a computer-generated Laplacian filter mask was prepared, as shown in Fig. 10.15(b). A photoreduced version of this mask was then placed in the filter plane of Fig. 10.12, yielding the enhanced image of Fig. 10.15(c). One clearly notes the fact that much of the smudging has effectively been removed. A less elegant filter is a simple high-pass filter achieved by placing an opaque dot (whose radius is proportional to the frequency band to be blocked) on axis in the frequency plane ($P2$). This results in the edge-enhanced image of Fig. 10.15(d). Note the sharp outlining of the edge ridges due to this high-pass filtering and the complete removal of the low-frequency smudge noticeable in both Figs. 10.15(a) and 10.15(c).

For details on other optical enhancement algorithms (e.g., Schlieren filters) performed in our laboratories, the reader is referred to Olimb/1986. The point we want to make here is that, while they may lack some of the precision and flexibility of digital filtering algorithms, optical processors are very fast, can be made compactly, and can produce enhancements adequate for many practical applications.

10.4 CONCLUSIONS

This chapter has provided an overview of some digital and optical modalities for enhancing degraded fingerprints. Enhancement is a first step toward further processing, leading to efficient encoding, storage, and recognition of latent prints. We believe that this remains a rich area for research, particularly in view of the rapid advances in the speed and processing capabilities of digital computers, particularly microcomputers, and in the capabilities of electro-optical signal processors. For example, one might use synthetic discriminant-function filters to detect and locate minutiae at arbitrary positions and rotations in a fingerprint and, thus, form a feature vector for later recognition tasks. Also, data-compression techniques such as Gaussian pyramid coding could be used for more efficient transmission and storage of fingerprints. Both of these operations could be accomplished with hybrid electro-optical systems that combine the flexibility of digital processing and the high throughput of optical processing. Given the potential payoff to society due to the application of such high-technology tools in criminalistics, one must ask whether we can afford not to actively pursue their application.

ACKNOWLEDGMENTS

This research was supported in part by ALM, Inc., and was performed in conjunction with the Center for Forensic Studies at Texas Tech University. Collaboration with Dr. Roland Menzel, Director of the Center for Forensic Studies, is acknowledged. Most of the research was carried out by our graduate students, notably Mr. Adonis Barsallo, Lt. Hal Olimb, and Mr. Jaw-Horng (Tony) Tarng. The assistance of Mrs. Pansy Burtis and Mr. Ravi Budruk with the typing of the manuscript and the figures, respectively, is gratefully acknowledged.

REFERENCES

Andrews, H. C., and Hunt, B. R. (1977). *Digital Restoration*, Prentice-Hall, Englewood Cliffs, New Jersey.

Andrews, H. C. (1976). Monochrome digital image enhancement, *Appl. Opt.*, *15*: 495.

Barsallo, A. E. (1985). *Digital Enhancement of Degraded Fingerprints*, M.S. Thesis, Dept. of Electrical Engineering, Texas Tech University, Lubbock, Texas.

Baxes, G. A. (1984). *Digital Image Processing, A Practical Primer*, Prentice-Hall, New York.

Burckhardt, C. B. (1970). A simplification of Lee's method of generating holograms by computer, *Appl. Opt*, *9*: 1949.

Chiralo, R. and Berdon, L. (1978). "Adaptive Digital Enhancement of Latent Fingerprints," Proceedings of the 1978 Carnahan Conference on Crime Countermeasures, Lexington, Kentucky.

Considine, P. S. and Gonsalves, R. A. (1978). Optical image enhancement and image restoration, *Topics in Applied Physics: Optical Data Processing* 12, (D. Casasent, ed. Springer-Verlag, Berlin Heidelberg.

Ekstrom, M. P., ed. (1984). *Digital Image Processing Techniques*, Academic Press, New York.

Federal Bureau of Investigation (FBI) (1963). *The Science of Fingerprints*, U. S. Government Printing Office.

Goodman, J. W. (1977). Operations achievable with coherent optical information processing systems, *Proc. IEEE*, *65*: 29–38.

Goodman, J. W. (1968). *Introduction to Fourier Optics*, McGraw-Hill Book Co., San Francisco.

Gonzales, R. C. and Wintz, P. (1987). *Digital Image Processing*, Second Edition, Addison Wesley, Reading, Massachusetts.

Horner, J. L., ed. (1987). *Optical Signal Processing*, Academic Press, San Diego.

Kasturi, R. and Walkup, J. F. (1986). Nonlinear image restoration in signal-dependent noise, *Advances in Computer Vision and Image Processing*, Vol. 2 (T. S. Huang, ed.) JAI Press, Greenwich, Connecticut, pp. 167–212.

Krile, T. F., Walkup, J. F., Barsallo, A., Olimb, H., and Tarng, J-H (1987). Applications of image processing in criminalistics, *Proc. Soc. Photo-Optical Instrumentation Engineers*, *743*: 191–197.

Lee, S. H. (19810. Coherent optical processing, *Topics in Applied Physics: Optical Information Processing*, Vol. 48 (S. H. Lee, ed.), Springer-Verlag, Berlin Heidelberg, pp. 60–65.

Lehar, A. F. and Stevens, R. J. (1984). Image processing systems for enhancement and deblurring of photographs, *Opt. Engr.*, *23*: 303.

Niblack, W. (1986). *An Introduction to Digital Image Processing*, Prentice-Hall International (UK), London.

Olimb, Hal E. (1986) *Optical Enhancement of Degraded Fingerprints*, M.S. Thesis, Dept. of Electrical Engineering, Texas Tech University, Lubbock, Texas.

Peli, E. (1987). Adaptive enhancement based on a visual model, *Opt. Engr.*, *26*: 655.

Pratt, W. (1978). *Digital Image Processing*, John Wiley & Sons, New York.

Rosenfeld, A. and Kak, A. (1976). *Digital Picture Processing*, Academic Press, New York.

Stark, Henry (1982). *Applications of Optical Fourier Transforms*, Academic Press, New York.

Tarng, J. W. (1987). *Advanced Techniques for Digital Image Processing*, M.S. Thesis, Dept. of Electrical Engineering, Texas Tech University, Lubbock, Texas.

11

The Digital Morphological Sampling Theorem

Robert M. Haralick and Xinhua Zhuang*

*University of Washington
Seattle, Washington*

Charlotte Lin and James Lee

*Boeing High Technology Center
Seattle, Washington*

11.1 INTRODUCTION

Morphological operations on images have relevance to conditioning, labeling, grouping, extracting, and matching image-processing operations. Thus, from low-level to intermediate to high-level vision, morphological techniques are important. Indeed, many successful machine-vision algorithms employed in industry on the factory floor, processing thousands of images per day in each application, are based on morphological techniques. Among the recent research papers on morphology are (Crimmons and Brown/1985, Lee et al./1987a, Haralick et al./1987b, Maragos and Schafer/1987). Serra/ 1982 constitutes a comprehensive reference.

Many well-known relationships worked out in the classical context of the convolution operation have morphological analogs. In this chapter, we introduce the digital morphological sampling theorem, which relates to morphology as the standard sampling theorem relates to signal processing

---

*Present affiliation: University of Missouri–Columbia, Columbia, Missouri

and communications. The sampling theorem permits the development of a precise, multiresolution approach to morphological processing.

Multiresolution techniques (Ahuja and Swamy/1984, Klinger/1984, Tanimoto/1982, Uhr/1983) have been useful for at least two fundamental reasons: (1) the representation they provide naturally permits a computational mechanism to focus on objects or features likely to be at least a given specified size (Crowley/1984, Miller and Stout/1983, Rosenfeld/1983, Witkin/1984), and (2) the computational mechanism can operate on only those resolution levels that just suffice for the detection and localization of objects or features of specified size, while significantly reducing the number of operations performed (Burt/1984, Dyer/1982, Lougheed and McCubbrey/1980).

The usual resolution hierarchy, called a pyramid, is produced by low-pass filtering and then sampling to generate the next lower resolution level of the hierarchy. The basis for a morphological pyramid requires a morphological sampling theorem that explains how an appropriately morphologically filtered and sampled image relates to the unsampled image. It must explain what kinds of shapes are preserved and what kinds are suppressed or eliminated. It must explain the relationship between performing a less costly morphological filtering operation on the sampled image and performing the more costly equivalent morphological filtering operations on the original image. It is just these issues that we address in this chapter.

We analyze the constraints on sampling and on image objects to speed up morphological operations without sacrificing accurate shape analysis. The results are shown to be true under reasonable morphological sampling conditions. Before sets are sampled, they must be morphologically simplified by an opening or a closing. Such sampled sets can be reconstructed in two ways, by either a closing or a dilation. In both reconstructions, the sampled reconstructed sets are equal to the sampled sets. A set contains its reconstruction by closing and is contained in its reconstruction by dilation; indeed, these are extremal bounding sets. That is, the largest set that downsamples to a given set is its reconstruction by dilation; the smallest is its reconstruction by closing. Furthermore, the distance from the maximal reconstruction to the minimal reconstruction is no more than the diameter of the reconstruction-structuring element. Morphological sampling thus provides reconstructions positioned only to within some spatial tolerance that depends on the sampling interval. This spatial limitation contrasts with the sampling-reconstruction process in signal processing, from which only those frequencies below the Nyquist frequency can be reconstructed.

All set morphological relationships are immediately generalizable to gray-scale morphology via the umbra homomorphism theorems. For gray-scale images, the bounds that the reconstruction establishes are bounds in a spatial sense.

In Sec. 11.2, we review the basic definitions and properties for binary morphology operations. In Sec. 11.3, we develop the morphological sampling theorem for binary morphology. The homomorphism theorem between binary and gray-scale morphology implies that each result in binary morphology has a corresponding result in gray-scale morphology. Sec. 11.4 develops these gray-scale generalizations. Sec. 11.5 discusses the computational advantages of operating on morphologically sampled images and shows how successively sampled images can be operated on in a resolution hierarchy called a pyramid. The final section summarizes the key points and conclusions.

11.2 PRELIMINARIES

Let E denote the set of numbers used to index a row or column position on a binary image. We assume that the addition and subtraction operations are defined on E. The binary image itself can then be thought of as a subset of $E \times E$. Pixels are in this subset if and only if they have the binary value one on the image. This correspondence permits us to work with sets rather than with image functions; indeed, with sets in E^N. The first two operations of mathematical morphology are the dual operations of dilation and erosion. The *dilation* of a set $A \subseteq E^N$ with a set $B \subseteq E^N$ is defined by

$$A \oplus B = \{x \mid \text{for some } a \in A \text{ and } b \in B, x = a + b\}.$$

The *erosion* of A by B is defined by

$$A \ominus B = \{x \mid \text{for every } b \in B, x + b \in A\}.$$

The careful reader should be aware that the symbol $\ominus$ used by (Serra/1982) does not designate erosion. Rather, it designates Minkowski subtraction.

For any set $A \subseteq E^N$ and $x \in E^N$, let A_x denote the *translation* of A by x;

$$A_x = \{y \mid \text{for some } a \in A, y = a + x\}.$$

For any set $A \subseteq E^N$, let $\check{A}$ denote the *reflection* of A about the origin;

$$\check{A} = \{x \mid \text{for some } a \in A, x = -a\}.$$

Relationships satisfied by dilation and erosion include the following:

$$A \oplus B = B \oplus A$$

$$(A \oplus B) \oplus C = A \oplus (B \oplus C) \qquad (A \ominus B) \ominus C = A \ominus (B \oplus C)$$

$$(A \cup B) \oplus C = (A \oplus C) \cup (B \oplus C) \qquad (A \cap B) \ominus C = (A \ominus C) \cap (B \ominus C)$$

$$A \oplus B = \bigcup_{b \in B} A_b \qquad A \ominus B = \bigcap_{b \in B} A_{-b}$$

$$A \subseteq B \Longrightarrow A \oplus C \subseteq B \oplus C \qquad A \subseteq B \Longrightarrow A \ominus C \subseteq B \ominus C$$

$$(A \cap B) \oplus C \subseteq (A \oplus C) \cap (B \oplus C) \qquad (A \cup B) \ominus C \supseteq (A \ominus C) \cup (B \ominus C)$$

$$(A \oplus B)^C = A^C \ominus \check{B} \qquad\qquad\qquad A \ominus (B \cup C) = (A \ominus B) \cap (A \ominus C)$$

In practice, dilations and erosions are usually employed in pairs, either dilation of an image followed by the erosion of the dilated result, or image erosion followed by dilation. In either case, the result of iteratively applied dilations and erosions is an elimination of specific image detail smaller than the structuring element, without the global geometric distortion of unsuppressed features. For example, opening an image with a disk-structuring element smooths the contour, breaks narrow isthmuses, and eliminates small islands and sharp peaks or capes. Closing an image with a disk-structuring element smooths the contours; fuses narrow breaks and long, thin gulfs; eliminates small holes; and fills gaps on the contours.

Of particular significance is the fact that image transformations employing iteratively applied dilations and erosions are idempotent; that is, their reapplication effects no further changes to the previously transformed result. The practical importance of idempotent transformations is that they comprise complete and closed stages of image-analysis algorithms, because shapes can be naturally described in terms of under what structuring elements they can be opened or can be closed and yet remain the same. Their functionality corresponds closely to the specification of a signal by its bandwidth. Morphologically filtering an image by an opening or closing operation corresponds to the ideal nonrealizable bandpass filters of conventional linear filtering. Once an image is ideally band-passed filtered, further ideal band-pass filtering does not alter the result.

These properties motivate the importance of opening and closing, concepts first studied by (Matheron/1975), who was interested in axiomatizing the concept of size. Both Matheron's/1975 definitions and Serra's/1982 definitions for opening and closing are identical to the ones given here, but their formulas appear different because they use the symbol $\ominus$ to mean Minkowski subtraction rather than erosion.

The morphological filtering operations of opening and closing are made up of dilation and erosion performed in different orders. The *opening* of A by B is defined by

$$A \circ B = (A \ominus B) \oplus B.$$

The *closing* of A by B is defined by

$$A \bullet B = (A \oplus B) \ominus B.$$

Opening and closing satisfy the following basic relationships:

$$(A \circ B) \circ B = A \circ B \qquad (A \bullet B) \bullet B = A \bullet B$$

$$A \circ B \subseteq A \qquad\qquad A \subseteq A \bullet B$$

$$A \subseteq B \Longrightarrow A \circ C \subseteq B \circ C \qquad A \subseteq B \Longrightarrow A \bullet C \subseteq B \bullet C$$

$$(A \circ B)^C = A^C \bullet \check{B} \qquad (A \bullet B)^C = A^C \circ \check{B}$$

The reason that openings and closings deal directly with shape properties is apparent from the following representation theorem for openings:

$$A \circ B = \left\{ x \mid \text{for some } y, \ x \in B_y \subseteq A \right\}.$$

A opened by B contains only those points of A that can be covered by some translation B_y, which is, in turn, entirely contained inside A. Thus, x is a member of the opening if it lies in some area inside A that entirely contains a translated copy of the shape B. In this sense, A opened by B is the set of all points of A that can participate in areas of A that match B. If B is a disk of diameter d, for example, then $A \circ B$ would be that part of A that in no place is narrower than d.

The duality relationship $(A \circ B)^C = A^C \bullet \check{B}$ between opening and closing implies a corresponding representation theorem for closing:

$$A \bullet B = \left\{ x \mid x \in \check{B}_y \text{ implies } \check{B}_y \cap A \neq \varnothing \right\}.$$

A closed by B consists of all those points x for which x being covered by some translation $\check{B}_y$ implies that $\check{B}_y$ "hits" or intersects some part of A. A more extensive discussion of these relationships can be found in (Haralick et al./1987b).

11.3 THE BINARY DIGITAL MORPHOLOGICAL SAMPLING THEOREM

The preliminary part of this section sets the stage, discussing the appropriate morphological simplifying and filtering to be done before sampling. Certain relationships must be satisfied between the sampling set and the structuring element used for reconstruction. The main body of the section discusses two kinds of reconstructions of the sampled images: a maximal reconstruction accomplished by dilation and a minimal reconstruction accomplished by closing. Fundamental set-bounding relationships are proved, which show that the closing reconstruction of a set must be contained in the set itself which, in turn, must be contained in its dilation reconstruction. The closing reconstruction differs from the dilation reconstruction by just a dilation by the reconstruction-structuring element, so the set-bound relationships translate to geometric-distance relationships. The section concludes by defining a suitable set-distance function that measures the distance between the sampled set and the morphologically filtered set. The distance between the minimal reconstruction and the maximal reconstruction, and the distance between the morphologically filtered set and either of its reconstructions, are all less than the sampling distance.

The first conceptual issue that arises in developing a morphological sampling theorem is how to remove small objects, object protrusions, object intrusions, and holes before sampling. It is exactly the presence of this kind of small detail before sampling that causes the sampled result to be unrepresentative of the original; just as in signal processing, the presence of frequencies higher than the Nyquist frequency causes the sampled signal to be unrepresentative of the original signal. This *aliasing* means that signals must be low-pass filtered before sampling. Likewise, in morphology, the sets must be morphologically filtered and simplified before sampling. Small objects and object protrusions can be eliminated by a suitable opening operation. Small object intrusions and holes can be eliminated by a suitable closing. Since opening and closing are duals, we develop our motivation by considering just the opening operation.

Opening a set F by a structuring element K to eliminate small details of F raises, in turn, the issue of how K should relate to the sampling set S. If the sample points of S are too finely spaced, little will be accomplished by the reduction in resolution. On the other hand, if S is too coarse relative to K, objects preserved in the opening may be missed by the sampling. S and K can be coordinated by demanding that there be a way to reconstruct the opened image from the sampled opened image. Of course, details smaller than K are removed by the opening and cannot be reconstructed.

One natural way to reconstruct a sampled opening is by dilation. If S and K were coordinated to make the reconstructed image (first opened, then sampled, and then dilated) the same as the opened image, we would have a morphological sampling theorem nearly identical to the standard sampling theorem of signal processing. However, morphology cannot provide a perfect reconstruction, as is illustrated by the following one-dimensional continuous-domain example.

Let the image F be the union of three topologically open intervals,

$$F = (3.1, 7.4) \cup (11.5, 11.6) \cup (18.9, 19.8),$$

where (x, y) denotes the topologically open interval between x and y. We can remove all details of less than length 2 by opening with the structuring element $K = (-1, 1)$ consisting of the topologically open interval from -1 to 1. Then, the opened image $F \circ K = (3.1, 7.4)$. What should the corresponding sample set be? Consider a sampling set $S = \{x \mid x \text{ an integer}\}$, with a sample spacing of unity; other spacings such as 0.2, 0.5, or 0.7 could illustrate the same sampling concept as well. The sampled opened image $(F \circ K) \cap S = \{4, 5, 6, 7\}$. Dilating by K to reconstruct the image produces $[(F \circ K) \cap S] \oplus K = (3, 8)$, an interval that properly contains $F \circ K$. The dilation fills in between the sample points, but cannot "know" to expand on the left end by a length of 0.9 and yet expand by 0.4 on the right end. However, the reconstruction is the largest one for which the sampled reconstruction $\{[(F \circ K) \cap S] \oplus K\} \cap S$ produces the sampled opening $(F \circ K) \cap S = \{4, 5, 6, 7\}$. This is easily seen in the example, because substituting the closed interval $[3, 8]$ for the open interval $(3, 8)$ produces the sampled closed interval $[3, 8] \cap S = \{3, 4, 5, 6, 7, 8, \}$, which properly contains $(F \circ K) \cap S = \{4, 5, 6, 7\}$.

The difficulty in reconstructing a sampled opened image morphologically can be understood in terms of the standard sampling theorem. Consider the case of a precise constant binary-valued image. The required morphological simplification means that details smaller than K have been removed from all objects on the opened image, but this removal does not band-limit the image. In fact, the opened image belongs to a special class of infinite bandwidth signals, wherein reconstructing the sampled opened image as specified by the standard sampling theorem cannot produce the kind of aliasing found in Moire patterns. The standard sampling theorem reconstruction produces a band-limited signal that passes through the sample points. Thus, the step-like patterns, like the open intervals of F, get reconstructed with ringing throughout and with overshoot and undershoot at step edges. By contrast, the morphological reconstruction cannot produce ringing, but the position of any step edge is uncertain within the sampling interval.

In the remainder of this section, we give a complete derivation of the results illustrated in the example. First, note that to use a structuring element K as a reconstruction kernel, K must be large enough to ensure that the dilation of the sampling set S by K covers the entire space E^N. For technical reasons apparent in the derivations, we also require that K be symmetric, $K = \check{K}$. In the standard sampling theorem, the period of the highest frequency present must be sampled at least twice in order to properly reconstruct the signal from its sampled form. In mathematical morphology, there is an analogous requirement. The sample spacing must be small enough that the diameter of K is just smaller than these two sample intervals. Hence, the diameter of K is large enough that it can contain two sample points, but not three sample points. We express this relationship by requiring that

$$x \in K_y \implies K_x \cap K_y \cap S \neq \emptyset$$
$$\text{and } K \cap S = \{0\}.$$

The first condition implies that the dilation of sample points fills the whole space; that is, $S \oplus K = E^N$ when K is not empty. If the points in the sampling set S are spaced no further than d apart, then the corresponding reconstruction kernel K could be the topologically open ball of radius $2d$. In this case, $x \in K_y \Rightarrow K_x \cap K_y \cap S \neq \emptyset$. Notice that two points that are d apart can lie on the diameter of d. But, since the ball is topologically open, the diameter cannot contain three points spaced d apart. Hence, the radius of K is just smaller than the sampling interval. Also notice that if a sample point falls in the center of K, K will not contain another sample point.

Why does the morphological sampling theorem we develop here pertain mainly to the digital domain? Consider the two-dimensional continuous case in which there is a regular square-grid sampling, with the sample interval in each direction being of length L. To guarantee that $K \cap S = \{0\}$, the biggest possible disk K is the open disk having radius L.

The difficulty occurs with the condition $x \in K_y \Rightarrow K_x \cap K_y \cap S \neq \emptyset$. Figure 11.1 shows a square whose length L side is the sampling interval. It also shows several translates of K and a disk of the radius L. Select two points that are no further from each other than distance L in the following way. Take one point x to be in the interior of one shaded region of Fig. 11.1. Take the other point y to be opposite it and interior to the other shaded region. With this selection, the distance between the two points is guaranteed to be less than L. Yet, it is apparent from the geometry that since none of the four open disks of diameter L can contain the two points that are distance L apart, the condition $x \in K_y \Rightarrow K_x \cap K_y \cap S \neq \emptyset$ cannot be satisfied. This is because each open disk represents exactly the set of points each having the property that, if an open disk were centered at the point, the open disk would contain a sample point. Hence, with the disk K being defined by the L_2 norm, there can be no morphological sampling theorem in the continuous case. In fact, the only norm by which K can be defined that yields a morphological sampling theorem in the continuous case is the L_∞ norm.

Because the shaded region in Fig. 11.2 is so narrow, this difficulty does not arise in the digital case. Suppose that the original domain is discrete with nearest points at distance 1 from each other. Then, the condition $x \in K_y \Rightarrow K_x \cap K_y \cap S \neq \emptyset$ is easily satisfied for any $L \in \{2, 3, 4, 5, 6, 7\}$, since in this case $L(1 - \frac{\sqrt{3}}{2}) < 1$. Figure 11.2 illustrates the case where the

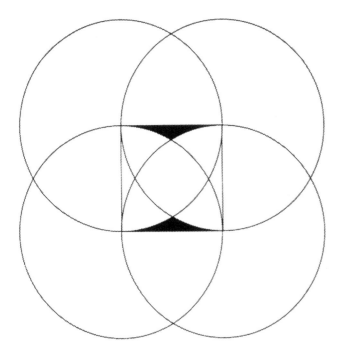

Figure 11.1 Illustration of how two points can be chosen no further apart than the sample distance, yet there is no sample point that is simultaneously less distant than the sample distance to each of them. Take the two points to be opposite each other and interior to one of the shaded regions. Consider the sample points to be the corners of the square.

sample interval L is 6. Notice that the distance between any pair of digital points, one from a region corresponding to one of the shaded regions of Fig. 11.1 and the other from the region opposite it, must be greater than L. Hence, for any two such points x and y, it is not the case that $x \in K_y$. So, the difficulty with $x \in K_y$ and $K_x \cap K_y \cap S = \varnothing$ cannot arise.

We now prove some propositions that lead to the binary morphological sampling theorem. In what follows, the set $F \subseteq E^N$, the reconstruction-structuring element, will be denoted by $K \subseteq E^N$, and the sampling set will be denoted by $S \subseteq E^N$. Although not necessary for every proposition, we assume that S and K obey the following five conditions:

1. $S = S \oplus S$
2. $S = \check{S}$
3. $K \cap S = \{0\}$
4. $K = \check{K}$

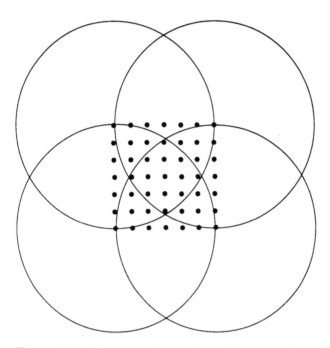

Figure 11.2 Illustration of how the condition $x \in K_y \Rightarrow K_x \cap K_y \cap S \neq \emptyset$ can be satisfied in the digital case where K is a circular disk. Notice that for any pair of digital points in the regions corresponding to the shaded region of Fig. 11.1, the distance between them is not less than the radius of the open disk K. Hence, the difficulty illustrated in Fig. 11.1 cannot arise.

> 5. $a \in K_b \Rightarrow K_a \cap K_b \cap S \neq \emptyset$

Figure 11.3 illustrates the S associated with a 3-to-1 down-sampling. Figure 11.4 illustrates a structuring element K satisfying conditions (3), (4), and (5). Since the dilation operation is commutative and associative, conditions (1) through (3) imply that the sampling set S with the dilation operation comprises an abelian group with the origin being its unit element. Thus, if $x \in S$, then $S_x = S$, and also, since $K \cap S = \{0\}$, $x \in S$ implies $K_x \cap S = \{x\}$. Both these facts are used in a number of the proofs to follow.

11.3.1 The Set-Bounding Relationships

It is obvious that since $0 \in K$, the reconstruction of a sampled set $F \cap S$ by dilation with K produces a superset of the sampled set $F \cap S$. That is, $F \cap S \subseteq (F \cap S) \oplus K$. The reconstruction by dilation is open, so that $[(F \cap S) \oplus K] \circ K = (F \cap S) \oplus K$. Moreover, as stated in the next

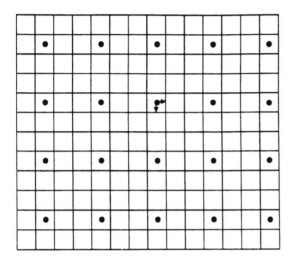

Figure 11.3 Illustration of sampling every third pixel by row and by column. The sampling set S is represented by all points that are shown as •.

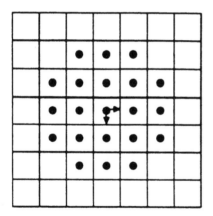

Figure 11.4 Illustration of a symmetric structuring element K that is a digital disk of radius $\sqrt{5}$. For the sampling set S of Fig. 11.3, $K \cap S = \{0\}$ and $x \in K_y$ implies $K_x \cap K_y \cap S \neq \emptyset$.

proposition, the erosion and dilation of the original image F by K bound the reconstructed sampled image.

(A) Proposition 1

Let $F, K, S \subseteq E^N$. Suppose $S \oplus K = E^N$ and $K = \check{K}$. Then, $F \ominus K \subseteq (F \cap S) \oplus K \subseteq F \oplus K$.

Proposition 1 shows that the reconstruction by dilation cannot be too far away from F, since the reconstruction is constrained to lie between F eroded by K and F dilated by K. Proposition 2 strengthens the closeness between F and the dilation reconstruction $(F \cap S) \oplus K$. Sampling F and sampling the dilation reconstruction of F produce identical results.

(B) Proposition 2

Let $F, K, S \subseteq E^N$. Suppose $K \cap S = \{0\}$. Then, $F \cap S = [(F \cap S) \oplus K] \cap S$.
  Proof:

$$[(F \cap S) \oplus K] \cap S = \left[\bigcup_{x \in F \cap S} K_x \right] \cap S$$

$$= \bigcup_{x \in F \cap S} K_x \cap S$$

$$= \bigcup_{x \in F \cap S} K_x \cap S_x$$

$$= \bigcup_{x \in F \cap S} (K \cap S)_x$$

$$= \bigcup_{x \in F \cap S} \{0\}_x$$

$$= F \cap S$$

From this result, it rapidly follows that sampling followed by a dilation reconstruction is an idempotent operation. That is, $([(F \cap S) \oplus K] \cap S) \oplus K = (F \cap S) \oplus K$.

Considering sampling followed by reconstruction as an operation, we discover that it is an increasing operation, distributed over union but not over intersection. That is:

1. $F_1 \subseteq F_2$ implies $(F_1 \cap S) \oplus K \subseteq (F_2 \cap S) \oplus K$
2. $[(F_1 \cup F_2) \cap S] \oplus K = [(F_1 \cap S) \oplus K] \cup [(F_2 \cap S) \oplus K]$
3. $[(F_1 \cap F_2) \cap S] \oplus K \subseteq [(F_1 \cap S) \oplus K] \cap [(F_2 \cap S) \oplus K]$

Proposition 3 states that the dilation reconstruction of a sampled F is always a superset of F opened by the reconstruction-structuring element K. Hence, if F is open under K, then F is contained in its dilation reconstruction.

(C) Proposition 3

Let $F, K, S \subseteq E^N$. Suppose $x \in K_y$ implies $K_x \cap K_y \cap S \neq \varnothing$. Then, $F \circ K \subseteq (F \cap S) \oplus K$.

Proof:

Let $x \in F \circ K$. Then, for some $y, x \in K_y \subseteq F$. Since $x \in K_y$, there exists $z \in K_x \cap K_y \cap S$. Now, $z \in K_y \subseteq F$ and $z \in S$ implies $z \in F \cap S$. Also, $z \in K_x$ implies that there exists $k \in K$ such that $z = x + k$. Then, $x = z - k$. Since $K = \check{K}, k \in K$ implies $-k \in K$. Now, since $z \in F \cap S$ and since $-k \in K$, $x = z - k \in (F \cap S) \oplus K$.

It quickly follows that $F \circ K \subseteq [(F \circ K) \cap S] \oplus K$. Thus, the reconstruction of the opened sampled image F is bounded by $F \circ K$ on the low side and $F \circ K$ dilated by K on the high side:

$$F \circ K \subseteq [(F \circ K) \cap S] \oplus K \subseteq (F \circ K) \oplus K.$$

If F is morphologically simplified and filtered so that $F = F \circ K$, then the previous bounds reduce to

$$F \subseteq (F \cap S) \oplus K \subseteq F \oplus K.$$

By reconsidering our example, $F = (3.1, 7.4) \cup (11.5, 11.6) \cup (18.9, 19.8)$, which is not open under $K = (-1, 1)$, we can see that such an F is not necessarily a lower bound for the reconstruction. In this case, $F \cap S = \{4, 5, 6, 7, 19\}$ and the reconstruction $(F \cap S) \oplus K = (3, 8) \cup (18, 20)$, which does not contain F. This suggests that the condition that F be open under K is essential in order to have $F \subseteq (F \cap S) \oplus K$.

We now show one last relation between the reconstruction $(F \cap S) \oplus K$ and F. The reconstruction $(F \cap S) \oplus K$ is the largest open set that, when sampled, produces $F \cap S$.

(D) Proposition 4

Let $A \subseteq E^N$ satisfy $A \cap S = F \cap S$ and $A = A \circ K$. Then, $A \supseteq (F \cap S) \oplus K$ implies $A = (F \cap S) \oplus K$.

Proof:

Suppose $A \supseteq (F \cap S) \oplus K$ and $A \cap S = F \cap S$ and $A = A \circ K$. Since $A \cap S = F \cap S$, $(A \cap S) \oplus K = (F \cap S) \oplus K$. But $A = A \circ K$ implies $A \subseteq (A \cap S) \oplus K = (F \cap S) \oplus K$. Now, $A \subseteq (F \cap S) \oplus K$, together with the supposition $A \supseteq (F \cap S) \oplus K$, implies $A = (F \cap S) \oplus K$.

Thus, we have established the maximality of the reconstruction $(F \cap S) \oplus K$ with respect to the two properties of being open and down-sampling to $F \cap S$. What about a minimal reconstruction? Certainly, we would expect a minimal reconstruction to be contained in the maximal reconstruction and to contain the sampled image. Since closing is extensive, we immediately have $F \cap S \subseteq (F \cap S) \bullet K$. Since $0 \in K$, erosion is an anti-extensive operation.

Hence, $(F \cap S) \bullet K = [(F \cap S) \oplus K] \ominus K \subseteq (F \cap S) \oplus K$. These relations suggest the possibility of a reconstruction by closing. The next proposition shows that a closing reconstruction has set bounds similar to those of the dilation reconstruction.

(E) Proposition 5

Let $F, K, S \subseteq E^N$. If $K = \check{K}$ and $x \in K_y$ implies $K_x \cap K_y \cap S \neq \varnothing$ and $0 \in K$, then $F \ominus K \subseteq (F \cap S) \bullet K \subseteq (F \cap S) \oplus K \subseteq F \oplus K$.

Proof:

Let $x \in F \ominus K$. Then, $K_x \subseteq F$. To show $x \in (F \cap S) \bullet K$, we will show that $x + k \in (F \cap S) \oplus K$ for every $k \in K$. So, let $k \in K$. Then, $x + k \in K_x$. But $x + k \in K_x$ implies $K_x \cap K_{x+k} \cap S \neq \varnothing$. Hence, there exists $s \in K_x \cap K_{x+k} \cap S$. Now, $s \in K_x$ and $K_x \subseteq F$ implies $s \in F$. Hence, $s \in F \cap S$. Furthermore, $s \in K_{x+k}$ implies there exists $k' \in K$, such that $s = x + k + k'$. Rearranging $s - k' = x + k$. But $K = \check{K}$, so $k' \in K$ implies $-k' \in K$. Now, $s \in F \cap S$ and $-k' \in K$ implies $s - k' \in (F \cap S) \oplus K$. Therefore, $F \ominus K \subseteq (F \cap S) \bullet K$. Since $0 \in K, (F \cap S) \bullet K \subseteq (F \cap S) \oplus K$. Since $F \cap S \subseteq F, (F \cap S) \oplus K \subseteq F \oplus K$.

For true reconstruction, the sampled reconstruction should be identical to the sampled image. Indeed, this is the case.

(F) Proposition 6

$[(F \cap S) \bullet K] \cap S = F \cap S$.

Consider our example $F = (3.1, 7.4) \cup (11.5, 11.6) \cup (18.9, 19.8)$, which is closed under $K = (-1, 1)$. If the sampling set S is the integers, then $F \cap S = \{4, 5, 6, 7, 19\}$. Closing $F \cap S$ with K can be visualized via the opening/closing duality $(F \cap S) \bullet K = [(F \cap S)^C \circ \check{K}]^C$. Opening the set $(F \cap S)^C$ with $\check{K} = K$ produces $(F \cap S)^C \circ K = \{x \neq 19 \mid x < 4 \text{ or } > 7\}$. Hence, $(F \cap S) \bullet K = [(F \cap S)^C \circ K]^C = \{x \mid x = 19 \text{ or } 4 \leq x \leq 7\}$, and sampling produces $[(F \cap S) \bullet K] \cap S = \{4, 5, 6, 7, 19\} = F \cap S$.

From the previous proposition, it rapidly follows that sampling followed by a reconstruction by closing is an idempotent operation. That is, $[(F \cap S) \bullet K] \cap S \bullet K = (F \cap S) \bullet K$.

A reconstruction by closing is obviously closed under K. Moreover, it can be quickly determined that

$F_1 \subseteq F_2$ implies $(F_1 \cap S) \bullet K \subseteq (F_2 \cap S) \bullet K$

$[(F_1 \cup F_2) \cap S] \bullet K \supseteq [(F_1 \cap S) \bullet K] \cup [(F_2 \cap S) \bullet K]$

$[(F_1 \cap F_2) \cap S] \bullet K \subseteq [(F_1 \cap S) \bullet K] \cap [(F_2 \cap S) \bullet K]$.

Furthermore, the closing reconstruction of a sampled F is always a subset of F closed by the reconstruction-structuring element K. That is, $(F \cap S) \bullet K \subseteq F \bullet K$, so that $[(F \bullet K) \cap S] \bullet K \subseteq F \bullet K$. Hence, a closing reconstruction of an image that is closed before sampling will be a subset of the closed image.

By considering a simple example, $F = \{0, 1\}$, which is not closed under $K = (-1, 1)$, we can see that F is not necessarily an upper bound for the reconstruction. In this case, $F \cap S = \{0, 1\} = F$ and the reconstruction $(F \cap S) \bullet K = F \bullet K = [0, 1]$, which properly contains F. This suggests that the condition that F be closed under K is essential in order to have $(F \cap S) \bullet K \subseteq F$.

We now show one last relation between the reconstruction $(F \cap S) \bullet K$ and F. The reconstruction $(F \cap S) \bullet K$ is the smallest closed set that, when sampled, produces $F \cap S$.

(G)   Proposition 7

Let $A \subseteq E^N$ satisfy $A \cap S = F \cap S$ and $A = A \bullet K$. Then, $A \subseteq (F \cap S) \bullet K$ implies $A = (F \cap S) \bullet K$.

11.3.2   Examples

To better illustrate the bounding relationships developed in the previous section between a set and its sample reconstructions, we show three simple examples. The domain of these examples is defined as $E \times E$, where E is the set of integers. The sample set S is chosen as the set of even numbers in both row and column directions. Thus,

$S = \{(r, c) \mid r \in E \text{ and is even}; c \in E \text{ and is even}\}.$

K is chosen as a box of size 3×3 whose center is defined as the origin. The sets S, K and the three example sets F_1, F_2, and F_3 are shown in Fig. 11.5. The sets F_1, F_2, and F_3 are 3×3 boxes having different origins, and the condition $F = F \circ K$ holds for all these example sets.

The results of $F \ominus K$, $F \cap S$, $(F \cap S) \bullet K$, $(F \cap S) \oplus K$, and $F \oplus K$ for sets F_1, F_2, and F_3 are shown in Figs. 11.6, 11.7, and 11.8, respectively.

(A)   Example 1

All the pixels contained in the vertical boundaries of F_1 have even column coordinates, and those in the horizontal boundaries of F_1 have even row coordinates. Since the sample set S consists of pairs of even numbers and F_1 is a 3×3 box, the set $F_1 \cap S$ consists of the four corner points of F_1 and is contained in the boundary set of F_1. Hence, the closing reconstruction of $F_1 \cap S$ recovers F_1, and the dilation reconstruction of $F_1 \cap S$ is equivalent to $F \oplus K$. In fact, the following two equalities hold only when (1) the sampling is every other row and column, (2) a set's vertical boundaries have even column coordinates, and (3) its horizontal boundaries have even row coordinates:

$(F \cap S) \bullet K = F$ and
$(F \cap S) \oplus K = F \oplus K.$

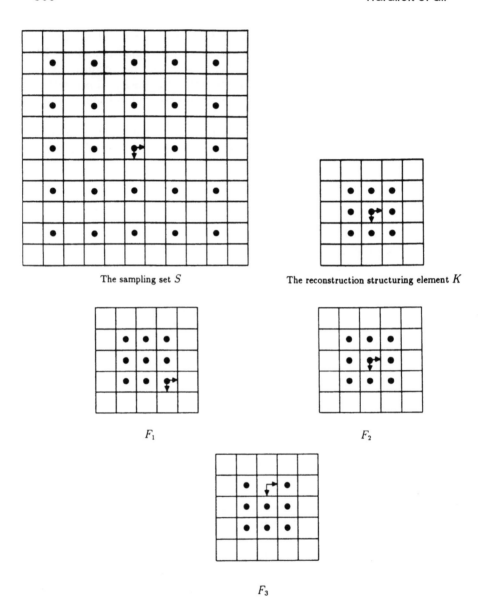

Figure 11.5 Illustration of a sampling set S, a reconstruction-structuring element K, and three sets, F_1, F_2, and F_3, each of which is open under K.

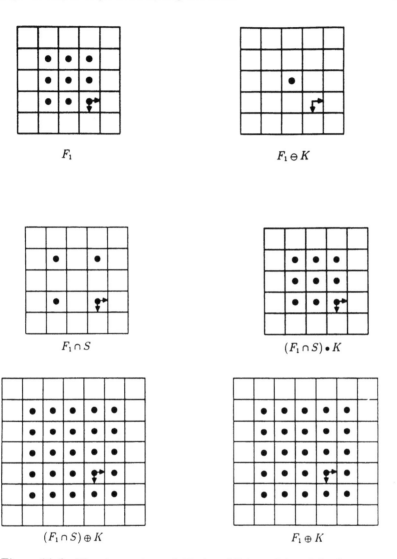

F_1

$F_1 \ominus K$

$F_1 \cap S$

$(F_1 \cap S) \bullet K$

$(F_1 \cap S) \oplus K$

$F_1 \oplus K$

Figure 11.6 How the erosion and dilation of F_1 bound the minimal reconstruction $(F_1 \cap S) \bullet K$ and the maximal reconstruction $(F_1 \cap S) \oplus K$, respectively, which in turn bound F_1 because F_1 is both open and closed under K.

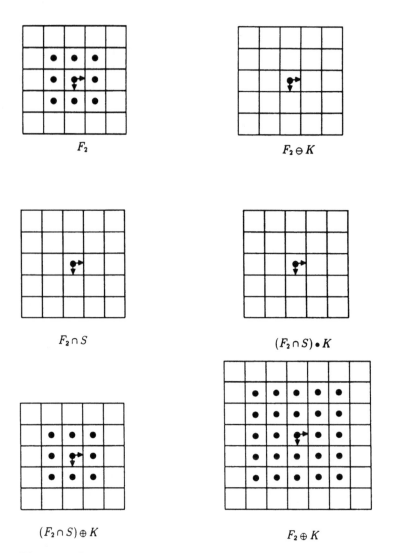

Figure 11.7 A second example of how the erosion and dilation of F_2 bound the minimal reconstruction $(F_2 \cap S) \bullet K$ and the maximal reconstruction $(F_2 \cap S) \oplus K$, respectively, which in turn bound F_2.

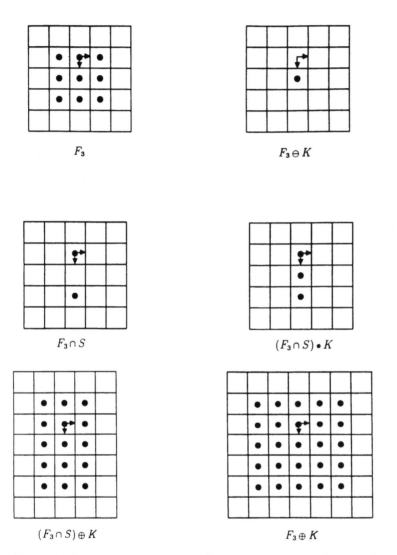

F_3

$F_3 \ominus K$

$F_3 \cap S$

$(F_3 \cap S) \bullet K$

$(F_3 \cap S) \oplus K$

$F_3 \oplus K$

Figure 11.8   A third example of how erosion and dilation of F_2 bound (in this case properly) the minimal reconstruction $(F_3 \cap S) \bullet K$ and the maximal reconstruction $(F_3 \cap S) \oplus K$, respectively, which in turn bound (in this case properly) F_3.

The bounding relationships for F_1, illustrated in Fig. 11.6, are

$$F_1 \ominus K \subseteq (F_1 \cap S) \bullet K = F_1 \subseteq (F_1 \cap S) \oplus K = F_1 \oplus K.$$

(B) Example 2

Since all pixels contained in the vertical boundaries of F_2 have odd column coordinates and those in the horizontal boundaries of F_2 have odd row coordinates and F_2 is a small 3×3 box, the set $F_2 \cap S$ does not contain any part of the boundary of F_2. Thus, the closing reconstruction of $F_2 \cap S$ equals $F_2 \ominus K$, and the dilation reconstruction of $F_2 \cap S$ is equivalent to F_2. Similar to Example 1, the following equalities hold only when the sampling is every other row and column and has its odd-column coordinates in its vertical boundaries and its odd-row coordinates in its horizontal boundaries:

$$F \ominus K = (F \cap S) \bullet K \text{ and}$$
$$F = (F \cap S) \oplus K.$$

The bounding relationships for F_2, illustrated in Fig. 11.7, are

$$F_2 \ominus K = (F_2 \cap S) \bullet K \subseteq F_2 = (F_2 \cap S) \oplus K \subseteq F_2 \oplus K.$$

(C) Example 3

The pixels contained in the vertical boundaries of F_3 have odd column coordinates, and the pixels in the horizontal boundaries of F_3 have even row coordinates. Hence, no equalities should exist in the bounding relationship. This is illustrated in Fig. 11.8. The bounding relationships for F_3 are

$$F_3 \ominus K \subseteq (F_3 \cap S) \bullet K \subseteq F_3 \subseteq (F_3 \cap S) \oplus K \subseteq F_3 \oplus K.$$

To show why the opening condition $F = F \circ K$ is needed for the bounding relationships involving F, we show an example set F_4 that deviates from the set F_3 by adding six extra points to it (see Fig. 11.9). The sample and reconstruction results of F_4, $F_4 \cap S$, $(F_4 \cap S) \bullet K$, and $(F_4 \cap S) \oplus K$ are exactly the same as the results for F_3. However, no bounding relationships between F_4 and its sample reconstructions are applicable. If we open F_4 by K, the bounding relationships exist because $F_4 \circ K = F_3$.

11.3.3   The Distance Relationships

We have established the maximality of the reconstruction $(F \cap S) \oplus K$ with respect to the property of being open and down-sampling to $F \cap S$, and the minimality of the reconstruction $(F \cap S) \bullet K$ with respect to the property of being closed and down-sampling to $F \cap S$. We now give a more precise characterization of how far $F \ominus K$ is from $F \bullet K$, how far $F \circ K$ is from $F \oplus K$, and how far $(F \cap S) \bullet K$ is from $(F \cap S) \oplus K$. This is important to know, since $F \ominus K \subseteq (F \cap S) \bullet K \subseteq F$ when $F = F \bullet K$; $F \subseteq (F \cap S) \oplus K \subseteq F \oplus K$ when

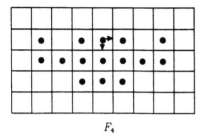

F_4

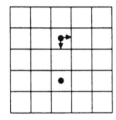

$F_4 \cap S$

Figure 11.9 A set F_4 that is not open under K. Its sampling $F_4 \cap S$ is identical to the sampling of F_3, yet the maximal reconstruction $(F_4 \cap S) \oplus K$ does not constitute an upper bound for F_4 as in the previous examples.

$F = F \circ K$; and $(F \cap S) \bullet K \subseteq F \subseteq (F \cap S) \oplus K$ when $F = F \circ K = F \bullet K$. Notice that in all three cases, the difference between the lower and the upper set bound is just a dilation by K. This motivates us to define a distance function to measure the distance between two sets and to work out the relation between the distance between a set and its dilation by K with the size of the set K. In this section, we show that with a suitable definition of distance, all these distances are less than the radius of K. Since K is related to the sampling distance, all the above-mentioned distances are less than the sampling interval.

For the size of a set B, denoted by $r(B)$, we use the radius of its circumscribing disk. Thus, $r(B) = \min_{x \in B} \max_{y \in B} \|x - y\|$. The more mathematically correct forms of *inf* for min and *sup* for max may be substituted when the space E is the real line. In this case, the proofs in this section require similar modifications. For a set A that contains a set B, a natural pseudodistance from A to B is defined by $\rho(A,B) = \max_{x \in A} \min_{y \in B} \|x - y\|$. Proposition 8 proves that (1) $\rho(A,B) \geq 0$, (2) $\rho(A,B) = 0$ implies $A \subseteq B$, and (3) $\rho(A,C) \leq \rho(A,B) + \rho(B,C) + r(B)$. The asymmetric relation (2) is weaker than the corresponding metric requirement that $\rho(A,B) = 0$ if and only if $A = B$, and relation (3) is weaker than the metric triangle inequality.

(A)   Proposition 8

1. $\rho(A,B) \geq 0$
2. $\rho(A,B) = 0$ if and only if $A \subseteq B$
3. $\rho(A,C) \leq \rho(A,B) + \rho(B,C) + r(B)$

Proof:

1. $\rho(A,B) \geq 0$, since $\rho(A,B) = \max_{x \in A} \min_{y \in B} \|x - y\|$ and $\|x - y\| \geq 0$.

2. Suppose $\rho(A,B) = 0$. Then $\max_{a \in A} \min_{b \in B} \|a - b\| = 0$. Since $\|a - b\| \geq 0$, $\max_{a \in A} \min_{b \in B} \|a - b\| = 0$ implies, for every $a \in A$, $\min_{b \in B} \|a - b\| = 0$. But $\|a - b\| = 0$ if and only if $a = b$. Hence, for every $a \in A$, there exists a $b \in B$ satisfying $a = b$; i.e., $A \subseteq B$. Suppose $A \subseteq B$. Then, for each $a \in A$, $\min_{b \in B} \|a - b\| = 0$. Hence, $\max_{a \in A} \min_{b \in B} \|a - b\| = 0$.

3.
$$\rho(A,C) = \max_{a \in A} \min_{c \in C} \|a - c\|$$

$$\leq \max_{a \in A} \min_{c \in C} \|a - b\| + \|b - c\| \text{ for every } b \in B$$

$$\leq \max_{a \in A} \left\{ \|a - b\| + \min_{c \in C} \|b - c\| \right\} \text{ for every } b \in B$$

$$\leq \max_{a \in A} \left\{ \|a - b\| + \max_{b' \in B} \min_{c \in C} \|b' - c\| \right\} \text{ for every } b \in B$$

$$\leq \rho(B,C) + \max_{a \in A} \|a - b\| \text{ for every } b \in B$$

But, $\max_{a \in A} \|a - b\| = \max_{a \in A} \|a - b' + b' - b\|$ for every $b, b' \in B$

$$\leq \max_{a \in A} \|a - b'\| + \|b' - b\| \text{ for every } b, b' \in B$$

Finally, $\max_{a \in A} \|a - b\| \leq \|b' - b\| + \max_{a \in A} \|a - b'\|$ for every $b, b' \in B$

Thus, $\max_{a \in A} \|a - b\| \leq \max_{b' \in B} \|b' - b\| + \min_{b' \in B} \max_{a \in A} \|a - b'\|$

$$\text{for every } b \in B$$

$$\leq \max_{b' \in B} \|b' - b\| + \rho(A,B)$$

Finally, $\rho(A,C) \leq \rho(B,C) + \max_{b' \in B} \|b' - b\|$ for every $b \in B$

$$\leq \rho(A,B) + \rho(B,C) + \min_{b \in B} \max_{b' \in B} \|b' - b\|$$

$$\leq \rho(A,B) + \rho(B,C) + r(B).$$

The pseudodistance ρ has a very direct interpretation. $\rho(A,B)$ is the radius of the smallest disk that, when used as a structuring element to dilate B, produces a result that contains A.

(B) Proposition 9

Let $disk(r) = \{x \mid \|x\| \leq r\}$ and $A, B \subseteq F^N$. Then, $\max_{a \in A} \min_{b \in B} \|a - b\| = inf\{r \mid A \subseteq B \oplus disk(r)\}$.

Proof:

Let $\rho_0 = \max_{a \in A} \min_{b \in B} \|a - b\|$ and $r_0 = inf\{r \mid A \subseteq B \oplus disk(r)\}$. Let $a \in A$ be given. Let $b_0 \in B$ satisfy $\|a - b_0\| = \min_{b \in B} \|a - b\|$. Now, $\rho_0 = \max_{x \in A} \min_{y \in B} \|x - y\| \geq \min_{b \in B} \|a - b\|$. Hence, $\rho_0 \geq \|a - b_0\|$, so that $a - b_0 \in disk(\rho_0)$. Now, $b_0 \in B$ and $a - b_0 \in disk(\rho_0)$ implies $a = b_0 + (a - b_0) \in B \oplus disk(\rho_0)$. Hence, $A \subseteq B \oplus disk(\rho_0)$. Since $r_0 = inf\{r \mid A \subseteq B \oplus disk(r)\}, r_0 \leq \rho_0$. Suppose $A \subseteq B \oplus disk(r_0)$. Then, $\max_{a \in A} \min_{b \in B \oplus disk(r_0)} \|a - b\| = 0$. Hence, $\max_{a \in A} \min_{b \in B} \min_{y \in disk(r_0)} \|a - b - y\| = 0$. But $\|(a - b) - y\| \geq \|a - b\| - \|y\|$. Therefore,

$$0 \geq \max_{a \in A} \min_{b \in B} \min_{y \in disk(r_0)} \|a - b\| - \|y\|$$

$$\geq \max_{a \in A} \min_{b \in B} \|a - b\| + \min_{y \in disk(r_0)} -\|y\|$$

$$\geq \max_{a \in A} \min_{b \in B} \|a - b\| - \max_{y \in disk(r_0)} \|y\|$$

Now, $\rho_0 = \max_{a \in A} \min_{b \in B} \|a - b\|$ and $r_0 = \max_{y \in disk(r_0)} \|y\|$ implies $0 \geq \rho_0 - r_0$, so that $r_0 \geq \rho_0$. Finally, $r_0 \leq \rho_0$ and $r_0 \geq \rho_0$ implies $r_0 = \rho_0$.

The pseudodistance ρ can be used as the basis for a true set metric by making it symmetric. We define the set metric $\rho_M(A,B) = \max\{\rho(A,B), \rho(B,A)\}$, also called the Hausdorf metric. The proof that ρ_M is indeed a metric follows rapidly after noting that $\rho_M(A,B) = inf\{r \mid A \subseteq B \oplus disk(r)$ and $B \subseteq A \oplus disk(r)\}$. This happens since:

$$\rho_M(A,B) = \max\{\rho(A,B), \rho(B,A)\}$$
$$= \max\{inf\{r \mid A \subseteq B \oplus disk(r)\} inf\{r \mid B \subseteq A \oplus disk(r)\}\}$$
$$= inf\{r \mid A \subseteq B \oplus disk(r) \text{ and } B \subseteq A \oplus disk(r)\}.$$

A strong relationship between the set distance and the dilation of sets must be developed to translate set-bounding relationships to distance bounding relationships. We show that $\rho(A \oplus B, C \oplus D) \leq \rho(A,C) + \rho(B,D)$

and then quickly extend the result to $\rho_M(A \oplus B, C \oplus D) \leq \rho_M(A, C) + \rho_M(B, D)$.

(C) Proposition 10

1. $\rho(A \oplus B, C \oplus D) \leq \rho(A, C) + \rho(B, D)$
2. $\rho_M(A \oplus B, C \oplus D) \leq \rho_M(A, C) + \rho_M(B, D)$

Proof:

1. $\rho(A \oplus B, C \oplus D) = \max\limits_{x \in A \oplus B} \min\limits_{y \in C \oplus D} \|x - y\|$

$$= \max\limits_{a \in A} \max\limits_{b \in B} \min\limits_{c \in C} \min\limits_{d \in D} \|a + b - c - d\|$$

$$\leq \max\limits_{a \in A} \max\limits_{b \in B} \min\limits_{d \in D} \min\limits_{c \in C} [\|a - c\| + \|b - d\|]$$

$$\leq \max\limits_{a \in A} \max\limits_{b \in B} \min\limits_{d \in D} \left[(\min\limits_{c \in C} \|a - c\|) + \|b - d\| \right]$$

$$\leq \max\limits_{a \in A} \min\limits_{c \in C} \|a - c\| + \max\limits_{b \in B} \min\limits_{d \in D} \|b - d\|$$

$$\leq \rho(A, C) + \rho(B, D)$$

2. $\rho_M(A \oplus B, C \oplus D) = \max\{\rho(A \oplus B, C \oplus D), \rho(C \oplus D, A \oplus B)\}$

$$\leq \max\{\rho(A, C) + \rho(B, D), \rho(C, A) + \rho(D, B)\}$$

$$\leq \max\{\rho(A, C), \rho(C, A)\}$$

$$+ \max\{\rho(B, D), \rho(D, B)\}$$

$$\leq \rho_M(A, C) + \rho_M(B, D)$$

From this last result, it is apparent that dilating two sets with the same structuring element cannot increase the distance between the sets. Dilation tends to suppress differences between sets, making them more similar. More precisely, if $B = D = K$, then $\rho_M(A \oplus K, C \oplus K) \leq \rho_M(A, C)$. It is also apparent that $\rho_M(A, A \oplus K) = \rho_M(A \oplus \{0\}, A \oplus K) \leq \rho_M(A, A) + \rho_M(\{0\}, K) = \rho_M(\{0\}, K) \leq \max\limits_{k \in K} \|k\|$. Indeed, since the reconstruction-structuring element $K = \check{K}$ and $\varnothing \in K$, the radius of the circumscribing disk is precisely $\max\limits_{k \in K} \|k\|$. Hence, the distance between A and $A \oplus K$ is no more than the radius of the circumscribing disk of K.

(D) Proposition 11

If $K = \check{K}$ and $0 \in K$, then $r(K) = \max\limits_{k \in K} \|k\|$.

Proof:

$$r(K) = \min\limits_{x \in K} \max\limits_{y \in K} \|x - y\| \leq \max\limits_{y \in K} \|0 - y\| = \max\limits_{y \in K} \|y\|$$

$$\text{and } \max_{y\in K} \|y\| = \frac{1}{2} \max_{y\in K} \|y - x + x + y\| \text{ for } x \in K$$

$$\leq \frac{1}{2} \left\{ \max_{y\in K} \|y - x\| + \max_{y\in K} \|x + y\| \right\} \text{ for } x \in K$$

$$\leq \frac{1}{2} \left\{ \max_{y\in K} \|x - y\| + \max_{y\in K} \|x - y\| \right\} \text{ for } x \in K$$

$$\leq \max_{y\in K} \|x - y\| \text{ for } x \in K$$

$$\leq \min_{x\in K} \max_{y\in K} \|x - y\| = r(K)$$

Since $\rho_M(A, A \oplus K) \leq \max_{k\in K} \|k\|$ and $\max_{k\in K} \|k\| = r(K)$, we have $\rho_M(A, A \oplus K) \leq r(K)$. Also, since $A \bullet K \supseteq A$, $\rho_M(A \bullet K, A) = \rho(A \bullet K, A)$. Since $0 \in K$, $A \bullet K \subseteq A \oplus K$. Hence, $\rho_M(A \bullet K, A) = \rho(A \bullet K, A) \leq \rho[(A \bullet K) \oplus K, A] = \rho(A \oplus K, A) \leq r(K)$.

It immediately follows that the distance between the minimal and maximal reconstructions, which differ only by a dilation by K, is no greater than the size of the reconstruction-structuring element; that is, $\rho_M[(F \cap S) \bullet K, (F \cap S) \oplus K] \leq r(K)$. When $F = F \circ K = F \bullet K$, $(F \cap S) \bullet K \subseteq F \subseteq (F \cap S) \oplus K$. Since the distance between the minimal and maximal reconstruction is no greater than $r(K)$, it is unsurprising that the distance between F and either of the reconstructions is no greater than $r(K)$.

(E) Proposition 12

If $A \subseteq B \subseteq C$, then (1) $\rho_M(A, B) \leq \rho_M(A, C)$, and (2) $\rho_M(B, C) \leq \rho_M(A, C)$.

Proof:

1. Since $A \subseteq B$, $\rho_M(A, B) = \rho(B, A)$, then

$$\rho(B, A) = \max_{b\in B} \min_{a\in A} \|b - a\|$$

$$\leq \max_{c\in C} \min_{a\in A} \|c - a\|, \text{ since } B \subseteq C$$

$$\leq \rho(C, A) = \rho_M(A, C), \text{ since } A \subseteq C.$$

2. The proof of (2) is similar to (1), with B taking the role of A and C taking the role of B.

Now, it immediately follows that if $F = F \circ K = F \bullet K$, $\rho_M(F, (F \cap S) \oplus K) \leq r(K)$ and $\rho_M[F, (F \cap S) \bullet K] \leq r(K)$.

These distance bounds can actually be shown under slightly less restrictive conditions. Suppose that $F = F \circ K$. Then it follows that $F \subseteq (F \cap S) \oplus K$. Since $F \cap S \subseteq F$, $(F \cap S) \oplus K \subseteq F \oplus K$. Hence, $F \subseteq (F \cap S) \oplus K \subseteq F \oplus K$. But $\rho_M(F, F \oplus K) \leq r(K)$. Hence, $\rho_M[F, (F \cap S) \oplus K] \leq r(K)$ and $\rho_M((F \cap S) \oplus K, F \oplus K) \leq r(K)$. It goes similarly with the closing reconstruction.

When the image F is open under K, the distance between F and its sampling $F \cap S$ can be no greater than $r(K)$. Why? It is certainly the case that $F \cap S \subseteq F \subseteq (F \cap S) \oplus K$. Hence, $\rho_M(F, F \cap S) \leq \rho_M[F \cap S, (F \cap S) \oplus K] \leq r(K)$.

If two sets are both open under the reconstruction-structuring element K, then the distance between the sets must be no greater than the distance between their samplings plus the size of K.

(F)   Proposition 13

If $A = A \circ K$ and $B = B \circ K$, then $\rho_M(A, B) \leq \rho_M(A \cap S, B \cap S) + r(K)$

Proof:

Consider $\rho(A, B)$. $\rho(A, B) \leq \rho(A, B \cap S)$. Since $A = A \circ K, A \subseteq (A \cap S) \oplus K$. Hence, $\rho(A, B) \leq \rho(A, B \cap S) \leq \rho[(A \cap S) \oplus K, B \cap S] \leq \rho(A \cap S, B \cap S) + r(K)$. Similarly, since $B = B \circ K$, $\rho(B, A) \leq \rho(B \cap S, A \cap S) + r(K)$.

Now,

$$\begin{aligned}
\rho_M(A, B) &= \max\{\rho(A, B), \rho(B, A)\} \\
&\leq \max\{\rho(A \cap S, B \cap S) + r(K), \rho(B \cap S, A \cap S) + r(K)\} \\
&= r(K) + \max\{\rho(A \cap S, B \cap S), \rho(B \cap S, A \cap S)\} \\
&= r(K) + \rho_M(A \cap S, B \cap S).
\end{aligned}$$

From this last result, it is easy to see that if F is closed under K, then the distance between F and its minimal reconstruction $(F \cap S) \bullet K$ is no greater than $r(K)$. Consider

$$\begin{aligned}
\rho_M[F, (F \cap S) \bullet K] &\leq \rho_M\{F \cap S, [(F \cap S) \bullet K] \cap S\} + r(K) \\
&= \rho_M(F \cap S, F \cap S) + r(K) = r(K)
\end{aligned}$$

These distance relationships mean that just as the standard sampling theorem cannot produce a reconstruction with frequencies higher than the Nyquist frequency, the morphological sampling theorem cannot produce a reconstruction whose positional accuracy is better than the radius of the circumscribing disk of the reconstruction-structuring element K. Since the diameter of this disk is just short of being large enough to contain two sample intervals, the morphological sampling theorem cannot produce a reconstruction whose positional accuracy is better than that of the sampling interval.

11.3.4   Examples

We use the example sets F_1, F_2, F_3, and F_4 in computing the distance between the original images and the sample reconstruction images. The values $\max_{y \in K} \|x - y\|$ for each $x \in K$ are shown in Fig. 11.10. The minimum

$2\sqrt{2}$	$\sqrt{5}$	$2\sqrt{2}$	
$\sqrt{5}$	$\sqrt{2}$	$\sqrt{5}$	
$2\sqrt{2}$	$\sqrt{5}$	$2\sqrt{2}$	

Figure 11.10 The $\max_{y \in K} \|x - y\|$ values for all $x \in K$, where K is the digital disk having radius $\sqrt{2}$.

value among them, $\sqrt{2}$, is the radius $r(K)$, since $r(K) = \min_{x \in K} \max_{y \in K} \|x - y\|$.

We now measure the distance between two sample reconstructions for all the example sets. To compute $\rho_M[(F_1 \cap S) \bullet K, (F_1 \cap S) \oplus K]$, we first compute $\rho[(F_1 \cap S) \oplus K, (F_1 \cap S) \bullet K]$ and $\rho[(F_1 \cap S) \bullet K, (F_1 \cap S) \oplus K]$. The values $\min_{y \in (F_1 \cap S) \bullet K} \|x - y\|$ for all $x \in (F_1 \cap S) \oplus K$ are shown in Fig. 11.11. The maximum value among them, $\sqrt{2}$, is the distance $\rho[(F_1 \cap S) \oplus K, (F_1 \cap S) \bullet K]$. Similarly, we can compute $\rho[(F_1 \cap S) \bullet K, (F_1 \cap S) \oplus K]$, which equals 0. Thus, $\rho_M[(F_1 \cap S) \bullet K, (F_1 \cap S) \oplus K]$ equals $\sqrt{2}$, which is exactly the radius $r(K)$.

$\sqrt{2}$	1	1	1	$\sqrt{2}$	
1	0	0	0	1	
1	0	0	0	1	
1	0	0	0	1	
$\sqrt{2}$	1	1	1	$\sqrt{2}$	

Figure 11.11 The $\min_{y \in (F_1 \cap S) \bullet K} \|x - y\|$ for all $x \in (F_1 \cap S) \oplus K$.

Similarly, the distances between two reconstructions for sets F_2, F_3, and F_4 can be measured, and they are all equal to $r(K)$.

What is the distance $\rho_M[F, (F \cap S) \oplus K]$ for the example sets? Since $F_1 = (F_1 \cap S) \bullet K$, $\rho_M[F_1, (F_1 \cap S) \oplus K] = \rho_M[(F_1 \cap S) \bullet K, (F_1 \cap S) \oplus K] = r(K)$. It is easy to see that $\rho_M[(F_2, (F_2 \cap S) \oplus K] = 0$ because $F_2 = (F_2 \cap S) \oplus K$. Figure 11.12 shows the values $\min_{y \in F_3} \|x - y\|$ for all $x \in (F_3 \cap S) \oplus K$, their maximum value being $\rho[(F_3 \cap S) \oplus K, F_3] = 1$. Since $F_3 \subseteq (F_3 \cap S) \oplus K$, $\rho[F_3, (F_3 \cap S) \oplus K]$ equals 0. Hence, $\rho_M[(F_3 \cap S) \oplus K, F_3] = 1 < r(K)$.

The distance $\rho[F_4, (F_4 \cap S) \oplus K]$ is interesting, since $F_4 \neq F_4 \circ K$. The $\min_{y \in (F_4 \cap S) \oplus K} \|x - y\|$ values for all $x \in F_4$ are shown in Fig. 11.13(a), their maximum value being $\rho[F_4, (F_4 \cap S) \oplus K] = 2$. The $\min_{y \in F_4} \|x - y\|$ values for all $x \in (F_4 \cap S) \oplus K$ are shown in Fig. 11.13(b); the maximum value is $\rho[(F_4 \cap S) \oplus K, F_4] = 1$. Thus, the distance $\rho_M[F_4, (F_4 \cap S) \oplus K]$ is equal to 2, which is greater than $r(K)$. This shows why the condition $F = F \circ K$ is required to bound the difference between F and its maximum reconstruction $(F \cap S) \oplus K$. Similarly, we find

$$\rho_M[(F_1 \cap S) \oplus K, F_1 \oplus K] = 0 < r(K)$$
$$\rho_M[(F_2 \cap S) \oplus K, F_2 \oplus K] = \sqrt{2} = r(K)$$
$$\rho_M[(F_3 \cap S) \oplus K, F_3 \oplus K] = 1 < r(K)$$
$$\rho_M[(F_4 \cap S) \oplus K, F_4 \oplus K] = 2 > r(K)$$

Note that since $F_4 \neq F_4 \circ K$, $\rho_M[(F_4 \cap S) \oplus K, F_4 \oplus K] \not\leq r(K)$. Using the minimum reconstruction, the positional accuracies for the example sets are:

$$\rho_M[F_1, (F_1 \cap S) \bullet K] = 0 < r(K)$$

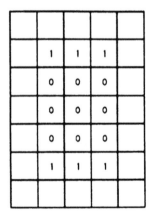

Figure 11.12   The $\min_{y \in F_3} \|x - y\|$ for each $x \in (F_3 \cap S) \oplus K$.

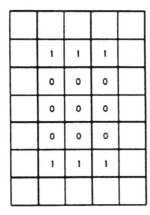

(a)

(b)

Figure 11.13   (a) Shows values for $\min_{y \in (F_4 \cap S) \oplus K} \|x - y\|$ for all $x \in F_4$; (b) shows values for $\min_{y \in F_4} \|x - y\|$ for all $x \in (F_4 \cap S) \oplus K$. The maximum among all these values is 2. Hence, $\rho_M[F_4 \cap S) \oplus k] = 2 > r(K)$.

$$\rho_M[F_2, (F_2 \cap S) \bullet K] = \sqrt{2} = r(K)$$
$$\rho_M[F_3, (F_3 \cap S) \bullet K] = 1 < r(K)$$
$$\rho_M[F_4, (F_4 \cap S) \bullet K] = 3 > r(K)$$

Also, since $F_4 \neq F_4 \bullet K$, $\rho_M[F_4, (F_4 \cap S) \bullet K] \not\leq r(K)$.

11.3.5 Binary Digital Morphological Sampling Theorem

This section summarizes the results developed in the previous sections. These results constitute the binary digital morphological sampling theorem.

Theorem 1: Binary Digital Morphological Sampling Theorem

Let $F, K, S \subseteq E^N$. Suppose that K and S satisfy the sampling conditions:

1. $S \oplus S = S$
2. $S = \check{S}$
3. $K \cap S = \{0\}$
4. $K = \check{K}$
5. $x \in K_y$ implies $K_x \cap K_y \cap S \neq \emptyset$

Then:

1. $F \cap S = [(F \cap S) \bullet K] \cap S$.
2. $F \cap S = [(F \cap S) \oplus K] \cap S$.
3. $(F \cap S) \bullet K \subseteq F \bullet K$.
4. $F \circ K \subseteq (F \cap S) \oplus K$.
5. If $F = F \circ K = F \bullet K$, then $(F \cap S) \bullet K \subseteq F \subseteq (F \cap S) \oplus K$.
6. If $A = A \bullet K$ and $A \cap S = F \cap S$, then $A \subseteq (F \cap S) \bullet K$ implies $A = (F \cap S) \bullet K$.
7. If $A = A \circ K$ and $A \cap S = F \cap S$, then $A \supseteq (F \cap S) \oplus K$ implies $A = (F \cap S) \oplus K$.
8. If $F = F \bullet K$, then $\rho_M[F, (F \cap S) \bullet K] \leq r(K)$.
9. If $F = F \circ K$, then $\rho_M[(F \cap S) \oplus K, F] \leq r(K)$.

11.4 THE GRAY-SCALE MORPHOLOGICAL SAMPLING THEOREM

In this section, we present the extension of the morphological sampling theorem from the binary case to the gray-scale case. We begin by reviewing some definitions and results (Haralick et al./1987).

We adopt the convention that the first $(N-1)$ coordinates of the N-tuples in a set $A \subseteq E^N$ constitute the spatial domain of A, and the Nth coordinate represents the surface; i.e., $A \subseteq E^{N-1} \times E$. For gray-scale images, $N = 3$ and an image is a function $f = E \times E \rightarrow E$. A set $A \subseteq E^{N-1} \times E$ is an *umbra* if and only if $(x,z) \in A$ implies that $(x,z) \in A$ for every $z \leq y$. The *top of A* is a function $T[A]$ mapping the spatial domain of A, $\{x \in E^{N-1}|$ for some $y \in E, (x,y) \in A\}$, to the *top surface* of A,

$$T[A](x) = \max \{y \in E \mid (x,y) \in A\}.$$

If $F \subseteq E^{N-1}$ and $f : F \rightarrow E$, the umbra of f is the set

$$U[f] = \{(x,y) \in F \times E \mid y \leq f(x)\}.$$

For $F, K \subseteq E^{N-1}, f : F \rightarrow E$, and $k : K \rightarrow E$, the dilation of f by k is the mapping $f \oplus k : F \oplus K \rightarrow E$ defined by

$$f \oplus k = T\{U[f] \oplus U[k]\},$$

from which it follows that

$$(f \oplus k)(x) = \max_{\substack{z \in K \\ x-z \in F}} \{f(x-z) + k(z)\}.$$

Similarly, the erosion of f by k maps $F \ominus K$ to E by

$$f \ominus k = T\{U[f] \ominus U[k]\},$$

from which it follows that

$$(f \ominus k)(x) = \min_{z \in K} \{f(x+z) - k(z)\}.$$

The umbra homomorphism theorem states that the operation of taking an umbra is a homomorphism from the gray-scale morphology to the binary morphology. That is, if $F, K \subseteq E^{N-1}, f : F \rightarrow E$, and $k : K \rightarrow E$, then

$U[f \oplus k] = U[f] \oplus U[k]$ and
$U[f \ominus k] = U[f] \ominus U[k]$.

Furthermore, the operations of max and union, and of min and intersection, are homomorphic under the umbra transformation. That is,

$U[\max \{f,k\}] = U[f] \cup U[k]$,
$U[\min \{f,k\}] = U[f] \cap U[k]$.

Two more notational conventions are needed before we begin developing the corresponding gray-scale morphological sampling results. If $S \subseteq E^{N-1}$ is a sampling set, and $f : F \rightarrow E$ is a gray-scale image, then the sampled version of f is the restriction of f to $F \cap S$ denoted by $f|_S$. Thus, $f|_S :$ $F \cap S \rightarrow E$ defined by $f|_S(x) = f(x)$ for $x \in F \cap S$. If $k : K \rightarrow E$, then $\check{k}$ is the reflection of k, defined by $\check{k} : \check{K} \rightarrow E$, where $\check{k}(x) = k(-x)$.

11.4.1   The Function-Bounding Relationships

The set-bounding relationships for the binary morphology have a direct correspondence to function-bounding relationships in gray-scale morphology. In this section, we develop the bounding relationships without spending much time or discussion, even though the extensions are somewhat more involved. The gray-scale analog to the relationship $F \ominus K \subseteq (F \cap S) \oplus K \subseteq F \oplus K$ is $f \ominus k \le f|_s \oplus k \le f \oplus k$, and it holds under very much the same conditions as the binary relationship holds. The only new requirement is for $k \ge 0$, which is stronger than the requirement that $0 \in U[k]$.

(A)   Proposition 14

Let $F, K, S \subseteq E^{N-1}$, $f : F \to E$, and $k : K \to E$. If $k = \check{k}$, $k \ge 0$, and $K \oplus S = E^{N-1}$, then $f \ominus k \le f|_s \oplus k \le f \oplus k$.

   Proof:

First we show that $f \ominus k \le f|_s \oplus k$. Let $x \in F \ominus K$. Then, $(f \ominus k)(x) = \min_{u \in K} \{f(x + u) - k(u)\}$. But $k = \check{k}$ and $k > 0$ imply $\min_{u \in K} \{f(x + u) - k(u)\} \le \min_{v \in K} \{f(x - v) + k(v)\}$. Now, $k = \check{k}$ implies $K = \check{K}$ and this, with $S \oplus K = E^{N-1}$, implies $F \ominus K \subseteq (F \cap S) \oplus K$. Thus, $x \in F \ominus K$ implies $x \in (F \cap S) \oplus K$. Hence, there exists a $w \in K$ such that $x - w \in F \cap S$. Because $\min_{z \in A} z \le \max_{z \in B} z$ when $B \cap A \ne \varnothing$, there results

$$(f \ominus k)(x) \le \min_{\substack{v \in K}} \{f(x - v) + k(v)\} \le \max_{\substack{v \in K \\ x - v \in F \cap S}} \{f(x - v) + k(v)\}$$

$$= \max_{\substack{v \in K \\ x - v \in F \cap S}} \{f|_s(x - v) + k(v)\} = (f|_s \oplus k)(x).$$

Next, we show that $f|_s \oplus k \le f \oplus k$. By the umbra homomorphism theorem, $U[f|_s \oplus k] = U[f|_s] \oplus U[k]$. But $U[f|_s] = U[f] \cap (S \times E)$. Hence, $U[f|_s \oplus k] = \{U[f] \cap (S \times E)\} \oplus U[k] \subseteq \{U[f] \oplus U[k]\} \cap \{(S \times E) \oplus U[k]\}$. But $(S \times E) \oplus U[k] = E^N$ if and only if $S \oplus K = E^{N-1}$. So, $U[f|_s \oplus k] \subseteq U[f] \oplus U[k] = U[f \oplus k]$ by the umbra homomorphism theorem. Hence, $T\{U[f|_s \oplus k]\} \le T\{U[f \oplus k]\}$, which, by definition of gray-scale dilation, implies $f|_s \oplus k \le f \oplus k$.

The analog of $F \cap S = [(F \cap S) \oplus K] \cap S$ is $f|_s = (f|_s \oplus k)|_s$. It holds under the condition that $k(0) = 0$.

(B)   Proposition 15

Let $F, K, S \subseteq E^{N-1}$, $f : F \to E$, $k : K \to E$, satisfy $S \oplus S = S$, $S = \check{S}$, $K \cap S = \{0\}$, $k = \check{k}$, and $k(0) = 0$. Then, $f|_s = (f|_s \oplus k)|_s$

   Proof:

Since $[(F \cap S) \oplus K] \cap S = F \cap S$, we need only to prove that for each $x \in F \cap S (f|_s \oplus k)|_s(x) = f(x)$. But $(f|_s \oplus k)|_s(x) = (f|_s \oplus k)(x)$. Since $x \in S$, $(f|_s \oplus k)|_s(x) = \max_{\substack{z \in K \\ x - z \in F \cap S}} \{f(x - z) + k(z)\}$. Furthermore, $x \in S$

and $x - z \in S$ imply $z = x - (x - z) \in S$, since $S \oplus S = S$ and $\check{S} = S$. Thus, $z = 0$ as is implied by $z \in K$, $z \in S$, and $K \cap S = \{0\}$. Finally, $(f|_s \oplus k)|_s(x) = f(x) + k(0) = f(x)$, since $k(0) = 0$.

To continue with the parallel development $f \circ k \leq f|_s \oplus k$, we first prove the stronger relation that for every $x \in F \circ K$, there exists an $s \in S \cap F$ such that $x \in K_s$ and $(f \circ k)(x) \leq f(s) + k(x - s)$. This result follows from the sampling condition $u \in K_v$ implies $S \cap K_u \cap K_v \neq \varnothing$ and a constraint on the structuring element $k : k(a) \leq k(a - b) + k(b)$ for every a, b satisfying $a \in K, b \in K$ and $a - b \in K$. This latter constraint is a new concept essential for the reconstruction-structuring element in the gray-scale morphology.

Before developing the proof for the inequality $(f \circ k)(x) \leq f(s) + k(x - s)$, it will be useful to explore the meaning of the inequality $k(a) \leq k(a - b) + k(b)$, since this is a constraint we have not had to deal with until now. The inequality $k(a) \leq k(a - b) + k(b)$, together with $k = \check{k}$, implies that $k(y) \geq 0$ for every $y \in K$. This can easily be seen by letting $a = x + y$ and $b = x$. This leads to $k(x + y) \leq k(x) + k(y)$. Then, let $a = x$ and $b = x + y$. This leads to $k(x) \leq k(y) + k(x + y)$. The two inequalities imply $k(x) \leq k(y) + k(x + y) \leq k(x) + 2k(y)$, from which $k(y) \geq 0$ quickly follows.

The inequality $k(a) \leq k(a - b) + k(b)$ also implies that for any integer $n \geq 2$, $k(nx) \leq nk(x)$ for every $x \in K$ satisfying $mx \in K$ for every m, $2 \leq m \leq n$. The proof is by induction. Taking $n = 2$, $a = 2x$ and $b = x$ establishes the base case $k(2x) \leq 2k(x)$. Suppose that for every m, $2 \leq m \leq n, k(ma) \leq mk(a)$ and $ma \in K$ and $a \in K$. Taking $a = (n + 1)x$ and $b = x$ produces $k(n + 1) \leq k(nx) + k(x)$. But $k(nx) \leq nk(x)$. Hence, $k((n + 1)x) \leq nk(x) + k(x) = (n + 1)k(x)$. Now, by induction, $k(nx) \leq nk(x)$ for every integer $n \geq 2$ satisfying $mx \in K$ for every m, $2 \leq m \leq n$.

Finally, notice that $k(a) \leq k(a - b) + k(b)$ for every $a, b \in K$ satisfying $a - b \in K$, and $K = \check{K}$, implies, as well, $k(b) \leq k(b - a) + k(a)$ for every $a, b \in K$ satisfying $a - b \in K$. Since $k = \check{k}$, we obtain $k(a - b) \geq \max \{k(a) - k(b), k(b) - k(a)\}$.

Constructing functions that satisfy the inequality is easy with the following procedure. Define $k(0) = 0$ and $k(1)$ to be any positive number. Suppose $k(m)$, $m = 0, \ldots, n$ have been defined. Take $k(n + 1)$ to be any number satisfying

$$\max_{\substack{u \\ 1 \leq u \leq n}} \{k(u) - k(n + 1 - u)\} \leq k(n + 1) \leq \min_{\substack{v \\ 1 \leq v \leq n}} \{k(v) + k(n + 1 - v)\}.$$

After k is defined for all nonnegative numbers in its domain, define $k(-n) = k(n)$ for $n \geq 0$.

The generating procedure works because $k = \check{k}$ and the inequality implies $k(x) - k(y - x) \leq k(y) \leq k(x) + k(y - x)$. Hence, $\max_{\substack{u \\ 1 \leq u \leq y}} \{k(u) - k(y - u) \leq k(y)\} \leq \min_{\substack{v \\ 1 \leq v \leq y}} \{k(v) + k(y - v)\}.$

(C)   Proposition 16

Let $F,K,S \subseteq E^{N-1}, f : F \to E, k : K \to E$. Suppose $u \in K_v$ implies $S \cap K_u \cap K_v \neq \varnothing$ and $k(a) \leq k(a - b) + k(b)$. Then, for every $x \in F \circ K$, there exists an $s \in S \cap F$ such that $x \in K_s$ and $(f \circ k)(x) \leq f(s) + k(x - s)$.

Proof:

Let $x \in F \circ K$. Then, $[x, (f \circ k)(x)] \in U[f \circ k]$. But $U[f \circ k] = U[f] \circ U[k]$. Hence, there exists $(u,v) \in E^{N-1} \times E$ such that $[x, (f \circ k)(x)] \in U[k]_{(u,v)} \subseteq U[f]$. Now $[x, (f \circ k)(x)] \in U[k]_{(u,v)}$ implies $[x, (f \circ k)(x)] - (u,v) \in U[k]$. So, $[x - u, (f \circ k)(x) - v] \in U[k]$, which implies $(f \circ k)(x) - v \leq k(x - u)$. Thus, $v \geq (f \circ k)(x) - k(x - u)$. But $U[k]_{(u,v)} \subseteq U[f]$ implies for every $a \in K$ $[a, k(a)] + (u,v) \in U[f]$. Hence, for every $a \in K, a + u \in F$ and $k(a) + v \leq f(a + u)$. Now $x \in K_u$ implies there exists $s \in K_x \cap K_u \cap S$, so that $s - u \in K$. And since $s - u \in K, k(s - u) + v \leq f(s - u + u) = f(s)$. Now $(f \circ k)(x) - k(x - u) \leq v$ and $v \leq f(s) - k(s - u)$ imply $(f \circ k)(x) - k(x - u) \leq f(s) - k(s - u)$. But $k(a) \leq k(a - b) + k(b)$ for every $a,b \in K$ satisfying $a - b \in K$. Letting $a = x - u$ and $b = s - u$, $a - b = x - s$, it is obvious that $x - u \in K, s - u \in K$, and $x - s \in K$. Hence, $k(x - u) - k(s - u) \leq k((x - u) - (s - u)) = k(x - s)$. Therefore, $(f \circ k)(x) \leq f(s) + k(x - s)$.

Corollary:

Let $F,K,S \subseteq E^{N-1}, f : F \to E$, and $k : K \to E$. Suppose $u \in K_v$ implies $S \cap K_u \cap K_v \neq \varnothing$ and $k(a) \leq k(a - b) + k(b)$. Then, $(f \circ k)(x) \leq (f|_s \oplus k)(x)$.

Having determined that $f \circ k \leq f|_s \oplus k$, the maximality of $f|_s \oplus k$ comes easily.

(D)   Proposition 17

Let $G,F,K,S \subseteq E^{N-1}, g : G \to E, k : K \to E$, and $f : F \to E$. If $k(a) \leq k(a - b) + k(b)$ for every $a \in K$ and $b \in K$ satisfying $a - b \in K$, and $x \in K_y$ implies $K_x \cap K_y \cap S \neq \varnothing$, then $g = g \circ k, g|_s = f|_s$, and $g \geq f|_s \oplus k$ implies $g = f|_s \oplus k$.

We continue our development with the bounding relations for the minimal reconstruction.

(E)   Proposition 18

Let $F,K,S \subseteq E^{N-1}, f : F \to E$, and $k : K \to E$. If $k = \check{k}, k(a) \leq k(a - b) + k(b)$ for every $a,b \in K$ satisfying $a - b \in K$, and $x \in K_y$ implies $K_x \cap K_y \cap S \neq \varnothing$, then for every $u \in K, f(x + z) - k(z) \leq (f|_s \oplus k)(x + u) - k(u)$ for each $z \in K$.

Proof:

Since $k(a) \leq k(a - b) + k(b)$, for every $a,b \in K$ satisfying $a - b \in K$, $k(z) \leq -k(u) + k(u - z)$ for every $u,z \in K$ satisfying $u - z \in K$. Hence,

$f(x + z) - k(z) \leq f(x + z) + k(u - z) - k(u)$ for every $u, z \in K$ satisfying $u - z \in K$. Making a change of variables $t = u - z$, there results

$$f(x + z) - k(z) \leq [f(x + u - t) + k(t)] - k(u).$$

Then, $f(x + z) - k(z) \leq [\max_{\substack{t \in K \\ x+u-t \in F \cap S \\ u-t \in K}} f|_s(x + u - t) + k(t)] - k(u)$

$$\leq (f|_s \oplus k)(x + u) - k(u)$$

for every $u \in K$.

(F) Proposition 19

Let $F, K, S \subseteq E^{N-1}, f : F \to E$, and $k : K \to E$. Suppose $u \in K_v$ implies $K_u \cap K_v \cap S \neq \emptyset, k = \check{k}$, and $k(a) \leq k(a - b) + k(b)$ for every $a, b \in K$ satisfying $a - b \in K$. Then, for every $x \in F \ominus K$, $(f \ominus k)(x) \leq (f|_s \bullet k)(x)$.

Proof:

Let $x \in F \ominus K$. Then, $(f \ominus k)(x) = \min_{z \in K} \{f(x + z) - k(z)\} \leq (f|_s \oplus k)(x + u) - k(u)$ for every $u \in K$. There results $(f \ominus k)(x) \leq \min_{u \in K} \{(f|_s \oplus k)(x + u) - k * u)\} = (f|_s \bullet k)(x)$.

(G) Proposition 20

Let $F, K, S \subseteq E^{N-1}$ and $f : F \to E, k : K \to E$. Then,

1. $f|_s = (f|_s \bullet k)|_s$.
2. $f|_s \bullet k \leq f \bullet k$.

Proof:

1. Since closing is an increasing operation, $f|_s \leq f|_s \bullet k$. Since dilation is an increasing operation, $f|_s \bullet k \leq (f|_s \bullet k) \oplus k = f|_s \oplus k$. Hence, $f|_s \leq (f|_s \bullet k)|_s \leq (f|_s \oplus k)|_s$. But $(f|_s \oplus k)|_s = f|_s$, and this proves $f|_s = (f|_s \bullet k)|_s$.
2. $U[f|_s \bullet k] = U[f|_s] \bullet U[k]$ by the umbra homomorphism theorem. Also, $U[f|_s] = U[f] \cap (S \times E)$, so that $U[f|_s] \subseteq U[f]$. Hence, $U[f|_s \bullet k] \subseteq U[f] \bullet U[k] = U[f \bullet k]$ by the umbra homomorphism theorem. This implies $f|_s \bullet k \leq f \bullet k$.

As before, the maximality comes easily.

(H) Proposition 21

Let $G, F, K, S \subseteq E^{N-1}, g : G \to E, f : F \to E$, and $k : K \to E$. Suppose $g|_s = f|_s$ and $g = g \bullet k$. Then, $g \leq f|_s \bullet k$ implies $g = f|_s \bullet k$.

11.4.2   The Gray-Scale Digital Morphologic Sampling Theorem

This section summarizes the results developed in the previous sections. These results constitute the gray-scale digital morphological sampling theorem.

Theorem 6: Gray-Scale Digital Morphological Sampling Theorem

Let $F, K, S \subseteq E^{N-1}$. Suppose K and S satisfy the sampling conditions:
1. $S \oplus S = S$.
2. $S = \check{S}$.
3. $K \cap S = \{0\}$.
4. $K = \check{K}$.
5. $x \in K_y$ implies $K_x \cap K_y \cap S \neq \varnothing$. Let $f : F \to E$ and $k : K \to E$. Suppose further that k satisfies
6. $k = \check{k}$.
7. $k(a) \leq k(a - b) + k(b)$ for every $a, b \in K$ satisfying $a - b \in K$.
8. $k(0) = 0$.
Then:
1. $f|_s = (f|_s \bullet k)|_s$.
2. $f|_s = (f|_s \oplus k)|_s$.
3. $f|_s \bullet k \leq f \bullet k$.
4. $f|_s \circ k \leq f \circ k$.
5. If $f = f \circ k = f \bullet k$, then $f|_s \bullet k \leq f \leq f|_s \oplus k$.
6. If $g = g \bullet k$ and $g|_s = f|_s$, then $g \leq f|_s \bullet k$ implies $g = f|_s \bullet k$.
7. If $g = g \circ k$ and $g|_s = f|_s$, then $g \geq f|_s \oplus k$ implies $g = f|_s \oplus k$.

11.4.3   Examples

Figure 11.14 illustrates an artificial example of opening a noisy 20×20 checkerboard image having 10×10 checks with a brick structuring element whose size is also 10×10. The gray-level difference between the bright and dark checks before the noise was added was 50. The noise is Gaussian noise with standard deviation 20. The example illustrates how, when the structuring element is matched to the structure of the signal, a surprising amount of noise reduction can take place.

The remaining examples illustrate how the morphological sampling theorem can lead to multiresolution processing techniques. The resolution hierarchy, called a pyramid, is produced typically by low-pass filtering and then sampling to generate the next lower resolution level. The

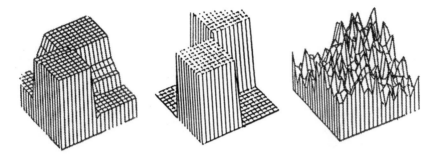

Figure 11.14   A 3-D plot illustrating the gray-scale opening operation: (a) shows a 20 × 20 checkerboard image with checker size 10 × 10 (the gray value of the bright checker is 100 and the gray value of the dark checker is 50), (b) shows the noisy image by adding a zero mean Gaussian noise of standard deviation 20 to (a), and (c) shows the result of opening (b) by a brick of size 10 × 10.

purpose of the low-pass filter is to remove from the higher resolution image those spatial frequencies that are higher than the Nyquist frequency corresponding to the sample spacing.   Figure 11.15 shows a five-level pyramid produced by pure sampling from a laser radar range map. The highest resolution image size is 256 × 256. A 2-pixel-wide line and a 4 × 4 box are placed intentionally at the upper right and upper left, respectively. Figure 11.16 shows the five-level pyramid produced by a 3 × 3 box filtering, followed by sampling to generate the next pyramid level.

Because multiresolution pyramids are used for the detection and identification of objects or features of at least a specified size, it is natural to ask if mathematical morphology (Matheron/1975, Serra/1972, Sternberg/1900, Haralick et al./1987a) might provide a better basis than do low-pass filters for constructing pyramids. This question is suggested by the fact that mathematical morphology deals directly with shape, whereas low-pass filtering techniques are based on linear combinations of sinusoidal waveforms, a representation far removed from shape.

Figure 11.17 shows a five-level morphological pyramid. At each level, the image is opened by a brick of size 3 × 3 and then sampled to generate the next lower resolution layer. Notice how the line in the upper right part of the image has been eliminated. Figure 11.18 shows a similar five-level morphological pyramid. In this pyramid, the image at each level has been opened by a 3 × 3 brick, sampled, and then reconstructed, using the maximal reconstruction. The next lower resolution layer is generated by sampling as before. The greater smoothness of the images in Fig. 11.17 over the corresponding images of Fig. 11.16 is precisely due to the smoothness

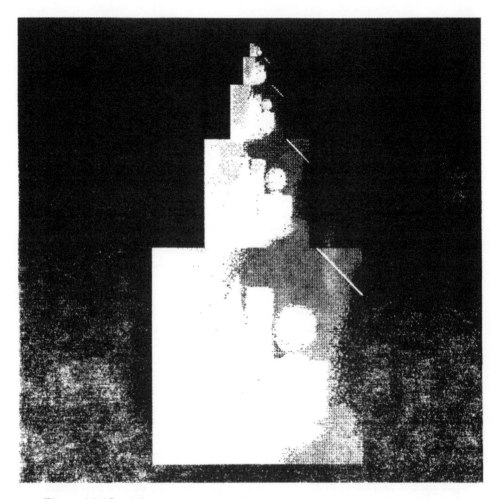

Figure 11.15 A five-level pyramid of laser radar range image produced by pure sampling. The highest resolution image size is 256 by 256. A 2-pixel-wide line segment and a 4 × 4 box are intentionally placed at the upper right and upper left of the image, respectively.

introduced by sampling and reconstructing. This illustrates that when images have been properly morphologically smoothed before sampling, the reconstructed images are indeed representative of the unsampled images.

Next, we describe a multiresolution matching scenario to integrate ground-truth maps with terrain region boundaries detected from registered

Figure 11.16 A five-level pyramid of the same image as in Fig. 11.15, produced by 3 × 3 box filtering and then sampling.

millimeter wave (MMW) radar polarimetric imagery acquired from the TABILS 5 section of the Air Force TABILS (target and background information library system) database to estimate position along a planned flight path.

The ground-truth database stores vertices of terrain region boundaries, along with region descriptions such as pine forest or wheat field. We

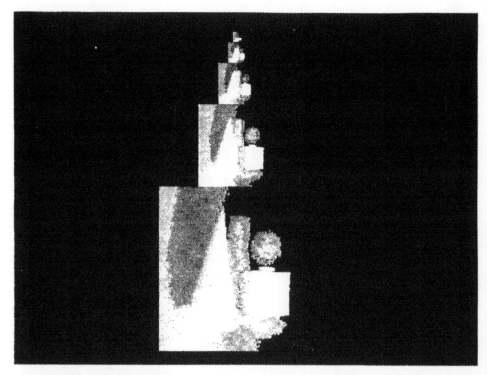

Figure 11.17 A five-level pyramid of the same image as in Fig. 11.16, produced by opening by a brick of 3 × 3 and then sampling to generate the next layer.

estimate flight position by matching detected boundaries in the incoming data to patterns in the ground truth. Figure 11.19(a) shows the boundary patterns for scene 2051A derived from the ground-truth vertices. The pattern in Fig. 11.19(b) corresponds to the current flight position, to match against ground truth. Large-scale boundary-pattern structures can estimate the displacement between detected and stored patterns over a large range at low accuracy and sampling density; smaller structures can be used over short ranges at increased accuracy and sampling. Therefore, we use a multiresolution coarse-to-fine matching strategy to meet the response-time needs of flight-path estimation.

Figure 11.19(c) illustrates morphological dilation pyramids for both the ground-truth boundaries and the detected edge patterns. Dilating the bi-

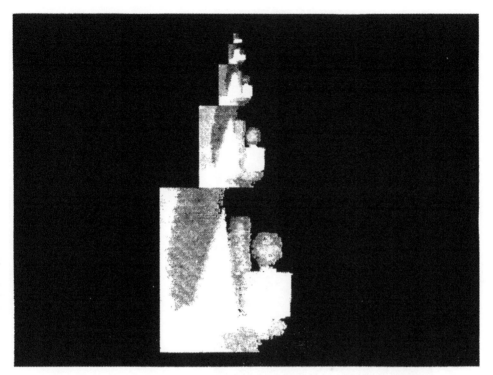

Figure 11.18 A five-level pyramid of the same image as in Fig. 11.15. In each level, the image is opened by a brick of 3 × 3 and then is sampled and reconstructed before it is down-sampled to generate the next level.

nary patterns by a 2 × 2 kernel before down-sampling (2 to 1 in both directions) retains the pattern points at coarse image resolution. Coarse-to-fine binary correlation on these pyramids (Lee et al./1987b) determines the best match: the flight position estimate is refined progressively, with estimates at one level restricting the search area at the next higher resolution.

The overall hierarchical correlation scheme outlined in Fig. 11.20 begins each stage with one pattern preshifted by its input offset. The pattern at the current resolution is shifted by various displacements, for which (binary) global-correlation error counts are determined. Each count sums over both

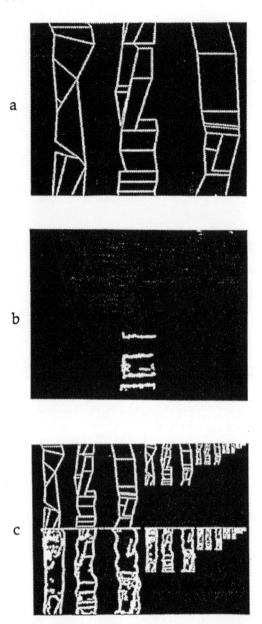

Figure 11.19 (a) Illustrates terrain boundary patterns for scene 2051 from the ground-truth vertex database, (b) illustrates detected region-boundary image corresponding to the current flight position, and (c) illustrates morphological dilation pyramids for the ground-truth boundary and detected edge patterns.

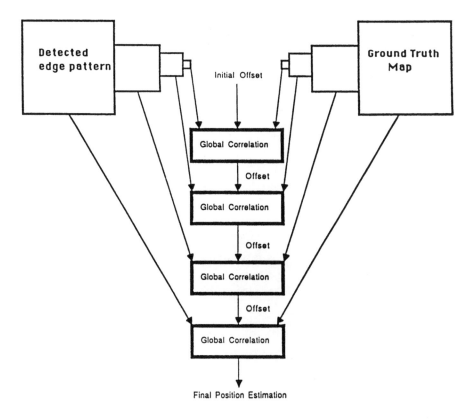

Figure 11.20   Illustration of multiresolution coarse-to-fine hierarchical correlation strategy for current flight position estimation.

the detected pattern points that are not ground-truth boundary points and the ground-truth boundary points (within the region enclosing the detected patterns) not corresponding to detected pattern points. The output offset of this stage, input to the next stage, equals the input offset plus the offset with the smallest error. We repeat to the finest image resolution, where the final offset specifies the estimated position of the current detected pattern in the ground-truth map. Figure 11.21 shows estimated positions overlaid in white on the ground truth for three detected pattern sections.

a

b

c

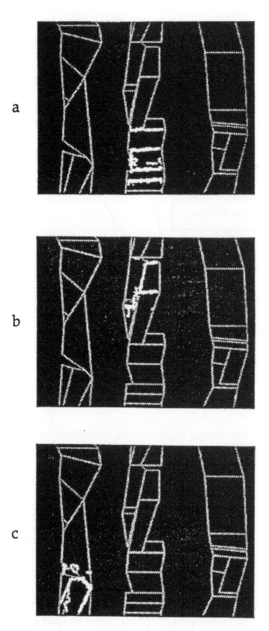

Figure 11.21   Illustrations of estimated positions overlaid on the ground-truth map for three sections of the detected edge patterns.

11.5 CONCLUSIONS

We have shown that before an image can be sampled, it must be morphologically simplified by an opening or a closing with the reconstruction-structuring element. A sampled image has a minimal and a maximal reconstruction. The minimal reconstruction is generated by closing the sampled image with the reconstruction-structuring element and is a valid reconstruction when the morphological simplification done before sampling is a closing. The maximal reconstruction is generated by dilating the sampled image with the reconstruction-structuring element and is a valid reconstruction when the morphological simplification done before sampling is an opening. The spatial meaning of the minimal and maximal reconstruction is direct. The minimal and maximal reconstructions delineate the spatial bounds within which the image event on the unsampled morphologically simplified image actually occurs. That is, the uncertainty due to sampling is precisely specified by the bounds given by the minimal and maximal reconstructions.

REFERENCES

Ahuja, N. and Swamy, S. (1984). Multiprocessor pyramid architectures for bottom-up image analysis, *IEEE Transactions on Pattern Analysis and Machine Intelligence, PAMI-6*: 463–475.

Burt, P. J. (1984). The pyramid as a structure for efficient computation. *Multiresolution Image Processing and Analysis* (A. Rosenfeld, ed.), Springer, Berlin, pp. 6–35.

Crimmins, T. R. and Brown, W. M. (1985). Image algebra and automation shape recognition, *IEEE Trans. on Aerospace and Electronic Systems, AES-21, no. 1*: 60–69.

Crowley, J. L. (1984). A multiresolution representation for shape, *Multiresolution Image Processing and Analysis* (A. Rosenfeld, ed.), Springer, Berlin, pp. 169–189.

Dyer, C. R. (1982). Pyramid algorithms and machines, *Multicomputers and Image Processing Algorithms and Programs* (K. Preston, Jr., and L. Uhr, eds.), Academic Press, New York, pp. 409–420.

Haralick, R. M., Lin, C., Lee, J., and Zhuang, X. (1987a). "Multiresolution Morphology," Proceedings of IEEE International Conference on Computer Vision, London.

Haralick, R. M., Sternberg, S. R., and Zhuang, X. (1987b). Image analysis using mathematical morphology, *IEEE Trans. on Pattern Analysis and Machine Intelligence, PAMI-9, no. 4*: 532–550.

Klinger, A. (1984). Multiresolution processing, *Multiresolution Image Processing and Analysis* (A. Rosenfeld, ed.), Springer, Berlin, pp. 86–100.

Lee, J., Haralick, R. M., and Shapiro, L. (1987a). Morphologic edge detection, *IEEE J. of Robotics and Automation, RA-2, no. 1*: 142–157.

Lee, J. S. J., Budak, P., Lin, C., and Haralick R. M. (1987b). "Adaptive Image Processing Techniques," Proceedings of SPIE Conference on Robotics '87, pp. 255–262.

Lougheed, R. M. and McCubbery, D. L. (1980). "The Cytocomputer: A Practical Pipelined Image Processor," Proceedings of 7th Annual International Symposium on Computer Architecture, pp. 1–7.

Maragos, P. and Schafer, R. W. (1987). Morphological filters (Part I: Their set-theoretic analysis and relations to linear shift-invariant filters; Part II: Their relations to median, order-statistic, and stack filters), *IEEE Trans. on Acoustics, Speech, and Signal Processing, ASSP-35*: 1153–1169.

Matheron, G. (1975). *Random Sets and Integral Geometry*, John Wiley & Sons, New York.

Miller, R. and Stout, Q. F. (1983). "Pyramid Computer Algorithms for Determining Geometric Properties of Images," Proceedings of Symposium on Computational Geometry, pp. 263–271.

Rosenfeld, A. (1983). Hierarchical representation: Computer representations of digital images and objects, *Fundamentals in Computer Vision—An Advanced Course* (O. D. Faugeras, ed.), Cambridge University Press, Cambridge, UK, pp. 315–324.

Serra, J. (1972). Stereology and structuring elements, *J. of Microscopy*: 93–103.

Tanimoto, S. L. (1982). Programming techniques for hierarchical parallel image processors, *Multicomputers and Image Processing Algorithms and Programs* (K. Preston, Jr., and L. Uhr, eds.), Academic Press, New York, pp. 431–429.

Uhr, L. (1983). Pyramid multicomputer structures and augmented pyramids, *Computing Structures for Image Processing*, Academic Press, London pp. 95–112.

Witkin, A. P. (1984). Scale space filtering: A new approach to multiscale description, *Image Understanding* (S. Ullman and W. Richards, eds.), Abex, Norwood, New Jersey, pp. 79–95.

Index

Biographies

Amir A. Amini is a member of the Institute of Electrical and Electronics Engineers. His research interests include computer vision and artificial intelligence. He received the B.S. degree in electrical engineering from the University of Massachusetts, Amherst, and M.S.E. and Ph.D. degrees in electrical engineering from the University of Michigan, Ann Arbor.

Sing-tze Bow is Professor and Director of the Laboratory of Image Processing and Pattern Recognition, Northern Illinois University, De Kalb and was formerly with Pennsylvania State University, University Park, Pennsylvania. The author of numerous technical publications, including *Pattern Recognition: Applications to Large Data-Set Problems* (Marcel Dekker, Inc.), Dr. Bow is a senior member of the Institute of Electrical and Electronics Engineers. He received the M.S.E.E. degree from the University of Washington, Seattle, and Ph.D. degree from Northwestern University, Evanston, Illinois.

Dragana Brzakovic is an Assistant Professor in the Department of Electrical and Computer Engineering, University of Tennessee, Knoxville. Her research interests include image understanding, knowledge-based pattern recognition, and applied artificial intelligence. She received the B.S. degree in electrical engineering from the University of Belgrade, Yugoslavia, and the M.E. and Ph.D. degrees in electrical engineering from the University of Florida, Gainesville.

Richard G. Casey is a researcher in visual pattern recognition, IBM Almaden Research Center, San Jose, California. His research interests include trainable recognition logic for small computers and the implementation of a self-learning technique for reading machine-printed text independent of typeface using cryptographic analysis. Dr. Casey received the B.E.E. degree from Manhattan College and M.S. and Eng.Sc.D. degrees from Columbia University, New York.

MASAKAZU EJIRI is a Senior Chief Research Scientist for the Central Research Laboratory, Hitachi Ltd., Tokyo, Japan. A Fellow of the Institute of Electrical and Electronics Engineers, his work involves control engineering, pattern recognition, robotics, machine vision, and artificial intelligence. He received the B.E. and Ph.D. degrees from Osaka University, Osaka, Japan.

RAFAEL C. GONZALEZ is President of the Perceptics Corporation and Professor of Electrical and Computer Engineering, University of Tennessee, Knoxville. The coauthor of several books, Professor Gonzalez is a Fellow of the Institute of Electrical and Electronics Engineers and an associate editor for the *Institute of Electrical and Electronics Engineers' Transactions on Systems, Man, and Cybernetics.* He received the B.S. degree in electrical engineering from the University of Miami, Florida, and the M.E. and the Ph.D. degrees in electrical engineering from the University of Florida, Gainesville.

ROBERT M. HARALICK is a Professor in the Department of Electrical Engineering, University of Washington, Seattle. A Fellow of the Institute of Electrical and Electronics Engineers, Professor Haralick also serves on the editorial board of the *Institute of Electrical and Electronics Engineers' Transactions on Pattern Analysis and Machine Intelligence.* He received the B.A. degree in mathematics, and the B.S., M.S., and Ph.D. degrees in electrical engineering from the University of Kansas, Lawrence.

LYNDON S. HIBBARD is with the Washington University School of Medicine, St. Louis, Missouri, and was a member of the Milton S. Hershey Medical Center, Pennsylvania State University College of Medicine, Hershey, Pennsylvania. He currently studies the application of imaging methods for metabolic and neurochemical mapping of the brain, reconstruction studies of brain microvasculature, and the display and interpretation of neuronal structures made from optical thin sections. He received the B.S. degree in chemistry from the University of Illinois, Urbana, and Ph.D. degree in physical chemistry from Michigan State University, East Lansing.

KAZUAKI IWAMURA is a member of the Central Research Laboratory, Hitachi Ltd., Tokyo, Japan. His research interests include drawing recognition techniques, neural network applications, and map-based information retrieval systems. A member of the Institute of Electronics, Information and Communication Engineers of Japan, he received the B.E. and M.E. degrees in mechanical engineering from Waseda University, Tokyo, Japan.

SHIGERU KAKUMOTO is a Senior Researcher for the Central Research Laboratory, Hitachi Ltd., Tokyo, Japan. His research areas include image processing, drawing recognition, and geographical information processing. A member of the Institute of Electronics, Information and Communication Engineers of Japan, He received the B.S. degree in biological science and B.E. degree in control engineering from Osaka University, Osaka, Japan.

THOMAS F. KRILE is a faculty member of Texas Tech University, Lubbock. His research areas include optical computing, neural networks, and image processing. A Fellow of the Optical Society of America, Dr. Krile received the B.S. and M.S. degrees in electrical engineering from the University of North Dakota, Grand Forks, and Ph.D. degree in electrical engineering from Purdue University, West Lafayette, Indiana.

JAMES LEE is a member of the Boeing High Technology Center. The author or coauthor of over 40 technical articles. Dr. Lee's research includes results on vision integrated with neural networks, adaptive image processing, and sensor fusion. He received the B.S. degree in control engineering from the National Chiao Tung University, M.S. degree in electrical engineering from the National Taiwan University, and Ph.D. degree from Virginia Polytechnic Institute and State University, Blacksburg.

CHARLOTTE LIN is a manager for the Boeing High Technology Center's Sensor Fusion Group. Previously, Dr. Lin developed image understanding for log interpretation and modeling/measurement methods for fluid dynamics, acoustics, and electromagnetics in reservoir rock. The author or coauthor of over 30 publications in applied research, Dr. Lin received the B.A. degree in liberal arts and M.A. and Ph.D. degrees in mathematics from Cornell University, Ithaca, New York.

SUNANDA MITRA is a faculty member of the Department of Electrical Engineering, Texas Tech University, Lubbock. Her research involves multisensor automatic target recognition, 3-D image reconstruction, and the analysis of medical imagery. She received the Ph.D. degree from Marburg University, Federal Republic of Germany.

TAKAFUMI MIYATAKE is a member of the Central Research Laboratory, Hitachi Ltd., Tokyo, Japan. His research areas include object recognition, picture processing, drawing recognition, and neural networks. A member of the Information Processing Society of Japan, he is a graduate of Hitachi Technical College.

SHIGERU SHIMADA is a member of the Central Research Laboratory, Hitachi Ltd., Tokyo, Japan. His research involves drawing recognition and knowledge-based systems. A member of the Information Processing Society of Japan and the Institute of Electronics, Information and Communication Engineers of Japan, he received the B.E. degree from Doshisha University, Kyoto, Japan, and M.S. degree from Nagoya Institute of Technology, Nagoya, Japan.

JOHN F. WALKUP is Professor of Electrical Engineering, Texas Tech University, Lubbock. A Fellow of the Institute of Electrical and Electronics Engineers, Dr. Walkup's research interests include optical signal processing, optical computing, digital image processing, and statistical optics. He received the Ph.D. degree in electrical engineering from Stanford University, Stanford, California.

TERRY E. WEYMOUTH is a faculty member of the Department of Electrical Engineering and Computer Science, University of Michigan, Ann Arbor. His research involves the application of blackboard architecture to machine vision problems, combining image events into relational structures that describe objects in a scene. Professor Weymouth received the Ph.D. degree from the University of Massachusetts, Amherst.

KWAN Y. WONG is a member of the IBM Almaden Research Center, San Jose, California. The author or coauthor of numerous technical publications, Dr. Wong is a senior member of the Institute of Electrical and Electronics Engineers. He received the B.E. and M.E. degrees in electrical engineering from the University of New South Wales, Sydney, Australia, and the Ph.D. degree from the University of California, Berkeley.

XINHUA ZHUANG is with the University of Missouri at Columbia and was a Visiting Professor at the University of Washington, Seattle. Previously, he served as Professor of Computer Science and Engineering, Zhejiang University, and as a research engineer for the Computing Technique Institute, Hangzhou, China. His research interests include pattern recognition, computer vision, and applied mathematics. He received undergraduate and graduate degrees in mathematics from Beijing University, China.

Milton Keynes UK
Ingram Content Group UK Ltd.
UKHW040711141024
449569UK00005B/96